Electricity from Renewable Resources

STATUS, PROSPECTS, AND IMPEDIMENTS

America's Energy Future Panel on Electricity from Renewable Resources

NATIONAL ACADEMY OF SCIENCES
NATIONAL ACADEMY OF ENGINEERING
NATIONAL RESEARCH COUNCIL
OF THE NATIONAL ACADEMIES

THE NATIONAL ACADEMIES PRESS
Washington, D.C.
www.nap.edu

THE NATIONAL ACADEMIES PRESS 500 Fifth Street, N.W. Washington, DC 20001

NOTICE: The project that is the subject of this report was approved by the Governing Board of the National Research Council, whose members are drawn from the councils of the National Academy of Sciences, the National Academy of Engineering, and the Institute of Medicine. The members of the panel responsible for the report were chosen for their special competences and with regard for appropriate balance.

Support for this project was provided by the Department of Energy under Grant Number DE-FG02-07-ER-15923 and by BP America, Dow Chemical Company Foundation, Fred Kavli and the Kavli Foundation, GE Energy, General Motors Corporation, Intel Corporation, and the W.M. Keck Foundation. Support was also provided by the Presidents' Circle Communications Initiative of the National Academies and by the National Academy of Sciences through the following endowed funds created to perpetually support the work of the National Research Council: Thomas Lincoln Casey Fund, Arthur L. Day Fund, W.K. Kellogg Foundation Fund, George and Cynthia Mitchell Endowment for Sustainability Science, and Frank Press Fund for Dissemination and Outreach. Any opinions, findings, conclusions, or recommendations expressed in this publication are those of the author(s) and do not necessarily reflect the views of the organizations that provided support for the project.

International Standard Book Number-13: 978-0-309-13708-9
International Standard Book Number-10: 0-309-13708-X
Library of Congress Control Number: 2009938602

Available in limited supply and free of charge from:

Board on Energy and Environmental Systems
National Research Council
500 Fifth Street, N.W.
Keck W917
Washington, DC 20001
202-334-3344

Additional copies of this report are available from the National Academies Press, 500 Fifth Street, N.W., Lockbox 285, Washington, DC 20055; (800) 624-6242 or (202) 334-3313 (in the Washington metropolitan area); Internet, http://www.nap.edu.

♲ Printed on recycled stock

Printed in the United States of America

THE NATIONAL ACADEMIES

Advisers to the Nation on Science, Engineering, and Medicine

The **National Academy of Sciences** is a private, nonprofit, self-perpetuating society of distinguished scholars engaged in scientific and engineering research, dedicated to the furtherance of science and technology and to their use for the general welfare. Upon the authority of the charter granted to it by the Congress in 1863, the Academy has a mandate that requires it to advise the federal government on scientific and technical matters. Dr. Ralph J. Cicerone is president of the National Academy of Sciences.

The **National Academy of Engineering** was established in 1964, under the charter of the National Academy of Sciences, as a parallel organization of outstanding engineers. It is autonomous in its administration and in the selection of its members, sharing with the National Academy of Sciences the responsibility for advising the federal government. The National Academy of Engineering also sponsors engineering programs aimed at meeting national needs, encourages education and research, and recognizes the superior achievements of engineers. Dr. Charles M. Vest is president of the National Academy of Engineering.

The **Institute of Medicine** was established in 1970 by the National Academy of Sciences to secure the services of eminent members of appropriate professions in the examination of policy matters pertaining to the health of the public. The Institute acts under the responsibility given to the National Academy of Sciences by its congressional charter to be an adviser to the federal government and, upon its own initiative, to identify issues of medical care, research, and education. Dr. Harvey V. Fineberg is president of the Institute of Medicine.

The **National Research Council** was organized by the National Academy of Sciences in 1916 to associate the broad community of science and technology with the Academy's purposes of furthering knowledge and advising the federal government. Functioning in accordance with general policies determined by the Academy, the Council has become the principal operating agency of both the National Academy of Sciences and the National Academy of Engineering in providing services to the government, the public, and the scientific and engineering communities. The Council is administered jointly by both Academies and the Institute of Medicine. Dr. Ralph J. Cicerone and Dr. Charles M. Vest are chair and vice chair, respectively, of the National Research Council.

www.national-academies.org

PANEL ON ELECTRICITY FROM RENEWABLE RESOURCES

LAWRENCE T. PAPAY, NAE,[1] Science Applications International Corporation (retired), *Chair*
ALLEN J. BARD, NAS,[2] University of Texas, Austin, *Vice Chair*
RAKESH AGRAWAL, NAE, Purdue University
WILLIAM CHAMEIDES, NAS, Duke University
JANE H. DAVIDSON, University of Minnesota, Minneapolis
J. MICHAEL DAVIS, Pacific Northwest National Laboratory
KELLY R. FLETCHER, General Electric
CHARLES F. GAY, Applied Materials, Inc.
CHARLES H. GOODMAN, Southern Company (retired)
SOSSINA M. HAILE, California Institute of Technology
NATHAN S. LEWIS, California Institute of Technology
KAREN L. PALMER, Resources for the Future
JEFFREY M. PETERSON, New York State Energy Research and Development Authority
KARL R. RABAGO, Austin Energy
CARL J. WEINBERG, Pacific Gas and Electric Company (retired)
KURT E. YEAGER, Galvin Electricity Initiative

America's Energy Future Project Director

PETER D. BLAIR, Executive Director, Division on Engineering and Physical Sciences

America's Energy Future Project Manager

JAMES ZUCCHETTO, Director, Board on Energy and Environmental Systems

Staff

K. JOHN HOLMES, Study Director
KATHERINE BITTNER, Senior Program Assistant (until July 2008)
LaNITA R. JONES, Program Associate
AMY HEE KIM, Christine Mirzayan Science and Technology Policy Graduate Fellow (until November 2008)
DOROTHY MILLER, Christine Mirzayan Science and Technology Policy Graduate Fellow (until August 2008)
JASON ORTEGO, Senior Program Assistant
STEPHANIE WOLAHAN, Christine Mirzayan Science and Technology Policy Graduate Fellow (until April 2009)
E. JONATHAN YANGER, Senior Program Assistant

---

[1]NAE, National Academy of Engineering.
[2]NAS, National Academy of Sciences.

Foreword

Energy, which has always played a critical role in our country's national security, economic prosperity, and environmental quality, has over the last two years been pushed to the forefront of national attention as a result of several factors:

- World demand for energy has increased steadily, especially in developing nations. China, for example, saw an extended period (prior to the current worldwide economic recession) of double-digit annual increases in economic growth and energy consumption.
- About 56 percent of the U.S. demand for oil is now met by depending on imports supplied by foreign sources, up from 40 percent in 1990.
- The long-term reliability of traditional sources of energy, especially oil, remains uncertain in the face of political instability and limitations on resources.
- Concerns are mounting about global climate change—a result, in large measure, of the fossil-fuel combustion that currently provides most of the world's energy.
- The volatility of energy prices has been unprecedented, climbing in mid-2008 to record levels and then dropping precipitously—in only a matter of months—in late 2008.
- Today, investments in the energy infrastructure and its needed technologies are modest; many alternative energy sources are receiving insufficient attention; and the nation's energy supply and distribution systems are increasingly vulnerable to natural disasters and acts of terrorism.

All of these factors are affected to a great degree by the policies of government, both here and abroad, but even with the most enlightened policies the overall energy enterprise, like a massive ship, will be slow to change course. Its complex mix of scientific, technical, economic, social, and political elements means that the necessary transformational change in how we generate, supply, distribute, and use energy will be an immense undertaking, requiring decades to complete.

To stimulate and inform a constructive national dialogue about our energy future, the National Academy of Sciences and the National Academy of Engineering initiated in 2007 a major study, "America's Energy Future: Technology Opportunities, Risks, and Tradeoffs." The America's Energy Future (AEF) project was initiated in anticipation of major legislative interest in energy policy in the U.S. Congress, and as the effort proceeded, it was endorsed by Senate Energy and Natural Resources Committee Chair Jeff Bingaman and former Ranking Member Pete Domenici.

The AEF project evaluates current contributions and the likely future impacts, including estimated costs, of existing and new energy technologies. It was planned to serve as a foundation for subsequent policy studies, at the academies and elsewhere, that will focus on energy research and development priorities, strategic energy technology development, and policy analysis.

The AEF project has produced a series of five reports, including this report on electricity from renewable resources, designed to inform key decisions as the nation begins this year a comprehensive examination of energy policy issues. Numerous studies conducted by diverse organizations have benefited the project, but many of those studies disagree about the potential of specific technologies, particularly those involving alternative sources of energy such as biomass, renewable resources for generation of electric power, advanced processes for generation from coal, and nuclear power. A key objective of the AEF series of reports is thus to help resolve conflicting analyses and to facilitate the charting of a new direction in the nation's energy enterprise.

The AEF project, outlined in Appendix A, included a study committee and three panels that together have produced an extensive analysis of energy technology options for consideration in an ongoing national dialogue. A milestone in the project was the March 2008 "National Academies Summit on America's Energy Future" at which principals of related recent studies provided input to the AEF study committee and helped to inform the panels' deliberations. A report chronicling the event, *The National Academies Summit on America's Energy Future:*

Summary of a Meeting (Washington, D.C.: The National Academies Press), was published in October 2008.

The AEF project was generously supported by the W.M. Keck Foundation, Fred Kavli and the Kavli Foundation, Intel Corporation, Dow Chemical Company Foundation, General Motors Corporation, GE Energy, BP America, the U.S. Department of Energy, and our own academies.

Ralph J. Cicerone, President
National Academy of Sciences
Chair, National Research Council

Charles M. Vest, President
National Academy of Engineering
Vice Chair, National Research Council

Preface

Shortly after the end of World War II, America's electricity use rose rapidly with the introduction of labor-saving appliances and tools in the home, the electrification of manufacturing processes and assembly lines in factories, and the increased distribution of refrigerated and frozen foods into markets. This unprecedented growth averaged almost 7 percent annually on a compound basis for two decades. Helping to fuel this growth was the lower price of electricity made possible by economies of scale achieved as new plants were built.

With the close of the 1960s and the start of the 1970s, a series of events changed the face of electric power economics and structure, and this process continues today. The 1970 National Environmental Policy Act (NEPA) and the creation of the U.S. Environmental Protection Agency (EPA) signaled that environmental considerations would be required for every decision regarding expansion, construction, and operation of electric power systems and components. In 1973 the Organization of the Petroleum Exporting Countries' oil embargo on the United States pointed out the vulnerability of the supply of transportation and boiler fuels. On the heels of the embargo, the United States experienced sharp increases in the cost of electricity due to the increased price of fuels. As the 1980s arrived, it became far more costly to construct large baseload power plants—particularly nuclear plants—because of lengthy approval processes and, post–Three Mile Island, reevaluation and redesign of nuclear safety systems.

The advent of deregulation due to legislation from 1978 onward meant that new project-financed independent power generators would look for least-cost options, which usually meant natural-gas-fired combined cycle power plants.

Based on a series of studies by the White House Office of Science and Technology Policy in the early 1970s, a few developers and utilities began to look into the possible use of renewable sources of energy for electric power production. In 1978, with the passage of the Public Utility Regulatory Policies Act (PURPA), small generation units and renewable resources were given special attention. The introduction of incentives such as tax credits at the federal and state level, as well as renewables portfolio standards (RPSs), spurred the development of renewable technologies. Growth in the 1980s and early 1990s was spotty, but the succeeding decade has seen a dramatic increase in renewable projects for electric power, particularly in wind and solar.

Today, there is a nexus of concerns about the U.S. energy portfolio: concerns about the environment, principally arising from climate change issues; concerns about energy security, principally due to the large amounts of oil imported from volatile parts of the world; and concerns about the economy, principally because of sharp increases in the price of oil, natural gas, and basic construction commodities. Collectively, these concerns beg the question of whether it is time for reevaluating and redesigning our electric infrastructure to extend energy efficiency to a much greater extent and use domestic, non-polluting, economically attractive energy sources. Thus, this provides the motivation for the continued but growing interest in renewable-based electric power.

Such concerns, consequently, have led to greater interest in renewable electric power. As part of the America's Energy Future (AEF) project initiated by the National Academy of Sciences and the National Academy of Engineering (Appendix A), the National Research Council convened the Panel on Electricity from Renewable Resources (Appendix B) to examine all the factors that must be considered if any renewable energy resource is to become a significant contributor to meeting U.S. energy needs (see Box P.1 for the full statement of task). Presented in this stand-alone report, the work of this independent panel also serves as input to the larger AEF study outlined in Appendix A.

This report of the panel considers resource bases, technologies, economics, environmental impacts, and deployment issues and also presents selected deployment scenarios and their impacts. The major focus is the relative near term, from the present to the year 2020. The report also considers, in less detail, the midterm between the years 2020 and 2035 and the long term beyond 2035. The goal of the report is to determine if renewable electric power technologies can make a significant (>20 percent) contribution to the total electric power needs of the United States and on what basis. It examines cost and deployment issues in detail.

This report is the result of considerable time and effort contributed by the panel members. Many issues needed a fair and honest discussion, and the panel members proved capable of the task. The panel in turn appreciates the dedicated

BOX P.1 **Task Statement for AEF Panel on Electricity from Renewable Resources**

This panel will examine the technical potential for electric power generation with alternative sources such as wind, solar-photovoltaic, geothermal, solar-thermal, hydroelectric, and other renewable resources. The panel will also consider the broader energy applications of renewables, especially low-temperature solar applications that may reduce electricity demands. The panel will evaluate technologies based on their estimated times to initial commercial deployment and will provide the following information for each:

- Initial deployment times <10 years: costs, performance, and impacts
- 10 to 25 years: barriers, implications for costs, and R&D challenges/needs
- >25 years: barriers and R&D challenges/needs, especially basic research needs.

The primary focus of the study will be on the quantitative characterization of technologies with initial deployment times <10 years. The panel will focus on those renewable resources that show the most promise for initial commercial development within a decade leading to substantial impact on the U.S. energy system, as well as consider the potential use of such technologies globally. In keeping with the charge to the overall scope of the America's Energy Future Study Committee, the panel will not recommend policy choices, but it will assess the state of development of technologies. In addition to a principal focus on renewable energy technologies for power generation, the panel will address the challenges of incorporating such technologies into the power grid, as well as the potential of improvements in the national electricity grid that could enable better and more extensive use of wind, solar-thermal, solar photovoltaics, and other renewable technologies.

and committed staff of the National Research Council, including K. John Holmes, study director and senior program officer with the Board on Energy and Environmental Systems (BEES); Amy Hee Kim, Dorothy Miller, and Stephanie Wolahan, all Christine Mirzayan Science and Technology Policy Graduate Fellows; James Zucchetto, director of BEES; Jonathan Yanger and Jason Ortego, senior program assistants; and Peter Blair, executive director of the Division on Engineering and Physical Sciences. Richard Sweeney of Resources for the Future also contributed to the economic analysis in Chapter 4 in his role as an unpaid consultant to the panel.

Lawrence T. Papay, *Chair*
Panel on Electricity from Renewable Resources

Acknowledgment of Reviewers

This report has been reviewed in draft form by individuals chosen for their diverse perspectives and technical expertise, in accordance with procedures approved by the Report Review Committee of the National Research Council (NRC). The purpose of this independent review is to provide candid and critical comments that will assist the institution in making its published report as sound as possible and to ensure that the report meets institutional standards for objectivity, evidence, and responsiveness to the study charge. The review comments and draft manuscript remain confidential to protect the integrity of the deliberative process. We wish to thank the following individuals for their review of this report:

Douglas M. Chapin, MPR Associates,
Paul DeCotis, State of New York,
Sam Fleming, Consultant,
Clark Gellings, Electric Power Research Institute,
Roy Gordon, Harvard University,
Narain Hingorani, Consultant,
Robert Hirsch, Management Information Services, Inc.,
Lester B. Lave, Carnegie Mellon University,
Timothy Mount, Cornell University,
Pedro Pizzaro, Southern California Edison,
Norman R. Scott, Cornell University,
Terrance Surles, Hawaii Natural Energy Institute, and
Jefferson Tester, Massachusetts Institute of Technology.

Although the reviewers listed above have provided many constructive comments and suggestions, they were not asked to endorse the conclusions or recommendations, nor did they see the final draft of the report before its release. The review of this report was overseen by Elisabeth M. Drake, Massachusetts Institute of Technology, and Robert A. Frosch, Harvard University. Appointed by the NRC, they were responsible for making certain that an independent examination of this report was carried out in accordance with institutional procedures and that all review comments were carefully considered. Responsibility for the final content of this report rests entirely with the authoring panel and the institution.

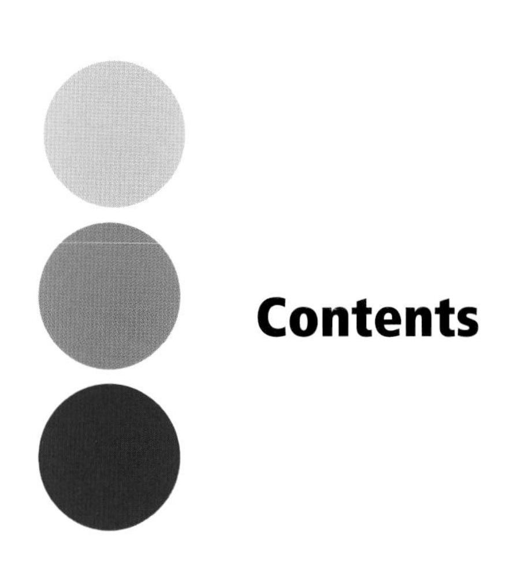

Contents

Appendixes

Summary

Renewable resources—the sun, wind, water, and biomass—were the first to be tapped to provide heat, light, and usable power. But throughout the 20th century and today, the dramatic increase in energy use for industrial, residential, transportation, and other purposes has been fueled largely by the energy stored in fossil fuels and, more recently, supplied by nuclear power. Linked to the exploitation and development of high-energy-density resources such as coal and oil at the scales required for powering the modern U.S. energy system are potentially significant environmental and other impacts. Concern about greenhouse gases released by the combustion of these fuels, for example, and awareness of eventual limits on the supply of fossil-fuel resources have strengthened interest in expanding the use of renewable energy resources. Escalations in energy prices, increasing worldwide demand for energy, and the need to ensure U.S. energy security have also combined to put energy in the headlines, increasing policy makers' interest in domestically produced renewable energy.

As part of the America's Energy Future (AEF) project initiated by the National Academy of Sciences and the National Academy of Engineering (Appendix A), the National Research Council convened the Panel on Electricity from Renewable Resources (Appendix B) to examine the technical potential for development and deployment of renewable electricity technologies. The full statement of task is provided in Box P.1 in the preface.

As a result of its study, the panel found that technologies for generation of electricity from renewable resources represent a significant opportunity—with attendant challenges—to provide low carbon dioxide (CO_2)–emitting electricity generation from resources available domestically and to generate new economic opportunities for the United States. Sufficient domestic renewable resources exist to allow renewable electricity to play a significant role in future electricity generation and thus help confront issues related to climate change, energy security, and the escalation of energy costs.

Generation of electricity from renewable resources has increased substantially over the past 20 years. As shown in Chapter 1, some sources have sustained a 20 percent or higher compound annual growth rate in capacity expansion and electricity generation over the past decade. However, non-hydroelectric renewable resources still provide only a small percentage of total U.S. electricity generation (about 2.5 percent of all electricity generated), even with these large recent growth rates. The most recent U.S. Energy Information Administration projections, which are presented in Chapter 1, indicate that under a "business as usual" scenario, the share of electricity generated from non-hydroelectric renewable resources in 2030 would be only 8 percent of the total U.S. electricity generated.

The panel concluded that sustained actions involving the coordination of policy, technology, and capital investment will be essential to achieving a greatly increased market penetration of renewable electricity. All three of these factors are important because improvements in the economics of renewable electricity generation, large increases in the scale and rate of deployment, and the establishment of consistent long-term policies are all required in order for renewables to make a material contribution to the nation's energy supply. Although continued technological advances are critical, the degree of penetration by renewable electricity will also be determined by actions that collectively center on sustainably improving the economic competiveness of electricity generated from renewable versus other resources and on policy initiatives that have a positive impact on competitive balance and the ease of deployment of renewable electricity.

CHALLENGES AND OPPORTUNITIES AHEAD FOR THE USE OF RENEWABLE ELECTRCITY

Immense challenges are presented by the need to reduce the vulnerabilities associated with climate change, energy supply interruptions, and volatile fossil-fuel markets. Reducing electric-sector CO_2 emissions by significant levels will require major changes in how we use and produce electricity. Cutting energy imports and substantially reducing our dependence on fossil fuels also will involve major changes. Reliance on a greater amount of renewable energy, particularly renewable electricity, can help address these challenges. Renewable energy is an attractive option because renewable resources available in the United States, taken collectively, can supply significantly greater amounts of electricity than the total current or projected domestic demand. These renewable resources are largely untapped today.

There are, however, important disadvantages to the use of non-hydropower renewables for electricity generation. The energy available from renewable resources is less concentrated than that provided by fossil-fuel or nuclear power, posing significant challenges to the development of renewable resources for electricity generation on a large scale. Generation must occur at the site of the resource and accommodate the temporal fluctuations characteristic of some non-hydropower renewable resources. At high penetrations of non-hydropower renewable sources, electricity system operators must deal with spatial and temporal constraints to integrating the generated electricity into the electric grid in ways that ensure a reliable, controllable supply of electricity. Large penetrations also will result in land-use requirements that in turn can lead to instances of local opposition to the siting of generation and transmission facilities.

In turn, the use of renewable electricity provides some significant advantages over the use of fossil-based electricity. Many types of renewable electricity-generating technologies can be developed and deployed in smaller increments, and constructed more rapidly, than large-scale fossil- or nuclear-based generation systems, thus allowing faster returns on capital investments. Generation of electricity from most renewable resources also reduces vulnerability to increases in the cost of fuels and mitigates many environmental impacts, such as those associated with atmospheric emissions of greenhouse gases and emissions of regulated air pollutants. Further, distributed renewable electricity generation located at or near the point of energy use, such as solar photovoltaic systems installed at residential, commercial, or industrial sites, can offer operational and economic benefits while increasing the robustness of the electricity system as a whole.

FINDINGS

Shown in bold text are the most critical elements of the panel's findings based on its consideration of the material presented in Chapters 2 through 7 of this report.

Timeframes and Prospects for Renewable Technologies

To better assess the prospects for individual renewable electricity technologies, the panel separated its consideration of these technologies and their characteristic costs, performance, and impacts into three time periods: an initial period that considers present technologies out to the year 2020; a second that considers current and potential renewable electricity technologies over the 2020 to 2035 time period; and a third period that looks at technologies beyond 2035.

For the time period from the present to 2020, there are no current technological constraints for wind, solar photovoltaics and concentrating solar power, conventional geothermal, and biopower technologies to accelerate deployment. The primary current barriers are the cost-competitiveness of the existing technologies relative to most other sources of electricity (with no costs assigned to carbon emissions or other currently unpriced externalities), the lack of sufficient transmission capacity to move electricity generated from renewable resources to distant demand centers, and the lack of sustained policies. Expanded research and development (R&D) is needed to realize continued improvements and further cost reductions for these technologies. Along with favorable policies, such improvements can greatly enhance renewable electricity's competitiveness and its level of deployment. Action now will set the stage for greater, more cost-effective penetration of renewable electricity in later time periods. **It is reasonable to envision that, collectively, non-hydropower renewable electricity could begin to provide a material contribution (i.e., reaching a level of 10 percent or more, with trends toward continued growth) to the nation's electricity generation in the period up to 2020 with such accelerated deployment.** Combined with hydropower, total renewable electricity could approach a contribution of 20 percent of U.S. electricity by the year 2020.

In the period from 2020 to 2035, it is reasonable to envision that continued and even further accelerated deployment could potentially result in non-hydroelectric renewables providing, collectively, 20 percent or more of domestic electricity generation by 2035. In the third timeframe, beyond 2035, continued development of renewable electricity technologies could potentially provide lower costs and result in further increases in the percentage of renewable electricity generated from renewable resources. However, achieving a predominant (i.e., >50 percent) level of renewable electricity penetration will require new scientific advances (e.g., in solar photovoltaics, other renewable electricity technologies, and storage technologies) and dramatic changes in how we generate, transmit, and use electricity. Scientific advances are anticipated to improve the cost, scalability, and performance of all renewable energy generation technologies. Moreover, some combination of intelligent, two-way electric grids; scalable and cost-effective methods for large-scale and distributed storage (either direct electricity energy storage or generation of chemical fuels); widespread implementation of rapidly dispatchable fossil-based electricity technologies; and greatly improved technologies for cost-effective long-distance electricity transmission will be required. Significant, sustained, and greatly expanded R&D focused on these technologies is also necessary if this vision is to be realized by 2035 and beyond.

Resource Base

Solar and wind renewable resources offer significantly larger total energy and electricity potential than do other domestic renewable resources. Although solar intensity varies across the nation, the land-based solar resource provides a yearly average of more than 5×10^{22} J (13.9 million TWh) and thus exceeds, by several thousand-fold, present annual U.S. electrical energy demand, which totals 1.4×10^{19} J (~4,000 TWh). Hence, at even modest conversion efficiency, solar energy is capable, in principle, of providing enormous amounts of electricity without stress to the resource base. The land-based wind resource is capable of providing at least 10–20 percent, and in some regions potentially higher percentages, of current electrical energy demand. Other (non-hydroelectric) renewable resources can contribute significantly to the electrical energy mix in some regions of the country.

Renewable resources are not distributed uniformly in the United States. Resources such as solar, wind, geothermal, tidal, wave, and biomass vary widely in space and time. **Thus, the potential to derive a given percentage of electricity from renewable resources will vary from location to location. Awareness of such factors is important in developing effective policies at the state and federal levels to promote the use of renewable resources for generation of electricity.**

Renewable Technologies

Over the first timeframe through 2020, wind, solar photovoltaics and concentrating solar power, conventional geothermal, and biomass technologies are technically ready for accelerated deployment. During this period, these technologies could potentially contribute a much greater share (up to about an additional 10 percent of electricity generation) of the U.S. electricity supply than they do today. Other technologies, including enhanced geothermal systems that mine the heat stored in deep low-permeability rock and hydrokinetic technologies that tap ocean tidal currents and wave energy, require further development before they can be considered viable entrants into the marketplace. The costs of already-developed renewable electricity technologies will likely be driven down through incremental improvements in technology, "learning curve" technology maturation, and manufacturing economies of scale. Despite short-term increases in cost over the past couple of years, in particular for wind turbines and solar photovoltaics, there have been substantial long-term decreases in the costs of these technologies, and recent cost increases due to manufacturing and materials shortages will be reduced if sustained growth in renewable sources spurs increased investment in them. In addition, support for basic and applied research is needed to drive continued technological advances and cost reductions for all renewable electricity technologies.

In contrast to fossil-based or nuclear energy, renewable energy resources are more widely distributed, and the technologies that convert these resources to useful energy must be located at the source of the energy. Further, extensive use of intermittent renewable resources such as wind and solar power to generate electricity must accommodate temporal variation in the availability of these resources. This variability requires special attention to system integration and transmission issues as the use of renewable electricity expands. Such considerations will become especially important at greater penetrations of renewable electricity in the domestic electricity generation mix. **A contemporaneous, unified intelligent electronic control and communications system overlaid on the entire electricity delivery infrastructure would enhance the viability and continued expansion of renewable electricity in the period from 2020 to 2035.** Such improvements in the intelligence of the transmission and distribution grid could enhance the whole electricity system's reliability and help facilitate integration of renewable electricity into that system, while reducing the need for backup power to support the enhanced utilization of renewable electricity.

In the third time period, 2035 and beyond, further expansion of renewable electricity is possible as advanced technologies are developed, and as existing technologies achieve lower costs and higher performance with the maturing of the technology and an increasing scale of deployment. Achieving a predominant (i.e., >50 percent) penetration of intermittent renewable resources such as wind and solar into the electricity marketplace, however, will require technologies that are largely unavailable or not yet developed today, such as large-scale and distributed cost-effective energy storage and new methods for cost-effective, long-distance electricity transmission. Finally, there might be further consideration of an integrated hydrogen and electricity transmission system such as the "SuperGrid" first championed by Chauncey Starr, though this concept is still considered high-risk.

Economics

A principal barrier to the widespread adoption of renewable electricity technologies is that electricity from renewables (except for electricity from large-scale hydropower) is more costly to produce than electricity from fossil fuels without an internalization of the costs of carbon emissions and other potential societal impacts. Policy incentives, such as renewable portfolio standards, the production tax credit, feed-in tariffs, and greenhouse gas controls, thus have been required, and for the foreseeable future will continue to be required, to drive further increases in the use of renewable sources of electricity.

Unlike some conventional energy resources, renewable electricity is considered manufactured energy, meaning that the largest proportion of costs, external energy, and materials inputs, as well as environmental impacts, occur during manufacturing and deployment rather than during operation. In general, the use of renewable resources for electricity generation involves trading the risks of future cost increases for fossil fuels and uncertainties over future costs of carbon controls for present fixed capital costs that typically are higher for use of renewable resources than for use of fossil fuels. Except for biopower, no fuel costs are associated with renewable electricity sources. Further, in contrast to coal and nuclear electricity plants, in which larger facilities tend to exhibit lower average costs of generation than do smaller plants, for renewable electricity the opportunities for achieving economies of scale are generally greater at the equipment manufacturing stage than at the generating site itself.

The future evolution of costs for generation of electricity from renewable resources will depend on continued technological progress and breakthroughs. It will also depend on the potential for policies to create greater penetration and to accelerate the scale of production—largely an issue of long-term policy stability and policy clarity. Markets will generally exploit the lowest-cost resource options first, and thus the costs of renewables may not decline in a smooth trajectory over time. For example, in the case of wind power, the lowest-cost resources are generally available at the most accessible sites in the highest wind class areas. Development of these prime resources will thus entail significant resource cost shifts as markets adjust to exploit next-tier resources. **At present, onshore wind is an economically favored option relative to other (non-hydroelectric) renewable**

resources, and hence wind power is expected to continue to grow rapidly if recent policy initiatives continue into the future.

Although some forecasts show that biopower will play an important role in meeting future renewable portfolio standards targets, the degree of competition with and recent mandates for use of liquid biofuels for providing transportation fuel and, of course, the use of biomass for food, agricultural feed, and other uses will impact the prospects for greater use of biomass in the electricity market. The future of distributed renewable electricity generation from sources such as residential photovoltaics will depend on how its costs compare to the retail price of power delivered to end users, on whether prices fully reflect variations in cost over the course of the day, and on whether the external costs of fossil-based electricity generation are increasingly incorporated into its price.

Formulation of robust predictions about whether the price of electricity will meet or exceed the price required for renewable sources to be profitable and what their resulting level of market penetration will be remains a difficult proposition. Comparisons between past forecasts of renewable electricity penetration and actual data show that, while renewable technologies generally have met forecasts of cost reductions, they have fallen short of deployment projections. Further, the profitability and penetration of electricity generated from renewable resources may be sensitive to investments in energy efficiency, especially if efficiency improvements are sufficient to meet growth in the demand for electricity or lower the market-clearing price of electricity. If the financial operating environment for fossil-fuel and other in-place sources of electricity remains unchanged, then the competitiveness of renewable electricity may be affected more than that of other electricity sources. However, at this time, the deployment of renewable electricity is being driven by tax policies, in particular by the renewable production tax credit, and by renewable portfolio standards.

Environmental Impacts

Renewable electricity technologies have inherently low life-cycle CO_2 emissions as compared to fossil-fuel-based electricity production, with most emissions occurring during manufacturing and deployment. Renewable electricity generation also involves inherently low or zero direct emissions of other regulated atmospheric pollutants, such as sulfur dioxide, nitrogen oxides, and mercury. Biopower is an exception because it produces NO_x emissions at levels similar to those associated with fossil-fuel power plants. **Renewable electricity technologies (except biopower,**

high-temperature concentrated solar power, and some geothermal technologies) also consume significantly less water and have much smaller impacts on water quality than do nuclear, natural-gas-, and coal-fired electricity generation technologies.

Because of the diffuse nature of renewable resources, the systems needed to capture energy and generate electricity (i.e., wind turbines and solar panels and concentrating systems) must be installed over large collection areas. Land is also required for the transmission lines needed to connect this generated power to the electricity system. But because of low levels of direct atmospheric emissions and water use, land-use impacts tend to remain localized and do not spread beyond the land areas directly used for deployment, especially at low levels of renewable electricity penetration. Moreover, some land that is affected by renewable technologies can also be used for other purposes, such as the use of land between wind turbines for agriculture.

However, at a high level of renewable technologies deployment, land-use and other local impacts would become quite important. Land-use impacts have caused, and will in the future cause, instances of local opposition to the siting of renewable electricity-generating facilities and associated transmission lines. State and local government entities typically have primary jurisdiction over the local deployment of electricity generation, transmission, and distribution facilities. Significant increases in the deployment of renewable electricity facilities will thus entail concomitant increases in the highly specific, administratively complex, environmental impact and siting review processes. While this situation is not unique to renewable electricity, nevertheless, a significant acceleration of its deployment will require some level of coordination and standardization of siting and impact assessment processes.

Deployment

Policy, technology, and capital are all critical for the deployment of renewable electricity. In addition to enhanced technological capabilities, adequate manufacturing capacity, predictable policy conditions, acceptable financial risks, and access to capital are all needed to greatly accelerate the deployment of renewable electricity. Improvements in the relative position of renewable electricity will require consistent and long-term commitments from policy makers and the public. Investments and market-facing research that focuses on market needs as opposed to technology needs are also required to enable business growth and market transformations.

Successful technology deployment in emerging energy sectors such as renewable electricity depends on sustained government policies, at both the project and the program level, and continued progress requires stable and orderly government participation. Uncertainty created when policies cycle on and off, as has been the case with the federal production tax credit, can hamper the development of new projects and reduce the number of market participants. Significant increases in renewable electricity generation will also be contingent on concomitant improvements in several areas, including the size and training of the workforce; the capabilities of the transmission and distribution grids; and the framework and regulations under which the systems are operated. As with other energy resources, the material deployment of renewable electricity will necessitate large and ongoing infusions of capital. However, renewable energy requires a greater allocation of capital to manufacturing and infrastructure requirements than do the conventional fossil-based energy technologies.

Integration of the intermittent characteristics of wind and solar power into the electricity system is critical for large-scale deployment of renewable electricity. Advanced storage technologies will play an important role in supporting the widespread deployment of intermittent renewable electric power above approximately 20 percent of electricity generation, although electricity storage is not necessary below 20 percent. Storage tied to renewable resources has three distinct purposes: (1) to increase the flexibility of the resources in providing power when the sun is not shining or the wind is not blowing, (2) to allow the use of energy on peak when its value is greatest, and (3) to facilitate increased use of the transmission line(s) that connect the resource to the grid. The last is particularly relevant if the resource is located far from the load centers or if the system output does not match peak load times well, as is often the case with wind power. However, wind power's development is occurring long before widespread storage will be economical. Although storage is not required for continued expansion of wind power, the inability to maximize the use of transmission corridors built to move wind resources to load centers represents an inefficient deployment of resources. Several parties are currently exploring the co-location of natural-gas-fired generation and other types of electricity generation with wind power generation to bridge this gap between storage technology and asset utilization. The co-siting of conventional dispatchable generation sources (such as natural-gas-fired combustion turbines or combined cycle plants) with renewable resources could serve as an interim mechanism to increase the value of renewable electric power until advanced storage technologies are technically feasible and economically attractive. The location of such natural-gas-fired generation could be at or near the wind resource, or at an appropriate site within the control area. Another possibility is the co-siting of two (or more) renewable resources, such as wind and solar resources, which might on average interact synergistically with respect to their temporal patterns of power generation and needs for transmission capacity.

Finally, it is important to note that the deployment needs and impacts from renewable electricity deployment are not evenly distributed regionally. Development of solar and wind power resources has been growing at an average annual rate of 20 percent and higher over the past decade. Overall electricity demand is forecasted to continue to grow at just under 1 percent annually until 2030, with the southeastern and southwestern regions of the United States expected to see most of this growth. Although some of this growth may correspond to areas where renewable resources are available, some of it will not, indicating the possible need for increases in electricity transmission capacity.

Scale of Deployment

An understanding of the scale of deployment necessary for renewable resources to make a material contribution to U.S. electricity generation is critical to assessing the potential for renewable electricity generation. Large increases over current levels of manufacturing, employment, investment, and installation will be required for non-hydropower renewable resources to move from single-digit- to double-digit-percentage contributions to U.S. electricity generation. The Department of Energy's study of 20 percent wind penetration by 2030 discussed in Chapter 7 demonstrates the challenges and potential opportunities—100,000 wind turbines would have to be installed; up to $100 billion worth of additional capital investments and transmission upgrades would be required; 140,000 jobs would have to be filled; and more than 800 million metric tons of CO_2 emissions would be eliminated. **In the panel's opinion, increasing manufacturing and installation capacity, employment, and financing to meet this goal by 2030 is doable, but the magnitude of the challenge is clear from the scale of such an effort.**

Integration of Renewable Electricity

The cost of new transmission and upgrades to the distribution system will be important factors when integrating increasing amounts of renewable electricity. The nation's electricity grid needs major improvements regardless of whether renewable electricity generation is increased. Such improvements would increase the reliability of the electricity transmission system and would reduce the losses incurred with all electricity sources. However, because a substantial fraction of new renewable electricity generation capacity would come from intermittent and/or distant sources, increases in transmission capacity and other grid improvements are critical for significant penetration of renewable electricity sources. According to the Department of Energy's study postulating 20 percent wind penetration, transmission could be the greatest obstacle to reaching the 20 percent wind generation level. **Transmission improvements can bring new resources into the electricity system, provide geographical diversity in the generation base, and allow improved access to regional wholesale electricity markets.** These benefits can also generally contribute positively to the reliability, stability, and security of the grid. **Improvements in the system's distribution of electricity are needed to maximize the benefits of two-way electricity flow and to implement time-of-day pricing. Such improvements would more efficiently integrate distributed renewable electricity sources, such as solar photovoltaics sited at residential and commercial units. A significant increase in renewable sources of power in the electricity system would also require fast-responding backup generation and/or storage capacity, such as that provided by natural gas combustion turbines, hydropower, or storage technologies.** Higher levels of penetration of intermittent renewables (above about 20 percent) would require batteries, compressed air energy storage, or other methods of storing energy such as conversion of excess generated electricity to chemical fuels. Improved meteorological forecasting could also facilitate increased integration of solar and wind power. Hence, though improvements in the grid and related technologies are necessary and valuable for other objectives, significant integration of renewable electricity will not occur without increases in transmission capacity as well as other grid management improvements.

FUTURE PROSPECTS FOR RENEWABLE ELECTRICITY

Currently, use of renewable resources for electricity generation generally incurs higher direct costs than those currently seen for fossil-based electricity generation, whose price does not now include the costs associated with carbon emissions and other unpriced externalities. Some form of market intervention or combination of incentives is thus required to enable renewable resources to contribute substantially to the national electrical energy generation mix. Sustained, consistent, long-term policies that provide for production tax credits, market incentives, streamlined permitting, and/or renewable portfolio standards are essential to support significant growth of the market for renewable electricity. With such policies and economic incentives in place, up to 20 percent of additional domestic electricity generation could come from non-hydropower renewable technologies within approximately the next 25 years.

In turn, significant technological and scientific barriers must be surmounted if renewables are to provide upward of 50 percent or more of domestic electricity generation in a reliable, controllable system that also has a low-carbon-emissions footprint. The barriers include those related to transmission as well as system integration and flexibility, including storage and other enabling technologies. Specifically, large-scale and distributed electrical energy storage, and/or large capacities for rapidly controllable low-carbon-emission generation, would be required to reach such a goal. Further, a systemwide intelligent, digitally controlled grid could reduce the need for backup power and storage and further facilitate the penetration of renewable electricity into the marketplace. Significant R&D is required now if such technologies are to be available in time to facilitate deployment of renewable electricity at a level of 50 percent or higher. Research is also needed to ensure that large-scale deployment of renewable electricity will not lead inadvertently to undesirable environmental consequences.

CRITICAL UNKNOWNS

The panel notes that many major unknowns will affect the future of electricity from renewable resources. Several are highlighted below.

- *Technologies*—The prospects for reducing manufacturing costs and improving the efficiencies of renewable electricity technologies, including the potential for solar photovoltaics to bring the installed system cost down to less than $1 per watt with at least 10 percent module and system efficiency to enable widespread deployment without subsidies;
- *Economics*—The price of electricity in the future, how prices will be structured, and the explicit or implicit price of CO_2 imposed by any future climate policy;
- *Policy*—The structure of renewables portfolio standards, tax policies (production and/or investment tax credits), and other policy initiatives directed at renewable electricity;

- *Biomass*—The contribution of biomass to electricity production versus the use of the biomass energy resource base for the production of liquid fuels;
- *Transmission*—The mechanisms and responsibilities for increases in transmission capacity and other upgrades for the electricity grid; and
- *Transportation*—The degree to which renewable electricity can influence the transportation sector and reduce dependence on imported oil and liquefied natural gas through, in the near term, charging vehicle batteries and, in the long term, producing non-petroleum-based fuels.

CONCLUSION

A future characterized by a large penetration of renewable electricity represents a paradigm shift from the current electricity generation, transmission, and distribution system. There are many reasons why renewable electricity represents such a shift, including the spatial distribution and intermittency of some renewable resources, and issues related to greatly increasing the scale of deployment. Wind and solar, renewable energy resources with the potential for large near-term growth in deployment, are intermittent resources that have some of their base located far from demand centers. The transformations required to incorporate a significant penetration of additional renewables include transformation in ancillary capabilities, especially the expansion of transmission and backup power resources, and deployment of technologies that improve grid intelligence and provide greater system flexibility. Further, supplying renewable resources on a scale that would make a major contribution to U.S. electricity generation would require vast investment in and deployment of manufacturing and human resources, as well as additional capital costs relative to those associated with current generating technologies that have no controls on greenhouse gas emissions. The realization of such a future would require a predictable policy environment and sufficient financial resources.

Nevertheless, the promise of renewable resources is that they offer significant potential for low-carbon generation of electricity from domestic sources of energy that are much less vulnerable to fuel cost increases than are other electricity sources. Overall success depends on having technology, capital, and policy working together to enable renewable electricity technologies to become a major contributor to America's energy future.

1 Introduction

The uses of energy have evolved as humans have changed patterns of energy consumption. Although renewable resources such as wind, water, and biomass were the first sources of energy tapped to provide heat, light, and usable power, it was the energy stored in fossil fuels and, more recently, nuclear power that fueled the tremendous expansion of the U.S. industrial, residential, and transportation sectors during the 20th century. But as fossil-fuel consumption has increased, a result of population growth and growth in our standard of living, so have the concerns over energy security and the negative impacts of greenhouse gases on the environment. Volatilities in foreign energy markets affecting fuel prices and availability have long raised the issue of domestic energy security. In addition, recent concerns over the limited supply of fossil fuels and the greenhouse gases released by fossil-fuel combustion have spurred efforts to utilize renewables resources—wind, sunlight, biomass, and geothermal heat—to meet U.S. energy demands. At this time, renewable sources of energy, or renewables, have enormous potential to reduce the negative impacts of energy use and to increase the domestic resource base. The fundamental challenge is collecting the energy in renewable resources and converting it to usable forms at the scales necessary to allow renewables to contribute significantly to domestic energy supply.

A central issue for future U.S. energy systems is the role that renewable resources will play in electricity generation. Renewable electricity presents a significant opportunity to provide domestically produced, low carbon dioxide (CO_2)–emitting power generation and concomitant economic opportunities. Although renewable electricity generation has increased over the past 20 years, the percentage of U.S. electricity generation from non-hydroelectric renewable sources remains small. Though continued technological advances are critical, economic, political, and deployment-related factors and public acceptance also are key factors in determining the contribution of renewable electricity. Meeting the opportunity that renewables offer to improve the environment and energy and economic security will require a huge scale-up in deployment and increased costs over current fossil-fuel generating technologies. Additional requirements include the capacity to more efficiently manufacture and deploy equipment for the generation of electricity from renewables and policies that have a positive impact on the competitiveness of renewables and the ease of integration of renewables into the electricity markets.

BACKGROUND

Recent History

Box 1.1 outlines a history of major policy milestones for renewables. Martinot et al. (2005) separate the history of non-hydropower renewables policy into three distinctive phases. In response to the oil crisis and price shocks in the late 1970s, significant federal research funding was directed toward development of multiple alternative sources of energy and toward renewable resources in particular. The PURPA era was inaugurated with the passage of the Public Utility Regulatory Policies Act (PURPA) of 1978, which required public utilities to purchase power from qualifying renewable and combined heat and power facilities. In addition, state tax incentives, such as those offered in California and Colorado, provided further impetus to increase the use of renewables.

A period of stagnation followed the late 1970s. Progress in the development of renewables slowed as energy prices declined. Financial incentives were cut, and the electric power sector entered a period of restructuring. The mid-1980s saw a decrease in real prices for natural gas (Figure 1.1), which spurred considerable growth in the development of natural-gas-fired electricity generation plants. In addition, the annual growth in electricity demand slowed from an average of 6 percent during the 1960s and 1970s to less than 3 percent in the 1980s (EIA, 2008a). This drop reduced the price for renewables paid under PURPA. Martinot et al. (2005) note that this period lasted from about 1990 to 1997, and only a very small amount of non-hydroelectric renewables development occurred during that period.

BOX 1.1 **Major Policy Milestones for Non-Hydropower Renewable Electricity**

1978	Public Utilities Regulatory Policy Act enacted, requiring public utilities to purchase power from qualifying renewable facilities.
1978	Energy Tax Act provided personal income tax credits and business tax credits for renewables.
1980	Federal R&D for renewable energy peaked at $1.3 billion ($3 billion in 2004 dollars).
1980	Windfall Profits Tax Act gave tax credits for alternative fuels production and alcohol fuel blending.
1992	California delayed property tax credits for solar thermal (also known as concentrating solar) power, which caused investment to stop.
1994	Federal production tax credit (PTC) for renewable electricity took effect as part of the Energy Policy Act of 1992.
1996	Net metering laws started to take effect in many states.
1997	States began establishing policies for renewables portfolio standards (RPSs) and public benefits funds (PBFs) as part of state electricity restructuring.
2000	Federal PTC expired in 1999 and was not renewed until late in the year, causing the wind industry to suffer a major downturn in 2000. The PTC also expired in 2002 and 2004, both times causing a major slowing in capacity additions.
2001	Some states began to mandate that utilities offer green power products to their customers.
2004	Five new states enacted RPSs in a single year, bringing the total to 18 states plus the District of Columbia; PBFs were operating in 15 states.
2005	Energy Policy Act extended the PTC for wind and biomass for 2 years and provided additional tax credits for other renewables, including solar, geothermal, and ocean energy.
2007	Energy Independence and Security Act of 2007 provided support for accelerating research and development on solar, geothermal, advanced hydropower, and electricity storage.
2008	27 states and the District of Columbia had enacted RPSs, and another 6 states had adopted goals for renewable electricity.
2008	Emergency Economic Stabilization Act extended the PTC for 1 year and the investment tax credit for residential and commercial solar through 2016.
2009	American Recovery and Reinvestment Act extended the PTC for wind through 2012 and the PTC for municipal solid waste, biopower, geothermal, hydrokinetic, and some hydropower through 2013. It also provided funding for research and updating of the electricity grid.

Source: Updated from Martinot et al. (2005).

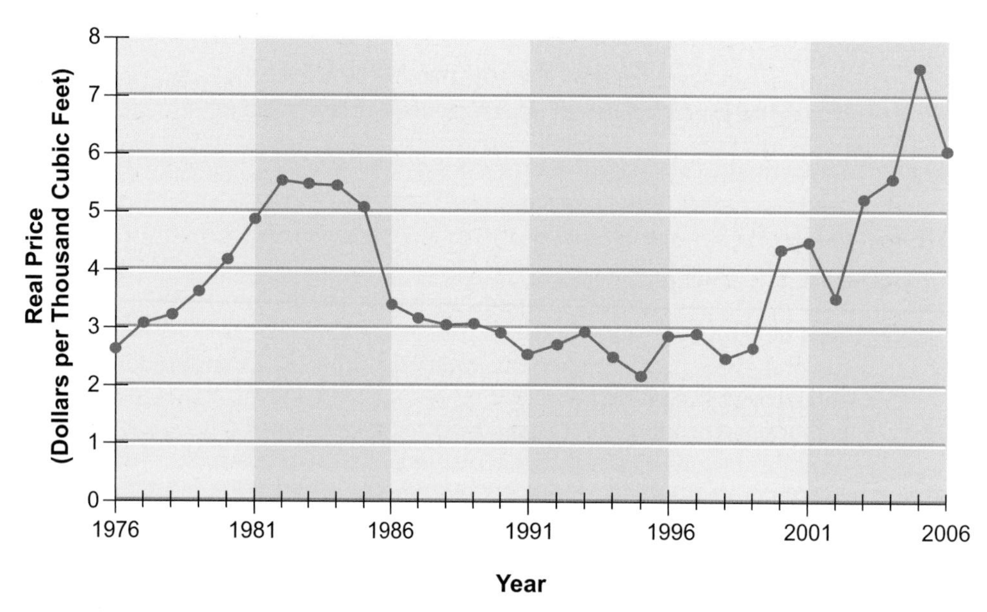

FIGURE 1.1 *Average price for natural gas for the electric power sector. Source: EIA, 2008a.*

Era of Strong Growth

Since the late 1990s, renewables have begun an era of strong growth in the United States, albeit from a small base. The amount of electricity produced from wind in particular began to increase, owing to advances in technology as well as favorable policies. Wind power electricity generation increased at a compounded annual growth rate of more than 20 percent from 1997 and 2006 and of more than 30 percent from 2004 to 2006 (EIA, 2008a). Solar photovoltaics (PV) have also seen similar growth rates in generation capacity in the United States. In 2008, non-hydropower renewables accounted for 3.4 percent of total electricity generation, up from 2.5 percent in 2007 (EIA, 2009). More details on the electricity capacity and the generation contributions from individual renewables are presented below in this chapter.

State Policies

Renewable Portfolio Standards

The generation of electricity from renewables has increased in part because of the effects of state-based policies adopted during the restructuring of many domestic electricity markets. One prominent policy mechanism for increasing the level of renewable electricity generation is the renewables portfolio standard (RPS), also known as the renewable energy standard. Typically, an RPS requires a specific percentage as the minimum share of the electricity produced (or sold) in a state that must be generated by some collection of eligible renewable technologies. The policies vary in a number of ways, such as the sources of renewables included; the form, timeline, and stringencies of the numerical goals; the extent to which utility-scale and end-use types of renewables are specified; and whether the goals include separate targets for particular renewable technologies.

As of 2008, 27 states and the District of Columba had RPSs, and another 6 states had voluntary programs (Figure 1.2). Wiser and Barbose (2008) estimate that full compliance with those RPSs will require an additional 60 GW of new

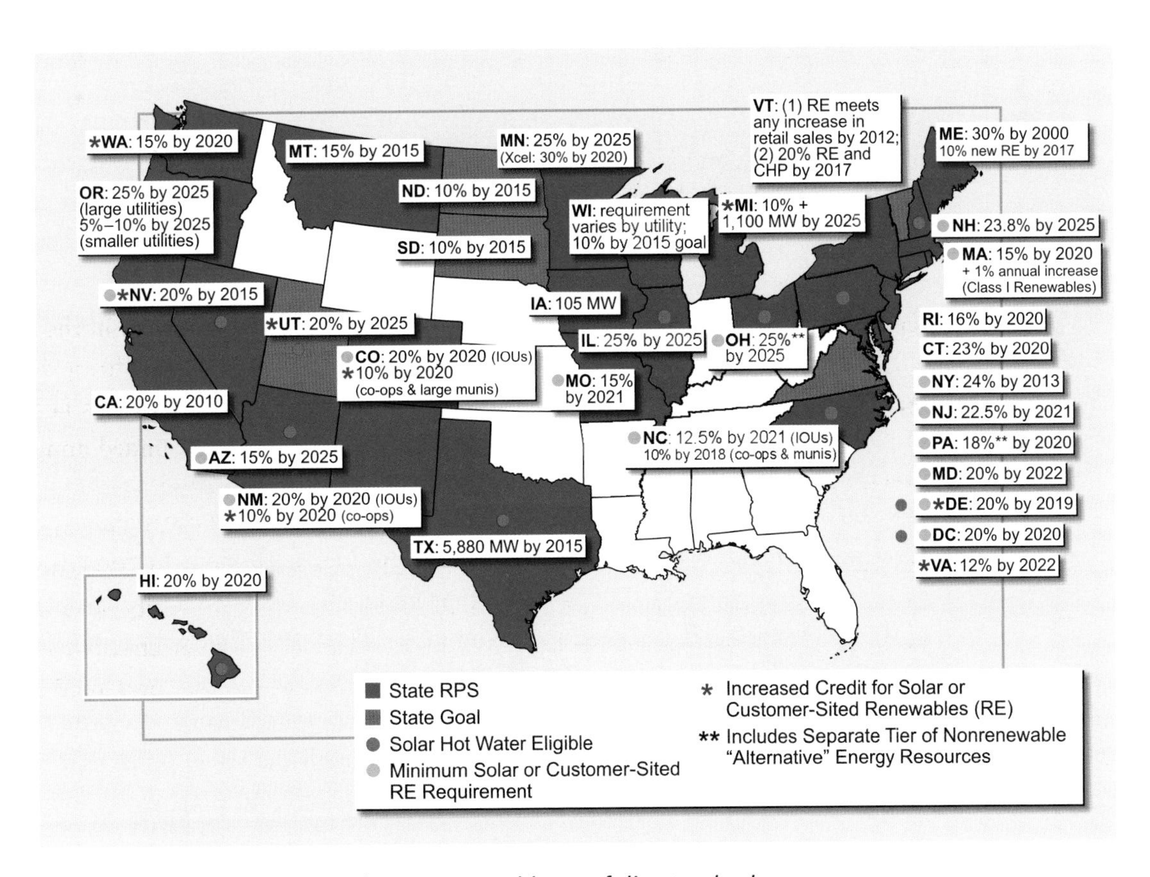

FIGURE 1.2 *Map of state renewable portfolio standards.*
Source: Database of State Incentives for Renewables and Efficiency, available at http://www.dsireusa.org. Courtesy of N.C. Solar Center at North Carolina State University and the Interstate Renewable Energy Council.

renewable electricity capacity by 2025. The actual RPS mandates vary from state to state. Maryland's RPS, for example, requires 9.5 percent renewable electricity by 2022, whereas California's requires 20 percent by 2010. Maine's original RPS required that 30 percent of all electricity be generated from renewable resources by 2000 and was later extended to require that new renewable energy capacity increase by 10 percent. Table D.1 in Appendix D shows details of these standards, including the timing for compliance, each standard's stringency, and the types of renewables covered. One element that varies among different standards is how each standard applies to specific sources of renewable energy.[1] Figure 1.3 shows the RPSs with specific requirements for electricity generation from solar and other distributed renewable resources.

Because of the variability in RPSs and the fact that they do not involve a direct cost, in contrast to the federal renewables production tax credits (PTCs; discussed below in the section titled "Federal Policies"), it is difficult to formulate a general assessment of the performance and electricity price impacts of state RPSs (Rickerson and Grace, 2007; Wiser and Barbose, 2008). Of the states that could be evaluated, Wiser and Barbose (2008) estimated that 9 of 14 were meeting their RPS requirements. However, state RPS policies are relatively recent and still evolving, and so experience with compliance remains limited. Two studies that have modeled the effectiveness of RPSs are Palmer and Burtraw (2005) and Dobesova et al. (2005). Palmer and Burtraw (2005) found that a national RPS was more cost-effective in promoting renewables than was a PTC or a carbon cap-and-trade policy that allocated allowances to all generators, including generators using renewables, on the basis of production costs. That study also found that the cost of implementing an RPS rose substantially when the standard for percentage of energy generated from renewables increased from 15 percent to 20 percent. Dobesova et al. (2005) found that under the Texas RPS the cost per ton of CO_2 emissions reduced was approximately the same as that with a pulverized coal plant with carbon capture and storage (CCS) or with a natural gas combined cycle plant with CCS, and was less cost-effective compared to an integrated coal gasification combined cycle plant with CCS (although the panel notes that no pulverized coal plants with CCS have been constructed and that cost estimates for

[1]A controversial aspect of some of the RPSs is the inclusion of some technologies not broadly accepted as renewable. For example, Pennsylvania includes waste coal in the state RPS. Ohio's Alternative Energy Resource Standard includes nuclear power and clean coal.

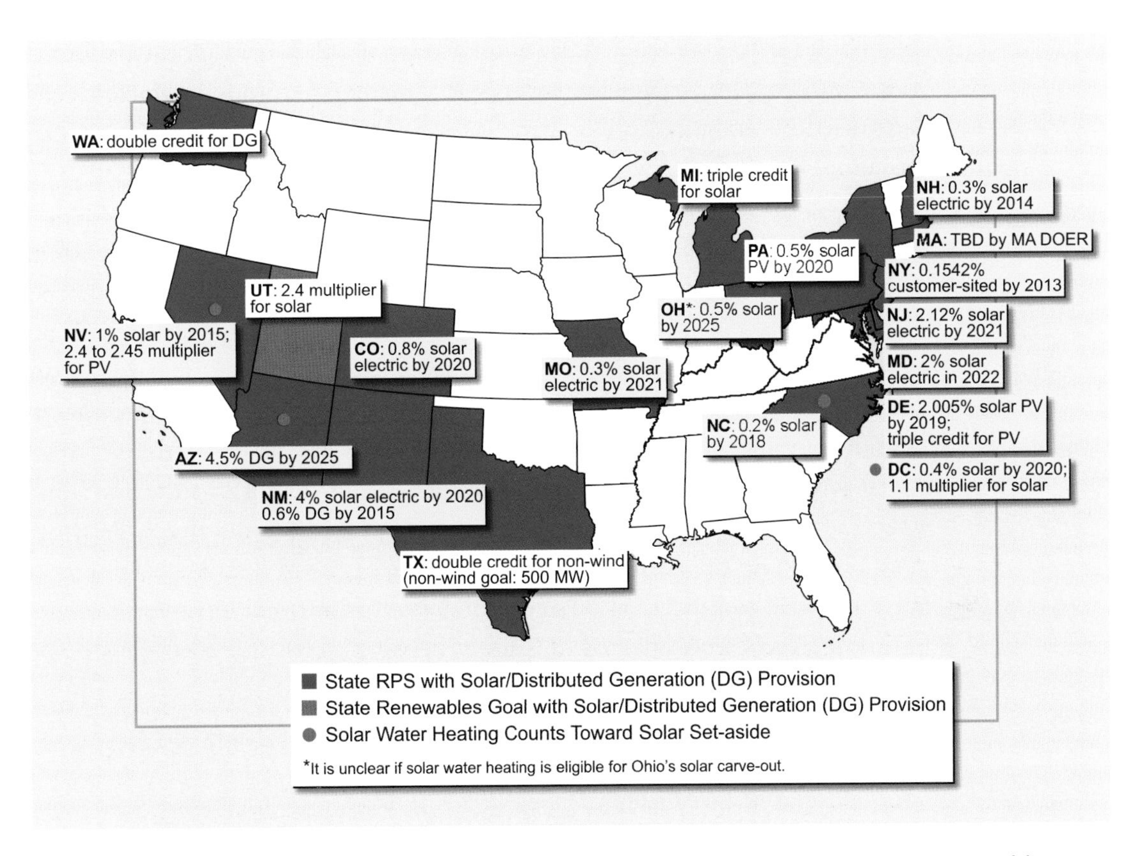

FIGURE 1.3 *Solar and distributed generation requirements within state renewables portfolio standards.*
Source: Database of State Incentives for Renewables and Efficiency, available at www.dsireusa.org. Courtesy of N.C. Solar Center at North Carolina State University and the Interstate Renewable Energy Council.

such facilities are thus highly speculative). Chapter 4 provides more details on the economic impacts of and market compliance strategies for RPSs.

Other State Policies

Other examples of state policies affecting renewable electricity generation include public benefit funds, net metering, green power purchasing agreements, tax credits, rebates, low-interest loans, and other financial incentives. Public benefit funds typically collect a small surcharge on electricity sales and specify that the funds so raised must be used for renewables. In 2004 such funds were investing more than \$300 million annually in renewable energy and are expected to collect more than

$4 billion for renewable energy cumulatively by 2017. An example from California is the program to subsidize rooftop PV systems for households and businesses, supported by the state's public benefit fund. Through California's Solar Initiative program, PV projects yielding 300 MW have been funded in 2007 and 2008 at a cost to California of $775 million in incentives, resulting in a total estimated project value of almost $5 billion considering private investments (CPUC, 2009). Net metering policies enable two-way power exchanges between a utility and individual homes and businesses—excess electricity generated by small renewable power systems installed in residences and businesses can be sold by the systems' owners back to the grid. Between 1996 and 2004, net metering policies were enacted in 33 states, bringing the total number of states with net metering to 39. Voluntary green power purchases allow consumers through a variety of state and utility programs to purchase electricity that comes from renewable resources. Between 1999 and 2004, more than 500 utilities in 34 states began to offer their retail customers the option to buy green power. Mandates that required utilities to offer green power products were enacted in 8 states between 2001 and 2007.[2]

Federal Policies

Production and Investment Tax Credits

Federal policies also contributed to the strong growth of renewables from the late 1990s onward. The major incentive for increasing electricity generation from renewable resources, particularly wind power, is the federal renewable electricity production tax credit. The PTC currently (in 2009) provides a 2.1¢ tax credit (originally passed as a 1.5¢ credit adjusted for inflation) for every kilowatt-hour of electricity generated in the first 10 years of the life of a private or investor-owned renewable electricity project. Originally established in the Energy Policy Act of 1992 for wind and closed-loop biomass plants brought on line between 1992 and 1993, respectively, the PTC was extended to January 1, 2002, and expanded to include poultry waste facilities in the Tax Relief Extension Act of 1999. The Economic Security and Recovery Act of 2001 included a 2-year extension of the PTC to 2004, and it was again extended in the Energy Policy Act of 2005 to apply through December 31, 2007. The PTC was extended further by the

---

[2]For information on the DOE Energy Efficiency and Renewable Energy (EERE) Green Power Network, see http://apps3.eere.energy.gov/greenpower.

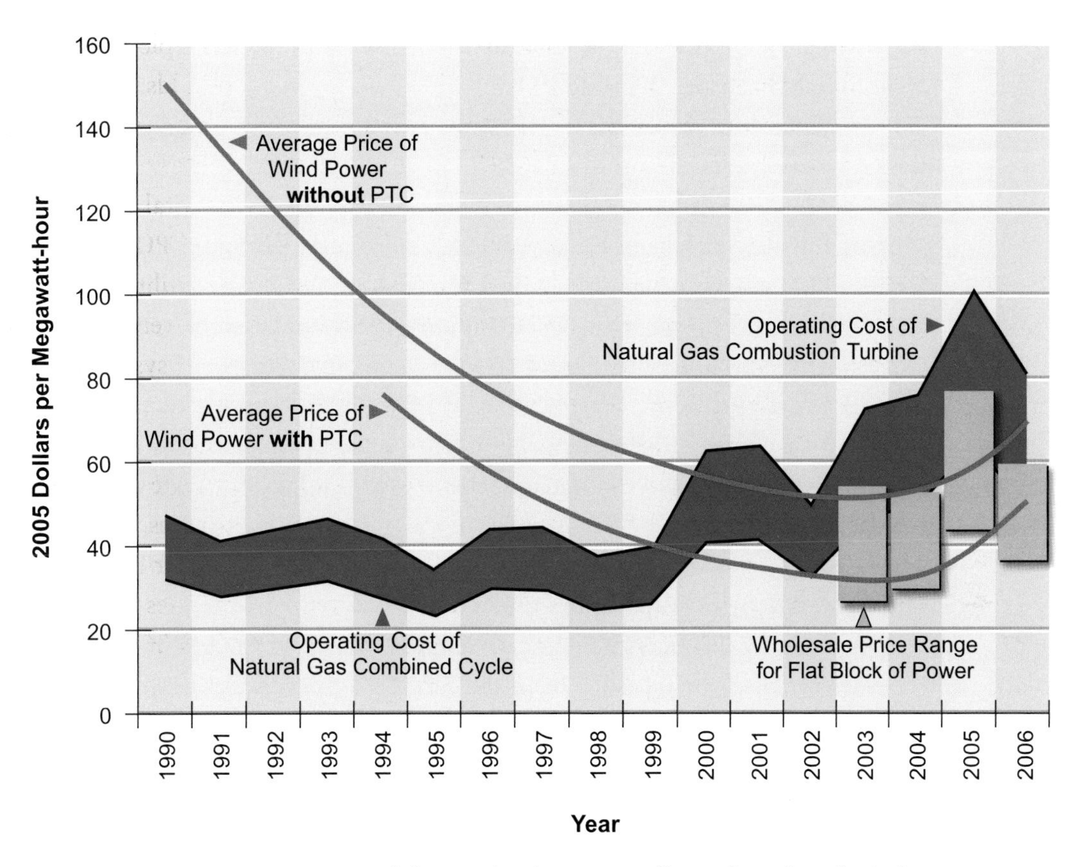

FIGURE 1.4 *Impacts of the production tax credit on the price of wind power compared to costs for natural-gas-fired electricity.*
Source: Wiser, 2008.

Tax Relief and Health Care Act of 2006 to apply through the end of 2008. The impact of the PTC on the competitiveness of wind power is shown in Figure 1.4.

Congress most recently extended the PTC and expanded incentives in the Emergency Economic Stabilization Act of 2008 and the American Recovery and Reinvestment Act (ARRA) of 2009. The 2008 bill added an 8-year extension (until 2016) of the 30 percent solar investment tax credit for commercial and residential installations and approved $800 million in bonds to help finance energy efficiency projects. The 2008 and 2009 bills together extend the PTC for wind through 2012 and the PTC for municipal solid waste, qualified hydropower, biomass, geothermal, and marine and hydrokinetic renewable energy facilities through 2013. Because of concerns that the current slowdown in business activity will reduce the capabilities of projects to raise investment capital, the ARRA allows owners of non-solar renewable energy facilities to elect a 30 percent investment tax credit rather than the PTC.

In contrast to the costs for RPSs, the costs of the PTC and other tax incentives for renewables are more straightforward to estimate, although there is some variability in the estimates.[3] The EIA estimates that the total federal subsidy and support for wind power in fiscal year 2007, primarily through the PTC, was $724 million, or approximately 2.3¢/kWh (EIA, 2008b). The estimate of the cost of the PTC alone ranges from $530 million to $660 million (EIA, 2008b). The Government Accountability Office (GAO) estimates that, from fiscal year 2002 through fiscal year 2007, revenue of $2.8 billion was foregone by the U.S. Treasury because of the Clean Renewable Energy bond tax credits, the exclusion of interest on energy facility bonds, and the new technology tax credits for renewable electricity production (the PTC) and renewable energy investment (GAO, 2007). The largest proportion of this expenditure was for the PTC and the much smaller renewable energy investment tax credit.

A study by GE Energy Financial Services examined the lifetime tax costs and revenues for the U.S. Treasury from the 5.2 GW of new wind power that came on line in 2007 (Taub, 2008). The study looked at both the costs of the PTC and the value of the accelerated depreciation allowed for wind power projects, and it offset those costs with revenues from increases in property taxes and other sources. It found that the lifetime costs of the PTC for the 5.2 GW of wind renewable electricity had a net present value in 2007 of $2.5 billion, which was offset by the estimated net present value of $2.75 billion obtained from taxes on the project and related economic activity. The largest source of revenue for the federal government from its investment in renewable electricity is the tax on project income, whereby the lifetime revenue stream is reduced to include the effect of 5-year Modified Accelerated Cost Recovery System depreciation.[4] Chapter 4 pro-

---

[3]Note that if the RPS policy includes tradable renewable energy credits (RECs, discussed in more detail in Chapter 4) then the price of the RECs provides a measure of the subsidy to renewable generators from the RPS program that is somewhat analogous to the cost to taxpayers of the PTC. However, not all RPS programs include tradable RECs. It should be noted that the real cost to the economy of either type of policy (RPS or PTC) is more complicated than either the cost of RECs or the value of the PTC.

[4]Several renewable technologies (wind, solar, geothermal, and small biomass generators) are also eligible for accelerated depreciation, which allows depreciation of their capital costs over 5 years instead of the 20-year lifetime depreciation for most fossil generators (15 years for new nuclear). This benefit allows project owners to reduce the taxes on income in the early years of

vides additional discussion of the PTC, including its impacts on new wind power generation.

Other Recent Initiatives

The ARRA offers other benefits for renewable electricity, including $2.5 billion for applied research, development, and deployment activities of the Department of Energy (DOE) Office of Energy Efficiency and Renewable Energy (EERE). This amount includes $800 million for the Biomass Program and $400 million for the Geothermal Technologies Program. Separate from the EERE portion is $400 million set-aside to establish the Advanced Research Projects Agency–Energy (ARPA–E) to support innovative energy research. The bill also includes $6 billion to support loan guarantees for renewable energy and electric transmission technologies, which is expected to guarantee more than $60 billion in loans. Finally, there is a significant focus on updating the nation's electrical grid. The ARRA budgeted a total of $11 billion to modernize the nation's electricity grid and required a study of the transmission issues facing renewable energy.

Current Policy Motivations

In the absence of a price on carbon, generating electricity from non-hydropower renewable resources generally is more expensive than generating electricity from coal, natural gas, or nuclear power at current costs. The exception recently has been wind power's competitiveness with electricity generated using natural gas. But there are other reasons that policy makers would choose to encourage research on, and development and deployment of, renewables. Greenhouse gas emissions from the combustion of fossil fuels are a growing concern. When burned to generate electricity, fossil fuels such as coal and to a lesser extent natural gas release large amounts of CO_2 and other greenhouse gases into the atmosphere. For example, according to the Energy Information Administration (EIA), energy-related CO_2 emissions from fossil fuel use in the United States amounted to almost 6000 million metric tons in 2007 (EIA, 2008a). The concentration of these gases in the atmosphere has very likely led to the increase in global average temperatures observed in recent decades. Increasing atmospheric concentrations of

---

operation. In addition, renewables may be eligible for a method of depreciation within the 5-year time period that allows depreciation of more than half of the investment value in the first 2 years of use.

CO_2 have been forecast to have a variety of impacts on the environment, including sea level rise, an increase in ocean acidification, and rapid changes in ecosystem ranges. In 2006, 69 percent of the electricity generated in the United States was produced by the combustion of fossil fuels (EIA, 2008a). Energy sources with low greenhouse gas emissions are an important component of strategies that aim to reduce or even maintain current levels of greenhouse gas emissions. Electricity generated from renewable resources in particular can contribute to this effort, because renewables can produce electricity without significant quantities of greenhouse gas emissions.

Another motivation for increasing the percentage of domestic electricity generated from renewables is energy security. Although 74 percent of the U.S. electricity generated from fossil fuels is produced from coal, an abundant resource in the United States, nearly all of the energy needed for the transportation sector is produced from oil (EIA, 2008a). Approximately 65 percent of the oil used in the United States is imported, often from politically unstable regions of the world (EIA, 2008a). Although this panel's report does not address the transportation sector, it is worth noting that with the advent of technologies such as electric and plug-in hybrid vehicles and concepts for using electricity from renewable resources to produce chemical fuels such as hydrogen, renewable electricity from a variety of sources has the potential in the long run to contribute to fueling the transportation sector.[5] Because the United States has some of world's most abundant solar, wind, biomass, and geothermal resources, renewables may help to secure supplies of domestic energy for all sectors.

Future Policy Era

Given the confluence of concerns over climate change and domestic energy security, as well as volatilities in energy prices, it is likely that over the next few years the United States will enter a new era of energy-related policymaking, including development of policies that will directly or indirectly affect production of electricity from renewable resources. Such concerns motivated the passage of the abovementioned Energy Independence and Security Act of 2007, which raised vehicle fuel economy standards for the first time in almost 30 years and mandated the use

---

[5]The use of domestic biomass to produce alternative liquid fuels for transportation is the subject of the report by the Panel on Alternative Liquid Transportation Fuels (NAS-NAE-NRC, 2009b). The relationship of that panel to the Panel on Electricity from Renewable Resources is discussed later in this chapter and is also shown in Appendix A.

of a large amount of biofuels for transportation; and the Emergency Economic Stabilization Act of 2008, which extended federal incentives for several kinds of renewable electricity and created additional incentives for solar and efficiency projects.

Several potential policy mechanisms might prove relevant to electricity generation from renewable resources. One such mechanism is a federal RPS, an approach that was considered for the Energy Independence and Security Act of 2007 but ultimately dropped from the legislation. There also have been recent initiatives with bipartisan support that have targeted U.S. greenhouse gas emissions. Policy options include a carbon tax or fee, under which electricity generators are required to pay a certain tax or fee per ton of CO_2 released to the atmosphere, and a cap-and-trade scheme, in which the government issues permits and sets a cap on the total amount of CO_2 that may be emitted. Under a cap-and-trade system, emitters could be allocated permits or required to purchase permits to cover their carbon emissions. Those who would need to increase their emissions might purchase credits either from those who have decreased their emissions or through some other market. The method of allocating permits can have major impacts on the deployment of renewables. Another method of distributing permits is for them to be auctioned to emitters and other participants. Another possible policy is the state or federal adoption of a carbon portfolio standard, which would require that all electricity suppliers meet an overall constraint on their carbon emissions rate. A carbon portfolio standard allows individual emitters to purchase low-carbon energy from any source and to seek out the lowest price. The role that renewable energy will play in any carbon regulatory system is unclear. Issues to be resolved include how RPSs are designed and integrated into cap-and-trade systems, whether generators using renewables will be issued allowances, and whether carbon caps will be sufficiently powerful to increase the markets for renewable energy in the near, mid, or long term.

CURRENT STATUS OF RENEWABLE ELECTRICITY GENERATION

U.S. Electricity Generation

The U.S. electricity sector generated 4.16 million GWh in 2007, almost 90 percent of which came from a combination of coal (49 percent), natural gas (21 percent), and nuclear (19 percent) facilities. Preliminary estimates for 2008 show a slight decline in total electricity generation to 4.12 million GWh (EIA, 2009). The compound annual growth rate for the 1999–2008 time period is about 1 percent (EIA, 2009).

The U.S. electricity sector's primary suppliers are more than 3000 utilities that operate under different market structures, depending on local and regional regulations.[6] In addition, more than 2000 other, non-utility, large power producers supply electricity to the grid. Traditionally, electricity was generated, transmitted, and distributed to users through vertically integrated utilities. However, efforts that began in the 1970s opened up electricity generation to more potential producers, and, since the early 1990s, many states have deregulated their electricity systems and have separated generation of electricity from its transmission and distribution. This shift has created different types of renewable electricity ownership structures and markets, which are described in more detail in Chapter 6. In general, the opening up of the electricity market can improve both the integration of renewables into the market and the ability to incorporate greater geographical diversity in the renewables mix.

In the late 1990s, the restructuring of the electricity sector led to a period of underinvestment in the electricity transmission system, principally due to uncertainty about the rate of return that would be allowed for investments in transmission (EPRI, 2004). This lapse created the present need to modernize the transmission and distribution system. It also has slowed the growth in transmission capacity needed to connect renewables. For example, California has 13,000 MW of potential solar projects waiting for approval to be connected to the grid as of January 2009 (AWEA/SEIA, 2009). As discussed above, the need to increase investment in the grid, including investments for renewables, began to be addressed in the ARRA of 2009.

U.S. Renewable Electricity

Renewables currently represent a small fraction of total U.S. electricity generation. The following statistics, including those for renewable electricity generation, come from the EIA (2008c). In total, renewable resources supplied 8.4 percent of

---

[6]The electric power sector includes electric utilities, independent power producers, and large commercial and industrial generators of electricity. A smaller amount of total electricity (approximately 4 percent) is generated by end users in the commercial, industrial, and residential sectors. Most of the end-user-generated electricity is consumed on-site, though a small amount may be sold to the electricity grid.

the total U.S. electricity generated, and non-hydroelectric renewables supplied 2.5 percent. Conventional hydroelectric power is the largest source of renewable electricity in the United States, generating 6.0 percent of the total electricity produced in 2007 by the U.S. electric power sector. Hydropower represents 71 percent of the electricity generated from renewable resources and in 2007 produced almost 250,000 GWh of electricity. Note that several state RPSs exclude hydropower as an acceptable renewable resource for meeting the state's target. Biomass electricity generation (biopower) is the second largest source, generating 55,000 GWh in 2007, corresponding to 16 percent of generation from renewables.[7] Biomass is unique because 52 percent of all biomass electricity generation comes from the industrial sector as opposed to the electric power sector.

Both hydropower and biomass have not grown much in terms of generation or generation capacity since 1990. Hydropower production, which is linked to widespread hydrologic conditions that can vary from year-to-year, dropped from a high of 356,000 GWh in 1997 to 216,000 GWh in 2001. Electricity generation from hydropower in 2007 was essentially the same as it was in 1992 (253,000 GWh), and hydropower generating capacity has remained generally constant since 1990. Electricity generation from biomass grew at an annual average rate of 1.1 percent from 1990 to 2006. Potential ecological concerns over existing hydropower plants, along with the 2007 Energy Independence and Security Act's mandates for biofuels for transportation, have led to uncertainty about whether either hydropower or biopower will yield greatly increased electrical generation in the foreseeable future.

Wind technology has progressed over the last two decades, and wind power has accounted for an increasing fraction of electricity generation in the United States. Although it now represents only about 1 percent of total U.S. electricity generation, wind power has grown at a 14 percent compound annual growth rate from 1990 to 2006 and at a 23 percent compound annual growth rate from 1997 to 2006. In 2007, wind power supplied more than 32,000 GWh of electricity, almost 5,500 GWh more than it had the year before (EIA, 2008a). EIA's preliminary estimate puts wind power electricity generation in 2008 at more than 52,000 GWh (EIA, 2009). An additional 5,200 MW of wind power generation capacity was installed in 2007, which represented 35 percent of all new generating capacity. Data for 2008 indicate that wind power generating capacity increased by more

---

[7]Biomass electricity generation includes electricity generated using wood and wood waste, municipal solid waste, landfill gases, sludge waste, and other biomass solids, liquids, and gases.

than 8,400 MW, breaking the record set in 2007 for largest annual installed wind power capacity (AWEA, 2008, 2009).[8] The growth in generating capacity was particularly strong in the western United States. Texas, the leader in U.S. wind power generation, added 2,760 MW of new capacity in 2008, for a total wind power generation capacity of 7,116 MW. Iowa more than doubled its wind capacity in 2008 by installing 1,517 MW on top of its 1,273 MW capacity existing at the end of 2007. Minnesota added 454 MW of new wind capacity, and Minnesota and Iowa were the states with the highest fraction of total electricity generation from wind power in 2007 (AWEA, 2008, 2009). However, there are issues that must be addressed related to the intermittency of wind as a renewable resource, such as the maintenance of a readily dispatchable source of power to compensate for times when wind power is not available. Issues related to intermittency and integrating renewables into the electricity grid are discussed further in later chapters, including Chapter 3 (technologies for grid integration), Chapter 4 (cost of renewables integration), and Chapter 6 (case studies of wind integration).

Concentrating solar power (CSP) and photovoltaic (PV) electricity generation by the electricity sector combined to supply 500 GWh in 2006 and 600 GWh in 2007, which constitutes 0.01 percent of total U.S. electricity generation. EIA data indicate that the compounded annual growth rate in net U.S. generation from solar was 1.5 percent from 1997 to 2007 (EIA, 2008a). That estimate, however, does not account for the growth in electricity generation by residential and other small PV installations, the sector that has displayed the highest growth rate for solar electricity.[9] Including these other sources, installations of grid-tied and off-grid solar PV in the United States have grown at a compounded annual growth rate of about 30 percent from 2000 to 2008 (Cornelius, 2007; Sherwood, 2008; SERI, 2009), although the total on-grid and off-grid generation capacity in 2008 is still fairly small (~1,000 MW).[10]

---

[8]If one assumes a 35 percent capacity factor—the fraction of time the technology is producing electricity or energy—the added total annual generation for 2008 would be more than 25,000 MWh.

[9]As noted by the EIA (2008a), electricity generation from CSP and PV was estimated for electric utilities, independent power producers, commercial electricity plants, and industrial plants only.

[10]For intermittent renewables such as solar and wind, quoting additions in generating capacity can be misleading since capacity factors—the fraction of time the technology is producing electricity or energy—can be low for renewables (approximately 10–25 percent for PV). However, for residential and other small PV installations that do not contribute electricity measured on the grid, capacity is a primary metric for assessing growth.

Geothermal heat represents the other major source of renewable electricity, generating 14,800 GWh of electricity in 2007 in the United States. According to EIA estimates (EIA, 2008a), electricity production from geothermal sources was larger than that from wind power as recently as 2003. However, the growth in geothermal electricity generation has been relatively flat since 1990, and geothermal electricity generation is now smaller than wind- and biomass-based U.S. electricity generation.

International Renewable Electricity

Renewable resources such as hydropower and geothermal energy have long been a major component of many countries' electricity sectors. Recently, electricity generation from solar and wind power has been expanding rapidly in parts of Europe and has also been emerging elsewhere. In particular, Germany and Spain have used aggressive feed-in tariffs to rapidly increase wind and solar electricity generation.[11] Because the tariff is resource-specific, solar PV can be as profitable to electricity generators as wind power. From 1998 to 2006 in Germany, the share of electricity generation from renewable resources increased from 4 percent to 14 percent—7 percent from wind, 1 percent from solar, and 6 percent from hydropower (Luther, 2008). In 2006, Germany produced approximately 31,000 GWh of electricity from wind and 2,200 GWh from solar PV (IEA, 2008). Spain produces 18 percent of its electricity demand from renewable resources, including 9.7 percent from hydropower and 7.6 percent from wind. Wind power in Spain generated more than 23,000 GWh of electricity in 2006, increasing from 6,500 GWh in 2000 (IEA, 2008). Denmark has the highest fraction of electricity generation from wind, 18.2 percent in 2005, for a total of 6,600 GWh. The high fraction from wind is aided by the interconnection of Denmark's power grid with that of Sweden, Norway, and Germany (Sharman, 2005): a large amount of available hydropower in Sweden and Norway can be adjusted rapidly to balance the variable output from Denmark's wind turbines.[12] The connection between these countries serves as an electricity sink at times of high wind generation and a source at

---

[11]The feed-in tariff is an electricity pricing law under which renewable electricity generators are paid at a set rate over a given period of time (Mendonca, 2007). The rates are differentiated by facility size and resource, and are set by a federal agency to ensure profitable operations.

[12]Because much of Denmark's electricity generation from wind replaces generation from hydropower, the benefits from reduced emissions of carbon and other pollutants are not as large as if wind power generation had replaced generation from fossil fuels.

times of low wind generation. In Spain, integration into the electricity grid of a sizable fraction of wind power is supported by a large excess-generation capacity that protects system reliability and by large hydroelectric plants that provide 18 percent of all generation capacity.

The growth of electricity generated from renewable resources, in particular in Europe and Asia, indicates increasing interest in moving away from carbon-based energy sources. Countries also view renewables in terms of their economic potential. Although its focus is electricity generation from renewable resources in the United States, this panel recognizes international activities in renewable electricity as important sources of experience that can benefit U.S. applications. International activities are also important because several of the companies involved in the development of domestic renewables projects or supplying the components for such projects are international companies. Thus, decisions on where to install wind power projects or where to locate manufacturing facilities are global decisions. For example, the wind turbine manufacturer with the largest U.S. market share is GE, but its share has decreased from 60 percent in 2005 to 44 percent in 2007, with a concomitant increase in the market share held by foreign-owned companies (DOE, 2008). Because of the cost of shipping wind turbines and the expected growth in installed capacity, several major global vendors have established new manufacturing or assembly facilities in the United States in conjunction with an increasingly stable regulatory environment. In terms of global manufacturing, almost 16 percent of wind turbines in 2006 were built in the United States; only Denmark, Germany, and Spain had a larger share of the manufacturing base (IEA, 2008). Thus, it is important to recognize that renewable electricity projects in the United States must compete in an international market for skilled labor, equipment, materials, and capital.

Private Investments

Private investment is essential for the deployment of renewable electricity on a scale that would significantly reduce carbon emissions and increase domestic production of low-carbon sources. Although federal funds can help enable basic research and development, renewable electricity must compete in the electricity market and must attract private capital to expand significantly. In 2007, $150 billion was invested in renewables worldwide, by many financial sectors, mostly in wind and solar PV. Figure 1.5, which indicates the level of investment in wind, solar, and biofuels projects in the United States since 2001, shows a 34-fold

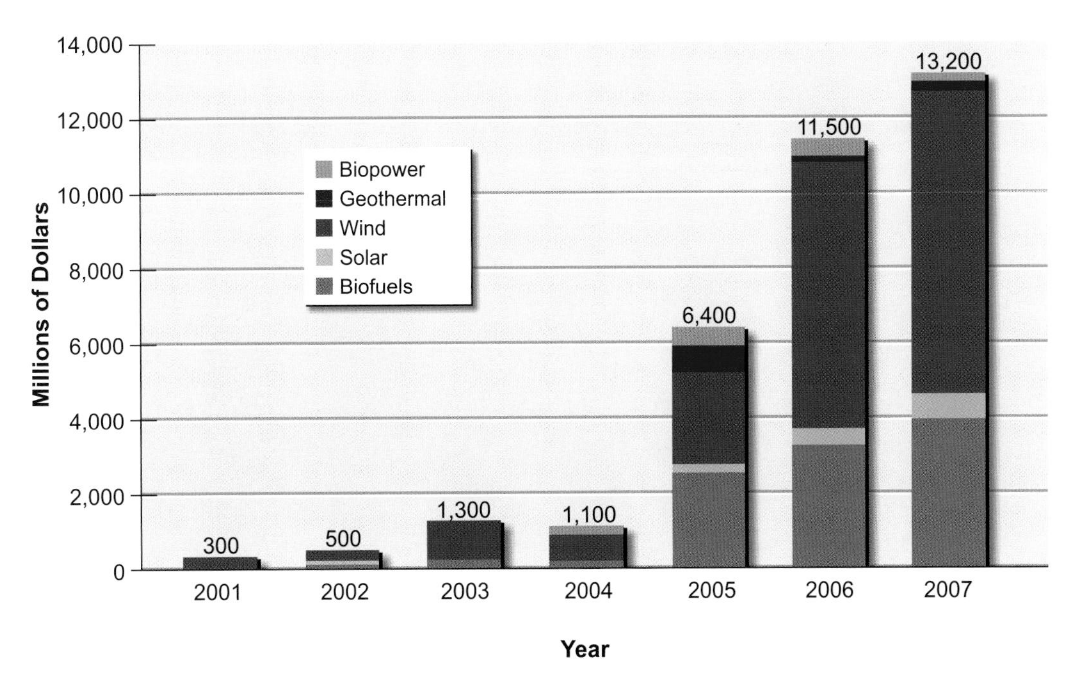

FIGURE 1.5 *Annual private investments in wind, biofuels, and solar power. Source: DOE/EERE, 2008.*

increase in investment, as reported by DOE/EERE (2008). This quotes information collected by New Energy Finance. Annual U.S. private investments have increased from $300 million in 2001 to $12 billion in 2007 (DOE/EERE, 2008). The largest, in wind power, totaled almost $8 billion in 2007. One forecast has investment in wind increasing to a cumulative total of $65 billion over the period from 2007 to 2015 (Emerging Energy Research, 2007).

Among the groups financing the clean technologies sector, venture capital firms have shown an especially strong interest. Representing a small fraction of all private investment, venture capital firms typically invest in small companies with high growth potential, such as start-up companies that are either too small to raise capital in public markets or too immature to obtain bank loans. Venture capital firms hope for large financial returns and successful exit events by going public or selling to large firms within a timeframe typically of 3–7 years. Investment numbers vary widely depending on who performs the analysis, but all sources have

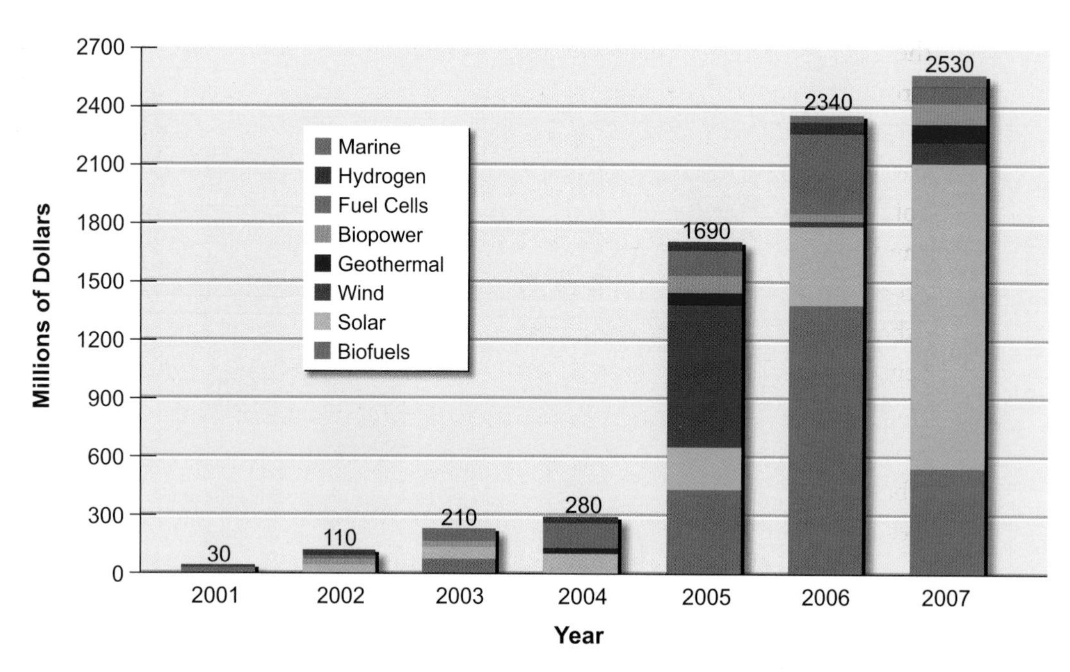

FIGURE 1.6 *Annual venture capital investment in wind, biofuels, and solar. Source: DOE/EERE, 2008.*

reported a sharp increase in venture capital investment in renewables.[13] Figure 1.6 shows that the venture capital investment in wind, solar, biofuels, and energy efficiency projects in the United States had increased 13-fold since 2001. According to the study by New Energy Finance, quoted by the DOE/EERE (2008), the two front-runners in recent years have been solar PV and energy efficiency technology companies, which each secured $1 billion in venture capital investment. This increasing trend of investment in clean energy projects continued in 2008, although recent constraints in credit have caused concern that investment capital for big renewable energy projects will tighten. A recent report by Dow Jones VentureSource found that, despite a 12 percent decrease in total venture capital investments in the second quarter of 2008, there was a strong increase in investment in energy and utility industries, with a total investment of $817 million, which represents an increase of 160 percent compared with the second quarter of 2007.[14] Of

[13]Investment keeps growing. Greentech Media. December 31, 2007. Available at http://www.greentechmedia.com/articles/the-green-year-in-review-444.html.

[14]Quarterly U.S. venture capital report. Dow Jones VentureSource. Available at http://www.venturecapital.dowjones.com.

the $817 million, $650 million was invested in renewable energy projects, with a strong focus on solar PV projects.

Some financial experts see a potential downside to venture capital firms' strong interest in renewable energy[15]—the timeframe in which start-ups can become profitable may not correlate well with the time required to make renewable energy companies commercially profitable. Programs such as the "entrepreneur-in-residence" program[16] between DOE and Kleiner, Perkins, Caufield, and Byers have been established as part of an effort to prevent this potential obstacle to investment, by using venture capital firms to help move clean energy technologies out of the national energy laboratories. The venture capital firms provide the early-stage investments to new start-up companies that are assisted by technology experts from the national laboratories. The program's objective is to increase the chances that new technologies will become commercially profitable.

REFERENCE CASE PROJECTION OF FUTURE RENEWABLE ELECTRICITY GENERATION IN THE UNITED STATES

Understanding how renewables fit into and compete in the wider electricity sector is critical for understanding the future of renewables and assessing the potential consequences of their large-scale deployment. One approach to understanding the electricity market—and thus gaining some perspective on the ability of renewable electricity technologies to compete with fossil-fuel and nuclear electricity—is offered by models, including energy-economic models. Such a perspective is important because the future of renewable electricity will depend largely on the ability of renewable electricity technologies to compete with fossil-fuel and nuclear electricity. It is also important to consider the extent to which a policy might affect energy demand. Models can demonstrate the potential impacts of demographic, economic, or regulatory factors on the use of renewable electricity within a framework that accounts for how such factors interrelate with use of all sources of electricity and with energy demand.

---

[15]"Dirty side to clean energy investing: Renewable investments have tripled since 2002, but is quick cash really what the sector needs?" CNN Money, March 27, 2007.

[16]National Laboratory Entrepreneur-in-Residence Program: Questions and Answers. DOE Energy Efficiency and Renewable Energy (EERE). Available at http://www1.eere.energy.gov/site_administration/entrepreneur.html.

However, such models are not predictors of the future, and hence the results of such models are not forecasts. Energy-economic models, as with all complex models, should not be confused with reality, or taken as prognosticators of the future (Holmes et al., 2009; NRC, 2007).

The EIA provides detailed projections of energy supply, demand, and prices through 2030, including for individual renewables within the electricity sector. Its most recent reference case is AEO 2009 Early Release (EIA, 2008d). The forecast is developed with the National Energy Modeling System (NEMS), an energy sector model with a high degree of detail that captures market feedbacks among various individual elements of the energy sector. AEO 2009 provides one scenario for the future of renewable electricity, albeit one used in a wide array of policy and technical settings. It assumes current policy conditions and thus does not take into account the potential for further energy- and climate-related initiatives. Updated annually, the EIA reference case is a moving reference, with the most recent forecast being more optimistic for renewables than was AEO 2008 (EIA, 2008d). It is important to note that the reference case estimate for renewable energy growth has changed significantly over the years, as Table 1.1 indicates.

In comparison with AEO 2008, AEO 2009 simulates an increase in the percentage of U.S. non-hydropower renewable electricity generation. As shown in Table 1.2, AEO 2008 estimated that by 2030 about 13 percent of all electricity generation would be from renewable resources, with only about 7 percent from non-hydropower renewables. AEO 2009 estimates that renewables will generate 14 percent of all U.S. electricity and that 8 percent will be generated from non-

TABLE 1.1 Predicted Annual Growth Rates of U.S. Non-hydropower Renewable Energy Generation

AEO Report Publication Year	Years	Predicted Annual Growth Rate (%)
2003	2001–2025	2.1
2004	2002–2025	4.2
2005	2003–2025	3.6
2006	2004–2025	4.2
2007	2005–2030	3.4
2008	2006–2030	5.1
2009	2007–2030	6.4

Source: EIA AEO reports published each year between 2003 and 2009. See also http://invisiblegreenhand.blogspot.com/2007/12/eia-2008-annual-energy-outlook.html.

TABLE 1.2 AEO 2009 Estimated Percentage of Overall U.S. Electricity Generation from Renewable Resources and Non-hydropower Renewable Resources, 2007–2030

	2007	2010	2020	2030
Total from renewable resources	8.5 (9.1)	10.7 (10.7)	13.3 (12.4)	14.1 (12.6)
Total from non-hydropower renewable resources	2.5 (2.8)	4.3 (3.9)	6.7 (6.1)	8.3 (6.8)

Note: The values estimated by AEO 2008 are shown in parentheses.
Source: EIA, 2008d,e.

hydropower renewables. Table 1.3 shows that AEO 2009 continues to see growth for both solar and wind, with solar growing at an annual average rate of more than 13 percent until 2030 and wind growing at almost 6 percent. Most of these values represent an increase over the estimates of AEO 2008, which simulated a smaller increase in the fraction of electricity generation from renewables and non-hydropower renewables. The main reason for the change in estimates between AEO 2008 and AEO 2009 is that additional state RPSs were taken into account in AEO 2009 that had not yet been passed when AEO 2008 was published. This difference demonstrates how reference case projections can change over time owing

TABLE 1.3 AEO 2009 Estimate of Electricity Generation from Renewable Resources (billion kilowatt-hours)

	Year				Annual Growth Rate
	2007	2010	2020	2030	2007–2030 (%)
Conventional hydropower	250 (260)	270 (293)	300 (301)	300 (301)	0.8 (0.6)
Geothermal heat	15 (16)	18 (18)	19 (24)	21 (31)	1.5 (2.9)
Municipal waste	16 (17)	21 (22)	22 (22)	23 (22)	1.5 (1.1)
Biomass	39 (41)	56 (53)	160 (135)	230 (172)	8.1 (6.4)
Solar (photovoltaic plus thermal)	1.3 (1.7)	3.9 (2.4)	18 (4.4)	23 (7.7)	13.3 (6.9)
Wind	32 (38)	81 (74)	94 (101)	130 (124)	6.2 (5.2)
Total from renewable resources	350 (380)	450 (461)	620 (587)	730 (658)	3.2 (2.5)
Total from non-hydropower	100 (110)	180 (169)	320 (286)	430 (356)	6.4 (5.1)
Total electricity generation (all sources)	4200 (4200)	4200 (4300)	4600 (4700)	5200 (5200)	0.9 (1.0)

Note: Data from AEO 2008 are shown in parentheses.
Source: EIA, 2008d,e.

to changes in policy and other factors. In addition, although both AEO 2008 and AEO 2009 predict significant growth in electricity generation from biomass, mandates under the Energy Independence and Security Act of 2007 have led to uncertainty about whether such growth will occur if the majority of the biomass resource base is devoted to the production of liquid fuels.

Overall, AEO 2009 estimates that electricity generation will rise at an annual growth rate of 0.9 percent, down from the 1.0 percent growth rate projected in AEO 2008. Table 1.4 indicates that this increase will not occur evenly across the United States and that growth in generation capacity within a region may not be the same as growth in electricity demand. AEO 2009 does not give projections at the state level but shows aggregated renewable electricity generation by region as a result of individual state RPSs, as seen in Figure 1.7. A significant portion of the qualifying renewables capacity in the Midwest, Northeast, Southwest, and Pacific Northwest is expected to come from wind. In the Mid-America Interconnected Network, 11,000 MW of wind capacity is expected in 2030, up from 220 MW in 2006. The majority of the new biomass capacity between 2006 and 2030 is

TABLE 1.4 AEO 2009 Estimated Annual Average Electricity Growth Rates from 2007 to 2030 by Region

	Growth in Electricity Demand (%)	Growth in Electricity Generation (%)
East Central Area Reliability Coordination (ECAR)	0.7 (0.7)	0.7 (0.6)
Electric Reliability Council of Texas (ERCOT)	1.1 (1.2)	1.1 (1.1)
Mid-Atlantic Area Council (MAAC)	0.9 (0.8)	1.0 (1.0)
Mid-America Interconnected Network (MAIN)	0.7 (0.6)	1.0 (0.8)
Mid-Continent Area Power Pool (MAPP)	0.7 (0.6)	1.6 (0.9)
Northeast Power Coordinating Council/NewYork (NY)	0.5 (0.5)	0.4 (0.5)
Northeast Power Coordinating Council/New England (NE)	0.6 (0.6)	1.0 (1.0)
Florida Reliability Coordinating Council (FL)	1.4 (1.6)	1.5 (2.2)
Southeastern Electric Reliability Council (SERC)	0.9 (1.2)	0.8 (0.9)
Southwest Power Pool (SPP)	0.9 (1.0)	0.4 (0.9)
Western Electricity Coordinating Council/ Northwest Power Pool Area (NWP)	1.0 (1.1)	0.9 (1.4)
Western Electricity Coordinating Council/ Rocky Mountain Power Area, Arizona, New Mexico, Southern Nevada Power Area (RA)	1.2 (1.5)	1.4 (1.5)
Western Electricity Coordinating Council/ California (CA)	0.9 (1.1)	1.2 (0.9)

Note: Data from AEO 2008 are shown in parentheses.
Source: EIA, 2008d,e.

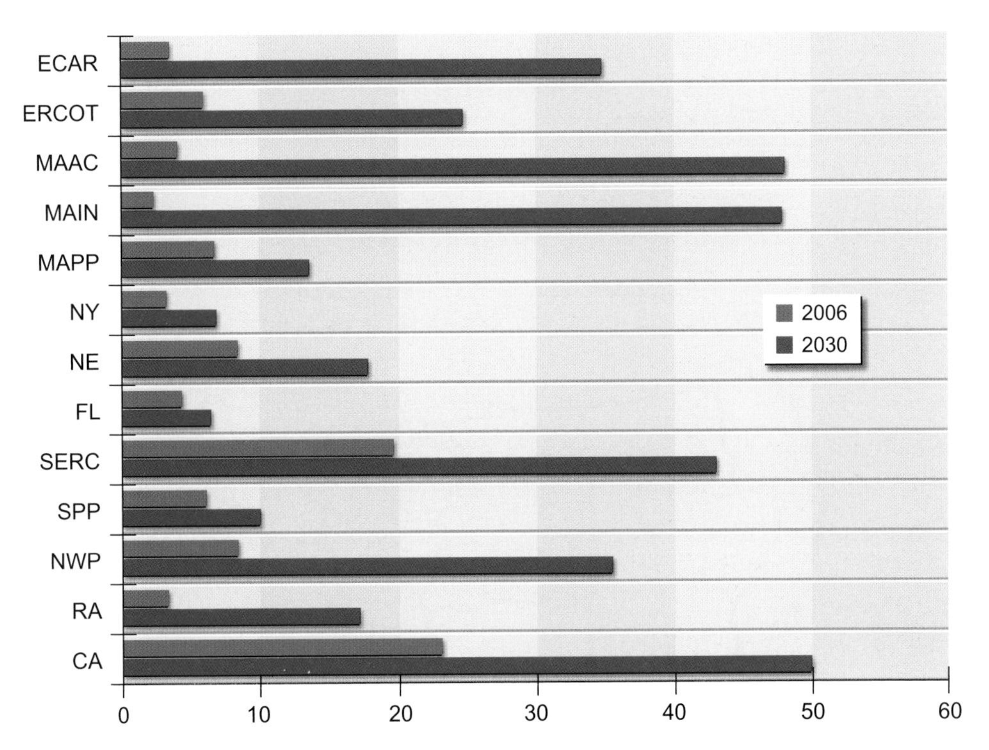

FIGURE 1.7 *Regional growth in nonhydroelectric renewable electricity generation, 2006–2030, in gigawatt-hours. Acronyms are defined in Table 1.4.*
Source: EIA, 2007.

projected to come from the Mid-Atlantic region (EIA, 2008d). Investment in solar power is expected to grow most significantly in Texas and California, especially given California's Solar Initiative (REPP, 2005). The regional distribution of the renewable resource base (see figures in Chapter 2) will be a guiding factor in the regional growth of renewable electricity generation. The existing regional variation in electricity generation can also be seen in Figure 1.8, which shows the different fuel mixes used for generating electricity in different parts of the country.

ISSUES OF SCALE

For electricity generation from renewable resources to fulfill a significant fraction of total U.S. electricity consumption, renewables need to be manufactured, deployed, and integrated into the electricity system on a much greater scale than they are today. Scaling up involves issues that go beyond the readiness of the

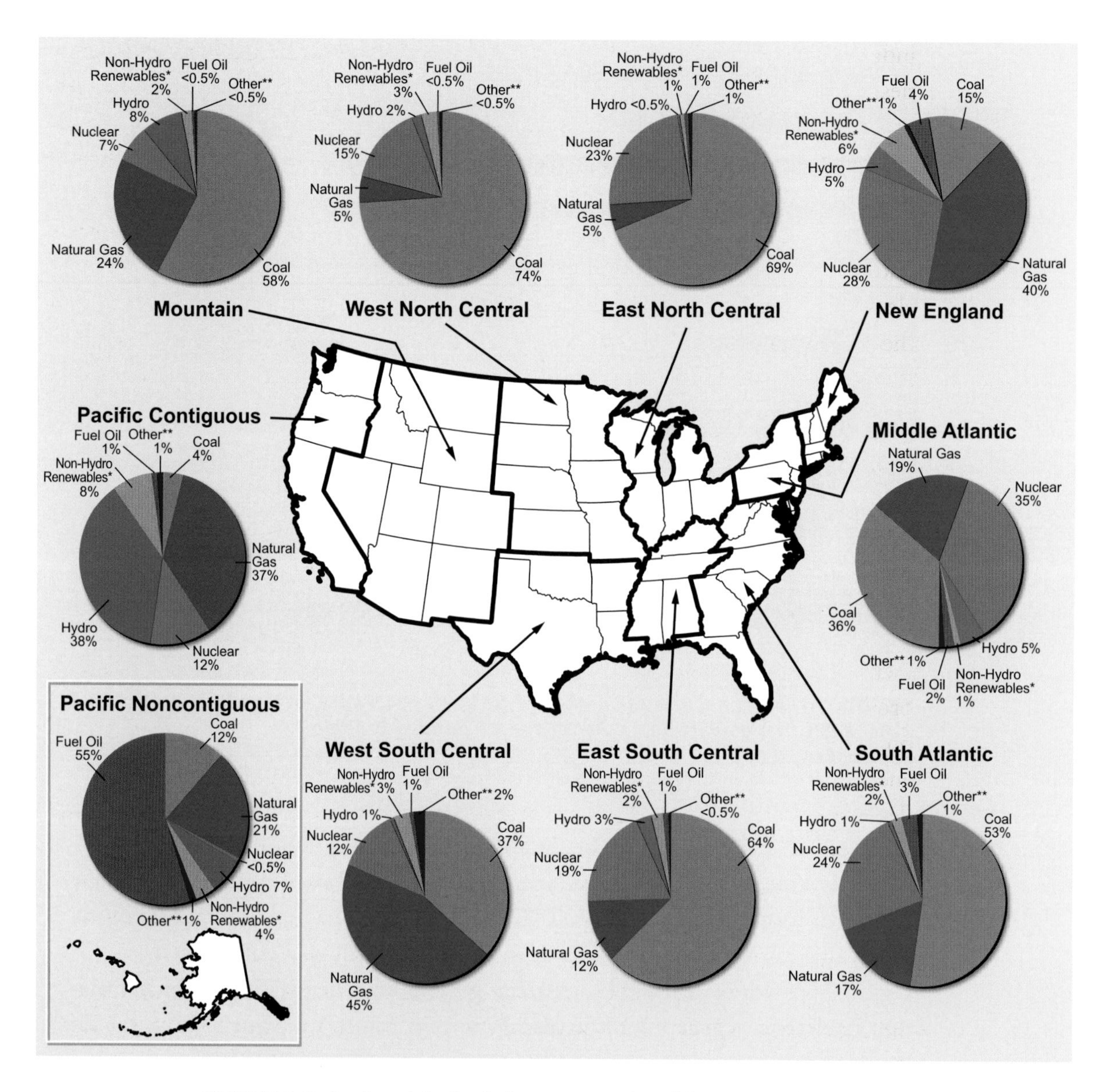

FIGURE 1.8 *Regional fuel mix for current electricity generation.*
Source: Edison Electric Institute, 2008.

individual renewable technologies, namely, issues related to manufacturing capacities, raw materials availability, workforce training and certification, and a host of other factors, including environmental effects. Issues that are related to the need to greatly expand the scale of renewable deployment will be discussed throughout the report. The final chapter of this report (Chapter 7) provides a quantitative discussion of the manufacturing, implementation, economics, and environmental issues and impacts associated with an increased level of deployment of renewable electricity. In general, the panel considers it critical that the reader have a sense of the scale issues associated with potentially achieving an aggressive but attainable level of renewable electricity deployment.

APPROACH AND SCOPE OF THIS REPORT

The panel's charge was to examine the technical potential for electric power generation from renewable resources such as wind, solar photovoltaic, geothermal, solar thermal, and hydroelectric power (see this report's preface for the full statement of task). In keeping with the overall plan for the America's Energy Future project (see Appendix A), the panel did not attempt to develop recommendations on policy choices but focused instead on characterizing the status of renewable energy technologies for power generation, especially technologies with initial deployment times of less than 10 years. In this report the panel also addresses the challenges of incorporating such technologies into the power grid; the potential for improvements in the electricity grid that could enable better and more extensive use of renewable technologies both in grid-scale applications and distributed at or near the customer's point of use; and potential storage needs.

The panel organizes its report around broad topics that are relevant for each individual source. Thus, the body of the report is organized around the topics of the resource bases, technologies, economics, impacts, and deployment. By necessity, much of the discussion addresses the technology readiness, costs, and impacts of individual renewable electricity sources. In this regard, the report's "storyline" could read like a puzzle, because each renewable (solar, wind, geothermal, biomass, and hydropower) has its own characteristic resource base, technology readiness, economics, and impacts. Solar electricity, for example, has the largest resource base and some well-developed technologies for tapping it but is still relatively expensive compared to other renewable electricity sources. However, the organization of the report emphasizes the degree to which these renewables share some common considerations. The report's discussion of the U.S. resource base (Chapter 2), technologies (Chapter 3), economics (Chapter 4), impacts (Chapter 5), and deployment (Chapter 6) is intended to present an integrated picture of renewables rather than snapshots of the individual renewable electricity sources. A quantitative discussion of issues related to accelerated deployment of renewables (Chapter 7) augments the more qualitative discussions presented in the preceding chapters.

The panel did not examine renewable energy for heating and hot water applications, which are considered in the upcoming report of the AEF Committee (NAS-NAE-NRC, 2009a). And although the panel devoted significant effort to considering the integration of renewables into the electricity grid, the full spectrum of issues and needs associated with the future of the electricity transmission and distribution systems falls under the purview of the Electric Power Transmission and Distribution subgroup of the AEF Committee (see Figure A.1 in Appendix A). The role that energy efficiency might play in the energy system and how efficiency might impact renewables are likewise not examined by this panel; they are addressed instead by the AEF Panel on Energy Efficiency in its upcoming report (NAS-NAE-NRC, 2009c). Similarly, the use of biofuels, such as corn and cellulosic ethanol, as alternative transportation fuels is not discussed by the present panel but instead is examined in the forthcoming report of the AEF Panel on Alternative Liquid Transportation Fuels (NAS-NAE-NRC, 2009b).

REFERENCES

AWEA (American Wind Energy Association). 2008. Wind Power Outlook 2008. Washington, D.C.

AWEA. 2009. Wind energy grows by record 8,300 MW in 2008. Press release, January 27. Washington, D.C.

AWEA/SEIA (AWEA/Solar Energy Industries Association). 2009. Green Power Superhighways: Building a Path to America's Clean Energy Future. Washington, D.C.

Cornelius, C. 2007. DOE solar energy technologies program. Presentation at the first meeting of the Panel on Electricity from Renewable Resources, September, 18, 2007. Washington, D.C.

CPUC (California Public Utilities Commission). 2009. California Solar Initiative CPUC Staff Progress Report. San Francisco.

Dobesova, K., J. Abt, and L. Lave. 2005. Are renewables portfolio standards cost-effective emissions abatement policy? Environmental Science and Technology 39:8578-8583.

DOE (U.S. Department of Energy). 2008. Annual Report on U.S. Wind Power Installation, Cost and Performance Trends: 2007. Washington, D.C.

DOE/EERE (DOE/Energy, Efficiency, and Renewable Energy). 2008. Renewable Energy Data Book. Washington, D.C.

EEI (Edison Electric Institute). 2008. Different regions of the country use different fuel mixes to generate electricity. Preliminary 2007 data. Available at http://www.eei.org/ourissues/ElectricityGeneration/FuelDiversity/Documents/diversity_map.pdf.

EIA (Energy Information Administration). 2008a. Annual Energy Review 2007. Washington, D.C.: U.S. Department of Energy, EIA.

EIA. 2008b. Federal Financial Interventions and Subsidies in Energy Markets 2007. Washington, D.C.: U.S. Department of Energy, EIA.

EIA. 2008c. Renewable Energy Annual, 2006. Washington, D.C.: U.S. Department of Energy, EIA.

EIA. 2008d. Annual Energy Outlook 2009 Early Release. DOE/EIA 0383(2009). Washington, D.C.: U.S. Department of Energy, EIA.

EIA. 2008e. Annual Energy Outlook 2008. DOE/EIA 0383(2008). Washington, D.C.: U.S. Department of Energy, EIA.

EIA. 2009. Electric Power Monthly. Washington, D.C.: U.S. Department of Energy, EIA.

Emerging Energy Research. 2007. U.S. Wind Power Markets and Strategies, 2007-2015. Cambridge, Mass.

EPRI (Electric Power Research Institute). 2004. Power Delivery System of the Future: A Preliminary Study of Costs and Benefits. Palo Alto, Calif.

GAO (General Accountability Office). 2007. Federal Electricity Subsidies: Information on Research Funding, Tax Expenditures, and Other Activities That Support Electricity Production. Washington, D.C.

Holmes, K.J., J.A. Graham, T. McKone, and C. Whipple. 2009. Regulatory models and the environment: Practice, pitfalls, and prospectus. Risk Analysis 29(9):159-170.

IEA (International Energy Agency). 2008. Key World Energy Statistics. Paris.

Luther, J. 2008. Renewable energy development in Germany. Presentation at the NRC Christine Mirzayan Fellows Seminar, March 5, 2008. Washington, D.C.

Martinot, E., R.Wiser, and J. Hamrin. 2005. Renewable Energy Policies and Markets in the United States. Prepared for the Energy Foundation's China Sustainable Energy Program, Center for Resource Solutions. San Francisco.

Mendonca, M. 2007. Feed-in Tariffs: Accelerating the Development of Renewable Energy. London: Earthscan.

NAS-NAE-NRC (National Academy of Sciences-National Academy of Engineering-National Research Council). 2009a. America's Energy Future: Technology and Transformation. Washington, D.C.: The National Academies Press.

NAS-NAE-NRC. 2009b. Liquid Transportation Fuels from Coal and Biomass: Technological Status, Costs, and Environmental Impacts. Washington, D.C.: The National Academies Press.

NAS-NAE-NRC. 2009c. Real Prospects for Energy Efficiency in the United States. Washington, D.C.: The National Academies Press.

NRC (National Research Council). 2007. Models in Environmental Regulatory Decision Making. Washington, D.C.: The National Academies Press.

Palmer, K., and D. Burtraw. 2005. Cost effectiveness of renewable energy policies. Energy Economics 27:873-894.

REPP (Renewable Energy Policy Project). 2005. Solar PV Development: Location of Economic Activity. Washington, D.C.

Rickerson, W., and R. Grace. 2007. The debate over fixed price incentives for renewable electricity in Europe and the United States: Fallout and future directions. White paper prepared for the Heinrich Böll Foundation. Washington, D.C.

SERI (Solar Energy Industries Association). 2009. U.S. Solar in Review 2008. Washington, D.C.

Sharman, H. 2005. Why wind power works for Denmark. Civil Engineering 158:66-72.

Sherwood, L. 2008. U.S. Solar Market Trends 2007. Latham, N.Y.: Interstate Renewable Energy Council.

Taub, S. 2008. Impact of 2007 Wind Farms on U.S. Treasury. Stamford, Conn.: GE Energy Financial Services.

Wiser, R. 2008. The development, deployment, and policy context of renewable electricity: A focus on wind. Presentation at the fourth meeting of the Panel on Electricity from Renewable Resources, March 11, 2008 Washington, D.C.

Wiser, R., and G. Barbose. 2008. Renewables Portfolio Standards in the United States: A Status Report with Data Through 2007. Berkeley, Calif.: Lawrence Berkeley National Laboratory.

Wiser, R., and M. Bolinger. 2008. Annual Report on U.S. Wind Power Installation, Cost and Performance Trends: 2007. DOE/GO-102008-2590. Washington, D.C.: U.S. Department of Energy.

2 Resource Base

The United States has a significant amount of renewable energy resources. This chapter details the resource base from wind, solar, geothermal, hydroelectric, and biomass sources of energy that could make a material contribution to the nation's electricity supply. Discussion of this resource base sets the stage for the scenarios of renewable energy deployment in Chapter 7.

Most renewable electricity generation must be located near the source of the renewable energy flux[1] being captured and converted into electricity. Hence, renewable energy sources are by nature local or regional, and those that may be unable to contribute significantly to total U.S. electricity generation could still contribute to a substantial share of the renewable-based electricity generated in regions where that specific type of renewable energy flux is abundant and well suited for development.

2007 BASELINE VALUES

In 2007 total U.S. electricity generation was 4.2 million GWh and peak generation capacity nationally was 998 GW (EIA, 2008); the average annual U.S. electric generation load in 2007 was thus approximately 480 GW. For reference, total U.S. primary energy consumption in 2007 was approximately 100 EJ. At approximately 35 percent generation efficiency, 42 EJ (corresponding to 11.7 million GWh at 100 percent generation efficiency) was used to provide the 4.2 million GWh of electricity generated in the United States in 2007.

[1]Energy flux is defined as the rate of energy transfer through a unit area.

WIND POWER

According to a study done by Pacific Northwest National Laboratory, the total estimated electric energy potential of wind for the continental United States is 11 million GWh per year from regions rated as Class 3 and higher[2] (Elliott et al., 1991)—a value greater than the 4.2 million GWh of electric energy generated in the United States in 2007. In energy units, 11 million GWh represents 40 EJ of energy, as compared to the 2007 domestic primary energy consumption of 100 EJ.

The domestic large-scale wind electric energy resource estimate of 11 million GWh is uncertain, however, and the actual wind resource could be higher or lower. One source of uncertainty is that the yearly wind electricity potential from the PNNL study was estimated from point-source measurements of the wind speed at a height of 50 m (Elliott et al., 1986). Modern wind turbines can have hub heights of 80 m or higher, where more wind energy resource is likely to be available. However, computer simulations of very-large-scale wind farm deployment show that an agglomeration of point-source wind speed data over large areas can significantly overestimate the actual wind energy resource base (Roy et al., 2004). Just as a large wind turbine will overshadow a wind turbine farther downwind, so a very extensive wind farm will also overshadow other wind farms downwind. Specifically, when the downwind length of the wind farm is comparable to, or larger than, the scale length of the atmosphere (approximately 50 km), then the point-source measurement extrapolation is no longer valid, and significantly overestimates the actual available wind energy resource (Keith et al., 2004).

Another consideration is that wind field deployment at levels needed to produce 5 million to 10 million GWh of electricity would entail extraction of a significant portion of the energy from the wind field of the continental United States for conversion into electric energy. Continental-scale simulations indicate that high levels of wind power extraction could, to various degrees, affect regional weather as well as climate. In addition to limiting the efficiency of large-scale wind farms, model calculations suggest that the extraction of wind energy from very-large-scale wind farms could have some measurable effect on weather and climate at the local or even continental and global scales (Roy et al., 2004; Keith et al., 2004).

More detailed meso-scale modeling and measurements are needed to clearly delineate the total U.S. extractable wind energy potential and the portion that can

[2]Wind class is a measure of wind power density, which is measured in watts per square meter and is a function of wind speed at a specific height.

be extracted without significant environmental impacts. Modeling activities are under way to determine the optimal distance between wind farms to minimize power loss (Frandsen et al., 2007). Assuming an estimated upper limit of 20 percent of the energy in the wind field for extraction, both regionally and on a continental scale, and a total U.S. onshore wind electricity value of 11 million GWh/yr, an upper value for the extractable wind electric potential would be about 2.2 million GWh/yr, equal to more than half of the electricity generated in 2007. This estimate assumes that large-scale wind farms are installed over all suitable Class 3 and higher wind speed areas in the continental United States, as mapped in Figure 2.1 (AWEA, 2007; DOE, 2008). The preceding analysis is limited to onshore wind energy resources.

Significant offshore wind energy resources also exist, and Europe has begun to develop its offshore resources. The available offshore wind capacity has been estimated at 907 GW for distances 5–50 nautical miles offshore (NREL, 2004a), which corresponds to 1.6 GWh/yr, assuming extraction of 20 percent of the energy in the wind field, i.e., almost 40 percent of 2007 U.S. electricity generation. The water at these locations varies from less than 30 meters to greater than 900 meters deep. Since a large percentage of the population lives along the coasts of the continental United States, offshore wind could be a renewable resource located close to population centers. These resources are also mapped in Figure 2.1 for the continental United States. Several states are now focusing wind development efforts on offshore wind resources, especially where onshore wind resources are well developed. However, offshore projects have been fraught with siting controversies, including the proposed development off Cape Cod, Massachusetts.

SOLAR POWER

The solar energy resource is extremely large. Taking 230 W/m^2 as a representative midlatitude, day/night average value for solar insolation[3] and 8×10^{12} m^2 as the area of the continental United States yields a yearly averaged, area-averaged, power generation potential of 1.84 million GW (Clean Edge, 2008). The solar resource thus provides annually to the continental United States the equivalent of

---

[3]Solar insolation is the amount of solar energy striking a flat surface per unit area per unit of time.

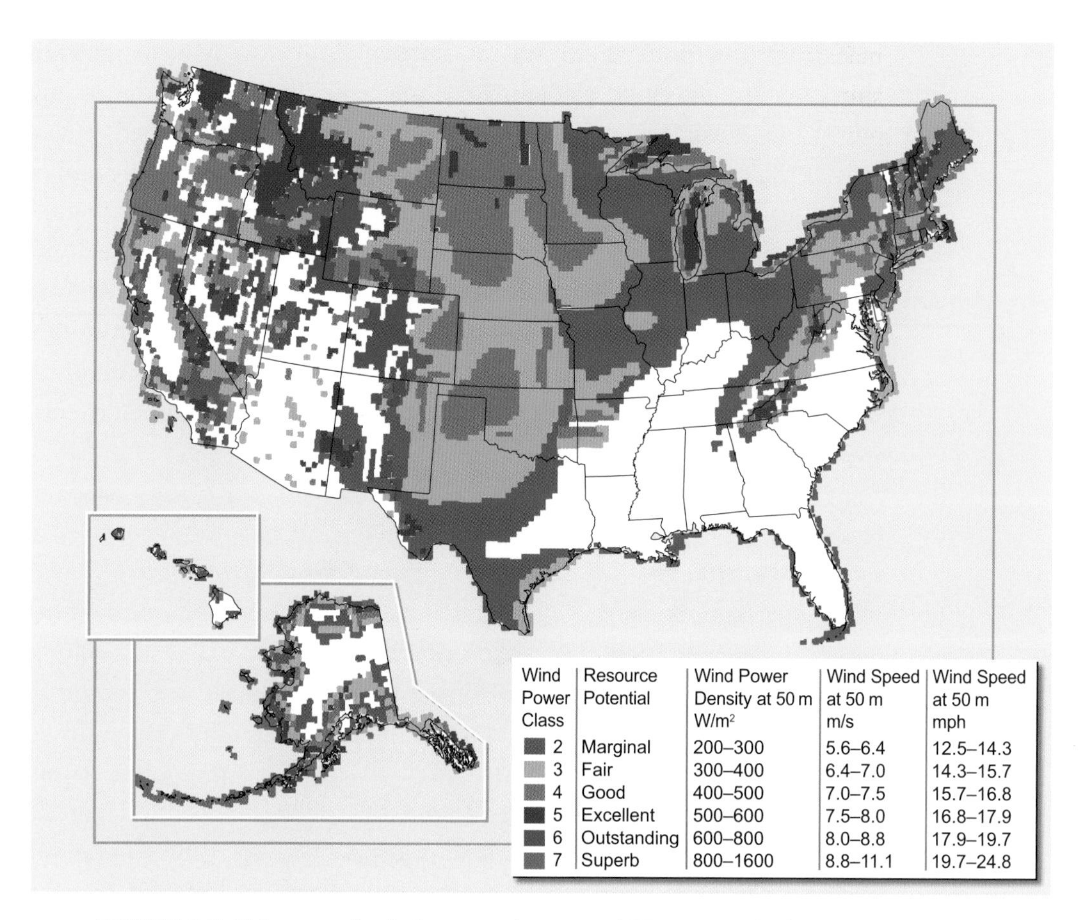

FIGURE 2.1 *U.S. map of wind power classes and 50-meter wind energy resource. Source: DOE, 2008.*

about 16 billion GWh of electric energy and, at a 10 percent average conversion efficiency, would therefore provide 1.6 billion GWh/yr of electricity. At a 10 percent conversion efficiency, coverage of 0.25 percent of the land area of the continental United States would be required to generate the 4.2 million GWh of electric energy generated domestically in 2007.

Solar Photovoltaic Power

Flat-plate photovoltaic (PV) arrays effectively use both direct and diffuse sunlight, thus enabling deployment over a larger geographic region than is possible with concentrated solar power. Although the yearly averaged total insolation varies significantly over the continental United States, the regional variation is approximately a factor of two, as shown in Figure 2.2. Estimates of the rooftop area suitable for installation of PV systems have been performed state-by-state for the whole United States. An analysis by the Energy Foundation and Navigant Consulting eliminated roofs on residences that were not generally facing southward and roofs that had too high a slope for routine installation of solar PV panels; considered the impacts of shading by trees, the presence of heating and air-conditioning units, and other obstacles on the remaining viable portion of the rooftops, but did not account for snow; and added suitable flat commercial building rooftop space to the total (Chaudhari et al., 2004). The analysis concluded that 22 percent of the available residential rooftop space, and 65 percent of commercial building rooftop space, was technically suitable for PV system installation. This total

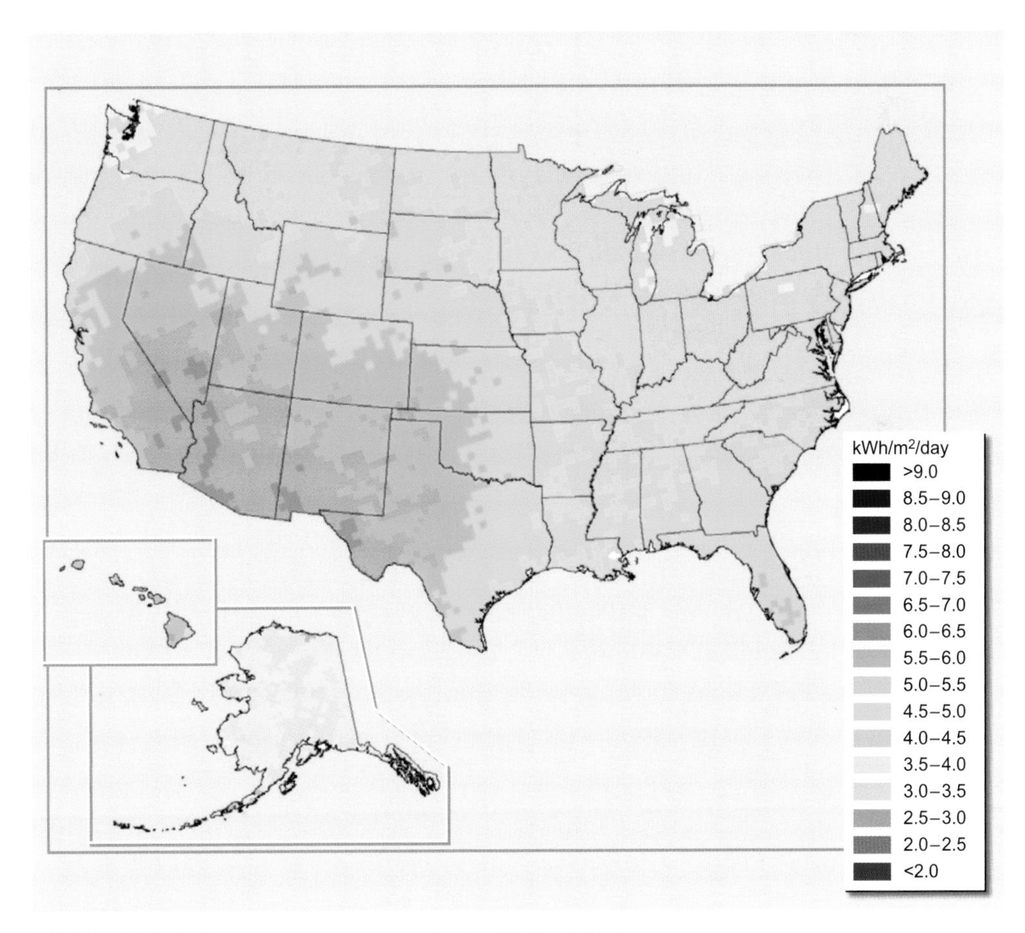

FIGURE 2.2 *Solar energy resources in the United States.*
Source: NREL, 2007.

rooftop area, along with state-by-state values for the average insolation, yielded a technical solar PV-based peak capacity of 1500–2000 GW at commercially available PV system conversion efficiencies of 10–15 percent. At an average 20 percent capacity factor, this peak-capacity value would thus result in the production of 13 million to 17.5 million GWh/yr of electric energy, still much larger than the 4.2 million GWh/yr of electricity generated in the United States in 2007. More conservative estimates indicate that existing suitable rooftop space could provide 0.9 million to1.5 million GWh/yr of PV-generated electricity (ASES, 2007). Clearly, with some (or perhaps no) amount of land set-aside for flat-plate PV-based solar electricity generation beyond that already available in existing rooftop areas, flat-plate solar PV has the potential to supply significantly more electricity than was generated in 2008 in the United States.

Concentrating Solar Power

Concentrating solar power (CSP) systems can only use the focusable, direct beam portion of incident sunlight and are thus limited to favored sites, primarily in the Southwest, that have abundant direct normal solar radiation. Figure 2.3 shows that despite variations in radiation intensity in the Southwest, all six states there have attractively high levels of insolation. A recent analysis by the Western Governors' Association identified suitable land area that has a high average insolation of more than 6.75 $kWm^{-2}day^{-1}$; it excluded land areas having a slope greater than 1 percent or a continuous area of smaller than 10 km^2, and national parks, nature reserves, and urban areas (WGA, 2006a). The analysis concluded that the Southwest has a concentrated solar power electricity peak generation capacity of 7000 GW. With an average annual capacity factor of 25–50 percent for CSP, depending on the thermal storage used for a plant, this land area could theoretically produce 15–30 million GWh of electric energy per year, again significantly more than the 4.2 million GWh total U.S. electricity supply in 2007.[4] Only a fraction of this land area at present could be developed economically for CSP-based electricity generation due to factors such as generation and transmission costs discussed in later chapters.

[4]See Figure 2 and Table 1 on page 83 of the ASES report (ASES, 2007).

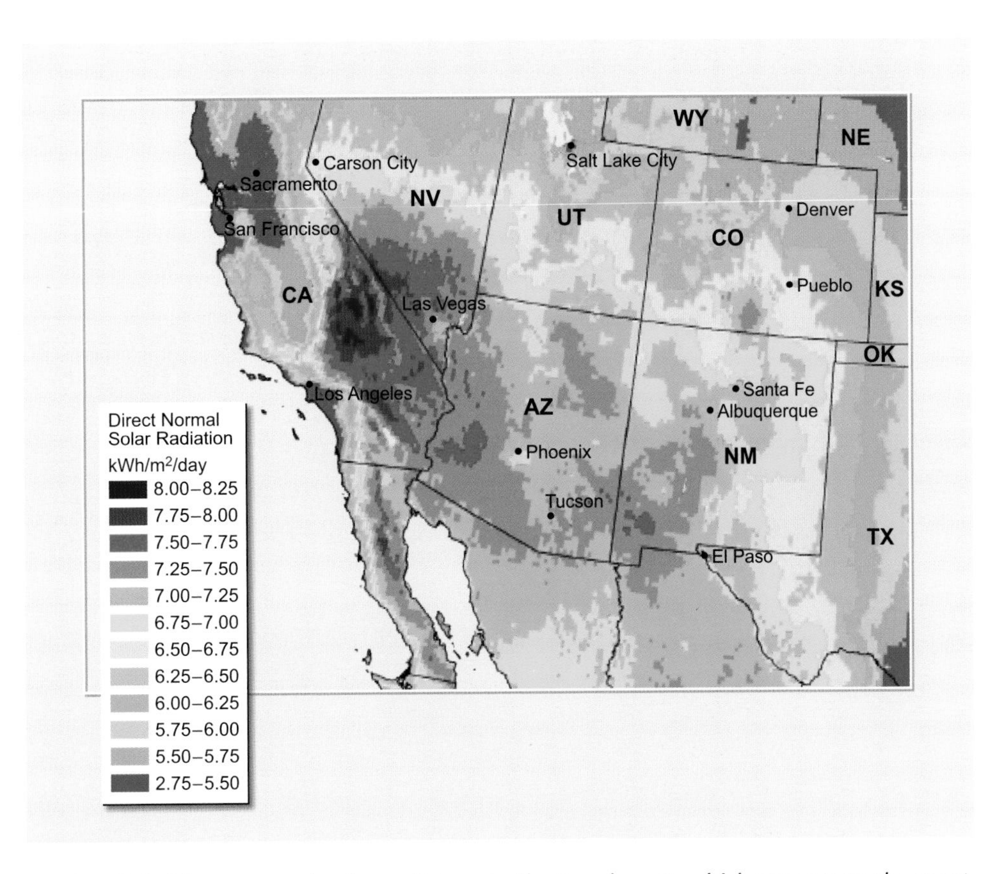

FIGURE 2.3 *Direct normal solar radiation in the Southwest, which represents the most suitable region for electricity generation from concentrated solar power.*
Source: National Renewable Energy Laboratory; reprinted in WGA, 2006a.

GEOTHERMAL POWER

Hydrothermal Energy

Geothermal energy exists as underground reservoirs of steam, hot water, and hot dry rocks in Earth's crust. Hydrothermal (sometimes referred to as conventional geothermal) electric generating facilities use hot water or steam extracted from these reservoirs and supply this energy to turbines to generate electricity. For reference, according to the U.S. Geological Survey (USGS, 1979), thermal energy stored as hydrothermal resources ranges between 2,500 EJ (0.67 billion GWh) and 9,700 EJ (2.7 billion GWh).

A regional study of known geothermal resources in the western United States found that 13 GW of electric power capacity exists in 140 hydrothermal sites identified in the region (Figure 2.4; WGA, 2006b). Of these 13 GW, the Western Governors' Association reported that 5.6 GW of capacity was considered viable for commercial development by 2015, which reflects the consensus of geothermal technology, development, and power-generating operations experts. Since hydrothermal facilities typically operate at 90 percent capacity during much of their operational life, the 13 GW from identified hydrothermal resources could provide up to 0.1 million GWh/yr of baseload electric energy. These same western states consumed slightly more than 1 million GWh/yr of electricity from 2000 through 2003 (WGA, 2006a). A nationwide assessment of the shallow hydrothermal resource base estimates an availability of 30 GW, with an additional 120 GW potential from unidentified hydrothermal resources that show no surface mani-

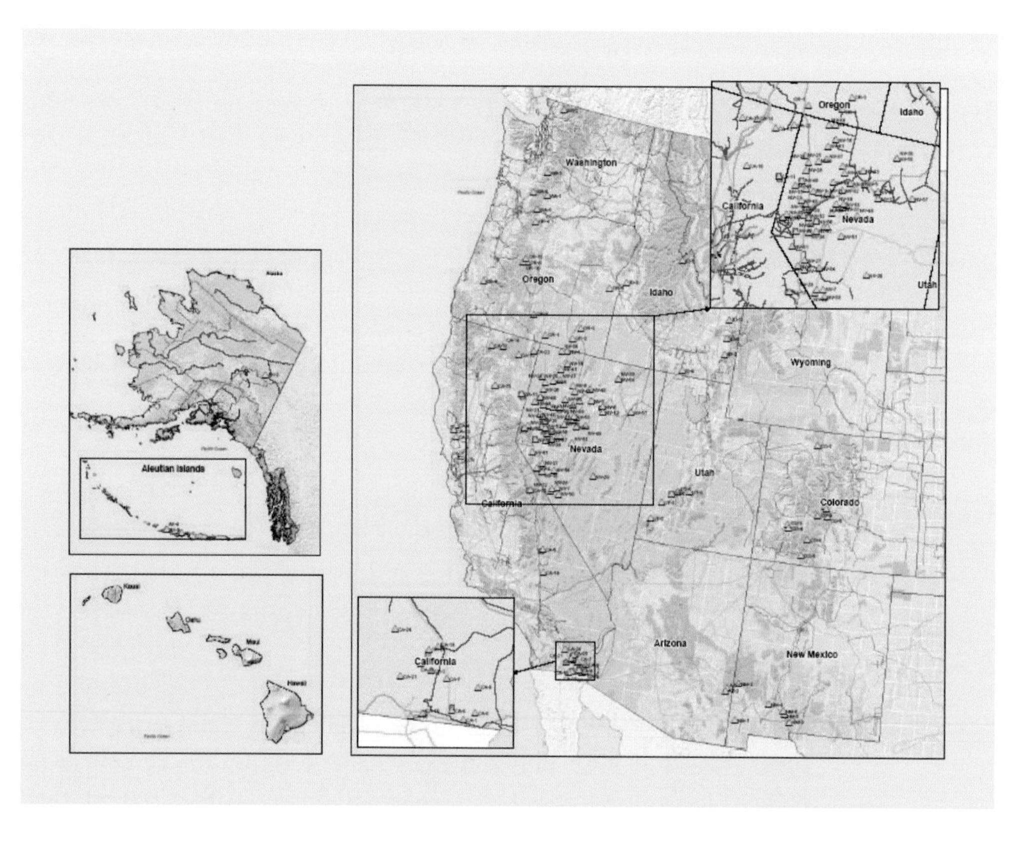

FIGURE 2.4 *Regional map of hydrothermal power sites resources identified by the Western Governors' Association.*
Source: National Renewable Energy Laboratory; reprinted in WGA, 2006a.

festations (NREL, 2006). The NREL study estimated that 10 GW could be developed by 2015.

Enhanced Geothermal Systems

Enhanced geothermal systems (EGSs) are engineered reservoirs created to extract heat from low-permeability and/or low-porosity geothermal resources, as defined by the Department of Energy. EGSs tap the vast heat resources available due to temperature gradients between the surface and depths of up to 10 km, as shown in the maps in Figure 2.5. The geothermal energy resource base located beneath the continental United States, defined as the total amount of heat trapped to 10 km depth, is estimated to be in excess of 13 million EJ (3.6 trillion GWh) (MIT, 2006). Figure 2.6 separates this heat content into a function of temperature and depth. The total heat stored is more than 130,000 times the total 2005 U.S. energy consumption of 106 EJ of energy. The extractable portion of this resource has been estimated at 200,000 EJ, i.e., about 2,000 times more than the primary energy consumed in the United States in 2005. At a conversion efficiency of 15 percent, a reasonable value in view of the typical ~200°C temperature difference between the temperature of the resource and the ambient temperature at the surface, the extractable geothermal resource could then, in principle, provide 30,000 EJ of electric energy.

In addition to the total amount of available energy, the rate at which it is extracted is also important. The mean geothermal heat flux over land at Earth's surface is approximately 60 mW/m^2 and in many areas is significantly less. An efficiency of 15 percent is estimated for electricity generation from this relatively low temperature heat in a turbine. Thus, on average, the extractable electric power density from the geothermal resource on a renewable basis is about 10 mW/m^2. At an extracted, producible electric power density of 10 mW/m^2, 100 GW of electric power (22 percent of the 2005 average U.S. electric load and 10 percent of the 2005 U.S. electric generation capacity) would thus require a minimum land area footprint of 1×10^{13} m^2.[5] For comparison, the land area of the continental United States is 8×10^{12} m^2, so the footprint needed to provide 20 percent of the 2005 average electric load from sustainably produced geothermal energy would exceed the total land area of the continental United States.

In practice, the in-place geothermal heat would be extracted at rates in excess

---

[5] 1×10^{11} W/(10^{-2} W m^{-2}) = 1×10^{13} m^2.

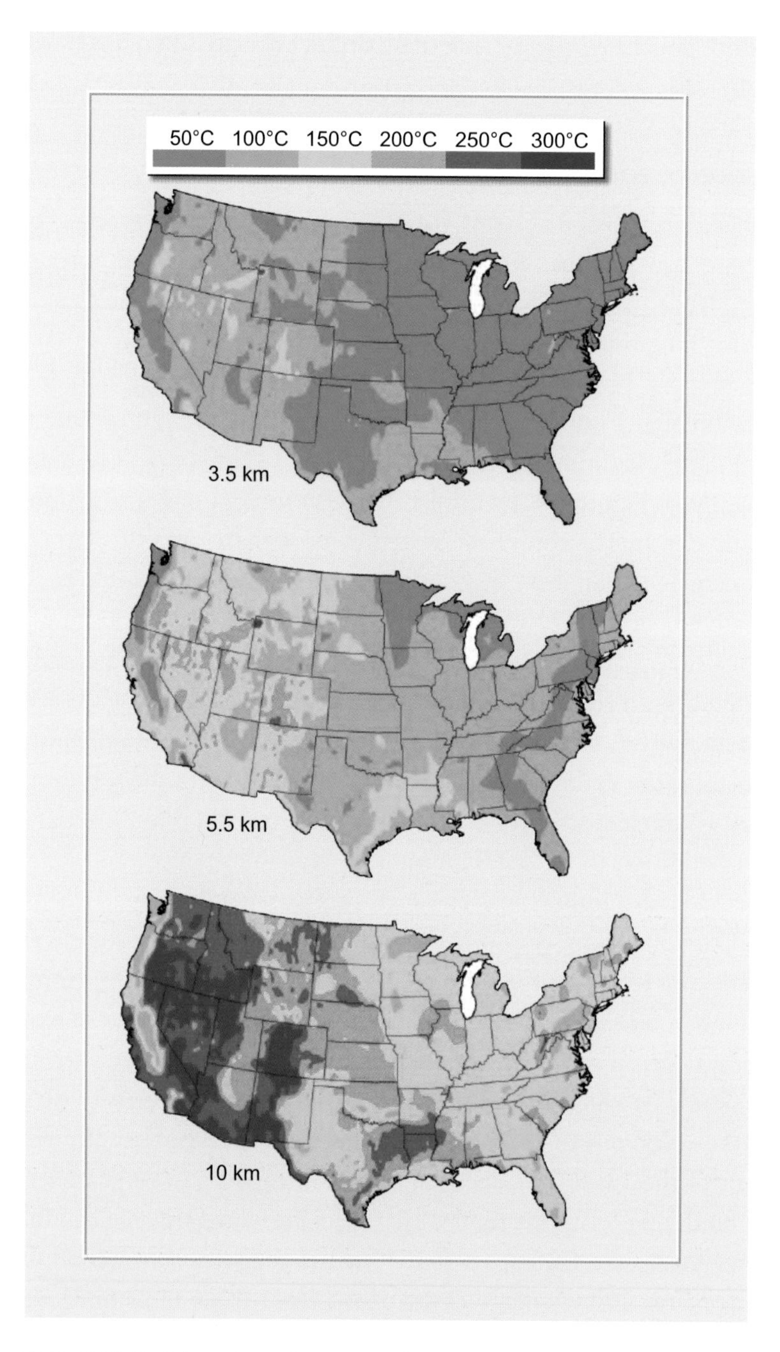

FIGURE 2.5 *Allocation of U.S. geothermal resources at 3.5 km (top panel), 5.5 km (middle panel), and 10 km (bottom panel) depths.*
Source: MIT, 2006. Copyright 2006 MIT.

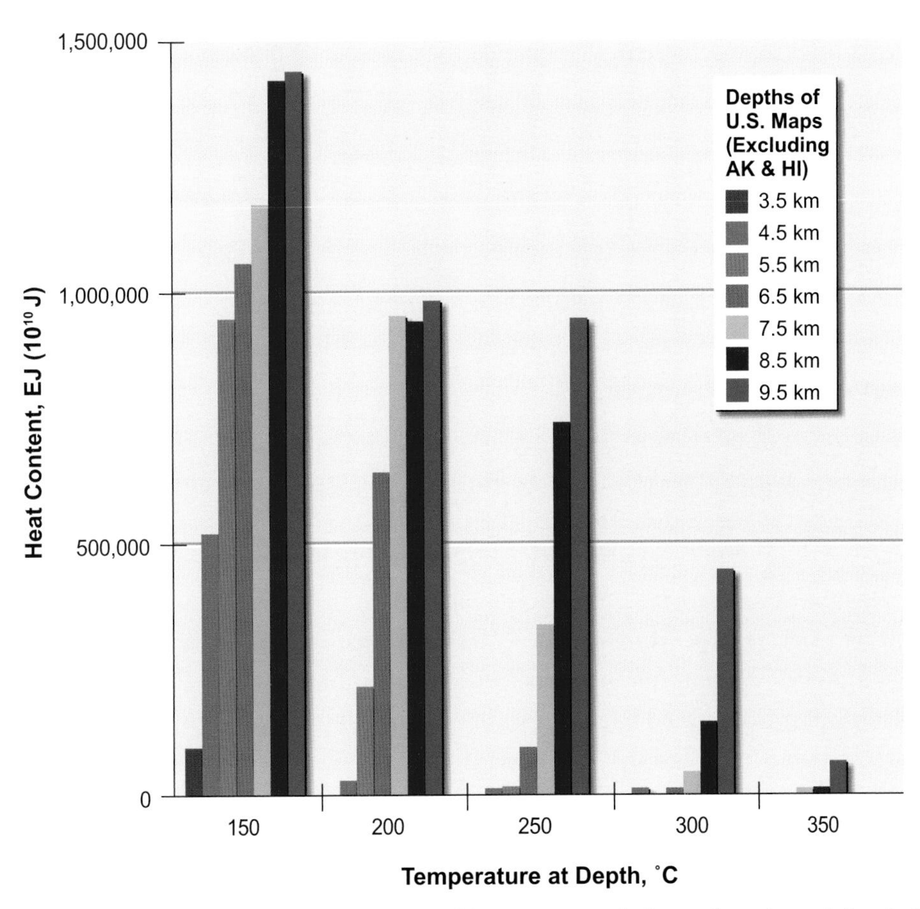

FIGURE 2.6 *Histogram of heat content (EJ) as a function of depth for slices 1 km thick. Source: MIT, 2006. Copyright 2006 MIT.*

of the natural geothermal heat flux; such extraction rates are not sustainable in the long term, because they would deplete the heat more rapidly than it would be restored by the natural geothermal flux. Such heat mining would reduce the land area needed to be tapped by allowing heat extraction to exceed the 10 mW/m^2 replacement rate. Indeed, a recent MIT report (2006) notes that some temperature drawdown should occur if such reservoirs are used most efficiently. In its analysis of the resource potential for EGS, the MIT report limited this heat mining by assuming that geothermal reservoirs would be abandoned when the temperature of the rocks fell by 10–15°C. Because heat extraction may not be uniform, the MIT report assumes that reservoirs would have a lifetime of 30 years, with periodic re-drilling, fracturing, and hydraulic simulation. The report estimates that reservoirs should be able to recover to their original temperature conditions within

100 years after abandonment. It contends that if only 10 percent or less of the stored heat is mined at any time, enhanced geothermal energy could be considered a renewable resource, because the huge resource base would support abandoning reservoirs for the 100-year period needed to restore the original temperature.

HYDROPOWER

Conventional Hydroelectricity

Conventional hydroelectricity generation in 2007 provided 0.25 million GWh. Hydroelectric generation capacity was 98 GW, representing about 9 percent of the total U.S. electric generation capacity (EIA, 2009).

Because use of the conventional hydroelectric resource is generally accepted to be near the resource base's maximum capacity in the United States, further growth will largely depend on non-conventional hydropower resources such as low-head power[6] and on microhydroelectric generation.[7] A 2004 DOE study of total U.S. water-flow-based energy resources, with emphasis on low-head/low-power resources, indicated that the total U.S. domestic hydropower resource capacity was 170 GW of electric power, of which 21 GW was from low-head/low-power, 26 GW was from high-head/low-power, and 123 GW was from high-head/high-power (DOE, 2004). These numbers represent only the identified resource base that was undeveloped and was not excluded from development. A subsequent study assessed this identified resource base for feasibility of development (DOE, 2006). After taking into consideration local land-use policies, local environmental concerns, site accessibility, and power transmission, the total potential domestic hydroelectric resource capacity was estimated to be 100 GW of electric power. This value was reduced to 30 GW of potential hydroelectric capacity after applying development criteria (DOE, 2006). A report from the Electric Power Research Institute (EPRI) determined that 10 GW of additional hydroelectric resource capacity could be developed by 2025 (EPRI, 2007). Of the 10 GW of potential capacity, 2.3 GW would result from capacity gains at existing hydroelectric facilities, 2.7 GW would come from small and low-power conventional hydropower

---

[6]Vertical difference of 100 feet or less in the upstream surface water elevation (headwater) and the downstream surface water elevation (tailwater) at a dam.

[7]Hydroelectric power installations that produce up to 100 kW of power.

facilities, and 5 GW would come from new hydropower generation at existing non-powered dams.

Hydrokinetic Power—Wave, Tide, and River Energy

Hydrokinetic energy is the energy associated with the flow of water, such as wave energy and the energy in water currents, including tides and rivers. As shown in Table 2.1, there is significant interest in developing such energy resources, based on permits filed with the Federal Energy Regulatory Commission. Permit activity is not a reliable predictor of future development of hydrokinetic resources, however, because often developers will apply before planning the facility or obtaining financing.

According to an EPRI report that assessed total U.S. wave energy potential (EPRI, 2005), all the wave energy in the coastal states of Washington and California combined could produce 0.44 million GWh/yr, and the wave energy from the Maine, New Hampshire, Massachusetts, Rhode Island, New York, and New Jersey coasts combined could produce 0.12 million GWh/yr (Figure 2.7). These values should be reduced by 10–15 percent to account for generation losses, resulting in a total electric generation potential of about 0.07 million GWh from the entire continental U.S. wave energy resource. Exhaustive use of the entire wave energy resource would therefore be required to produce less than 2 percent of the 4.2 million GWh of the electricity generated in the United States in 2007.

The largest U.S. wave resource lies off southern Alaska, which has an estimated resource base of 1.25 million GWh/yr, as shown in Figure 2.7 (EPRI, 2005). Extraction of this total amount of energy would involve tapping wave energy flows over relatively large areas of ocean, and the EPRI report also does

TABLE 2.1 Permit Activity for Hydrokinetic Resources (in megawatts of proposed capacity)

	Issued	Pending
Wave	170–330	1270–2150
Current	1025–3350	270–375
Tide	140–285	445
Inland	100	3550

Source: Federal Energy Regulatory Commission; presented in Miles, 2008.

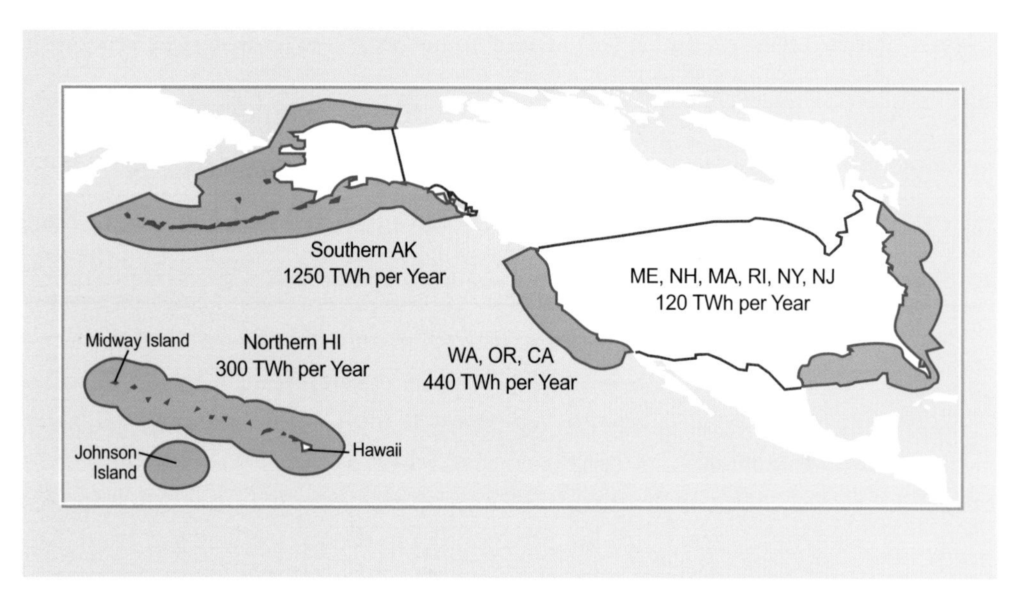

FIGURE 2.7 *U.S. wave energy resources for the continental United States. Source: EPRI, 2005.*

not indicate how the electric energy over such a large area of the ocean would be collected or transmitted to consumers in the lower 48 states.

The 2005 EPRI study also looked at tidal energy from a series of sites identified in Alaska, Washington, California, Massachusetts, Maine, New Brunswick, and Nova Scotia (EPRI, 2005). The total combined resource was estimated to have an annual average electric capacity potential of 152 MW, which corresponds to an annualized electric energy production of 1300 GWh/yr (EPRI, 2005)—enough to provide, if the stated resource were used in whole, 0.03 percent of the 2005 domestic generated electric energy.

EPRI's study of the electric energy potential in river currents yielded a value of 0.11 million GWh/year (EPRI, 2005). Thus, development of the entire U.S. river current electricity potential would be required to produce 0.1 million GWh/yr, which would represent less than 3 percent of the 2005 domestic electric energy production.

BIOPOWER

Biomass Resource Base

The USDA/DOE billion-ton study (2005) identified the potential for use of 1.3 billion dry tons (1 dry ton = 1,000 kg) per year of biomass without adversely affecting food production. The area involved in producing this resource base comprises 448 million acres (1.8×10^{12} m^2) of agricultural land (consisting of both cropland and pasture), which is 23 percent of the land area of the continental United States, and 672 million acres of forestland (2.7×10^{12} m^2), representing 34 percent of the land area of the continental United States (USDA/DOE, 2005). Agricultural land totaled 455 million acres in 1997, the year of the most recent complete inventory of land use. Hence, the total land area assumed to be used for such biomass farms is just over 57 percent of the total land area of the lower 48 states.

The amount of biomass sustainably removed from domestic agricultural lands and forestlands is 190 million dry tons annually, with about 142 million dry tons coming from forestland and the remainder coming from croplands. Only about 20 percent of this biomass is now in use. The USDA/DOE report projected that approximately 370 billion tons (double the present biomass production) could be made available sustainably for biomass uses from 672 million acres of forestland. To accomplish this would require a variety of methods, including using wood for electric power generation instead of burning that wood for forest management (as is done at the present time), using pulp residues, and logging residues.

The USDA/DOE report also projected that agricultural lands (cropland, idle cropland, and cropland pasture), which produce approximately 50 million tons per year for biomass uses, have the potential, within 35 to 40 years, to yield nearly 1 billion dry tons of biomass. This represents a 20-fold increase in the sustainable biomass yield relative to the present value. Of this projected 1 billion dry tons that might be available in 35–40 years, 300–400 million tons would come from crop residues and 350 million tons would result from the substitution of high-yield perennial biomass crops for other land uses on at least 40 million acres of land.

The geographical distribution of the biomass resource base shown in Figure 2.8 comes from Milbrandt (2005), which estimated a lower overall biomass resource base than does the USDA/DOE report. This is because the billion-ton study estimated future potential biomass resources in the country, while Milbrandt evaluated currently available biomass resources (though it considers a case study

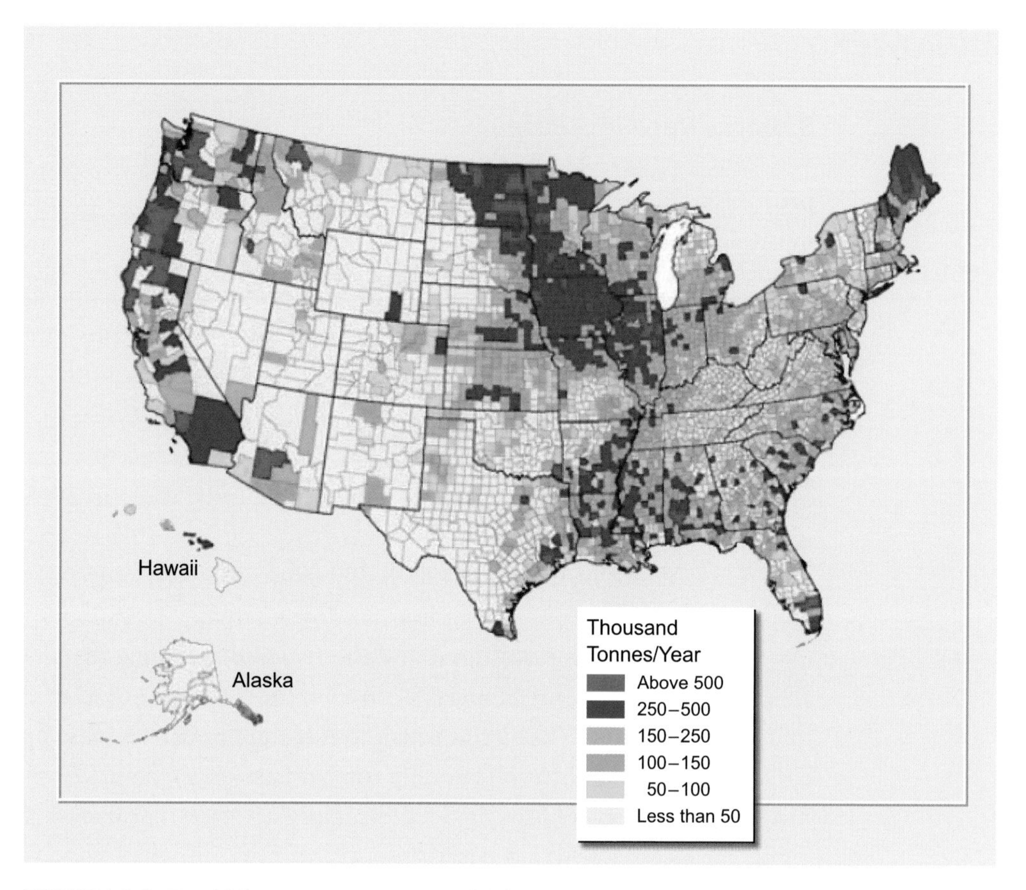

FIGURE 2.8 *Total biomass resources available in the United States, by county. Source: Milbrandt, 2005.*

of switchgrass on Conservation Reserve Program lands). The resource assessment is performed at a county level and it includes (1) residues from agriculture and forestry, (2) urban wood (secondary mill residues, MSW wood, utility tree trimming, and construction/demolition wood), and (3) methane emissions from manure management, landfills, and domestic wastewater treatment.

According to the USDA/DOE study, providing 1.3 billion dry tons per year of biomass would require increasing the yields of corn, wheat, and other small grains by 50 percent; doubling residue-to-grain ratios for soybeans; developing more efficient residue-harvesting equipment; managing cropland with no-till cultivation; growing perennial crops whose output is primarily dedicated to energy purposes on 55 million acres of cropland, idle cropland, and cropland pasture; using animal manure in excess of what can be applied on-farm for soil improvement; and using a larger fraction of other secondary and tertiary residues for biomass production. Attaining these levels of crop yield increases and collection would require research and new technologies such as genetic engineering to increase production. The ~50 million acres devoted to high-yield perennials were projected to have an average annual crop yield of approximately 8 dry tons/acre, in order to provide ~400 million dry tons of biomass annually from that portion of land. Supporting the billion-ton estimate was the assumption that agricultural lands in the United States could potentially provide in excess of 1 billion dry tons of sustainably collectable biomass, while continuing to meet food feed and export demands. This estimate included 446 million dry tons of crop residues (for example, more than 250 million tons from corn stover, as compared to the present value of 75 million tons annually), 377 million dry tons of perennial crops,[8] 87 million dry tons of grains used for biofuels, and 87 million dry tons of animal manure, process residues, and other residues generated in the consumption of food products.

The forthcoming report of the America's Energy Future (AEF) Panel on Alternative Liquid Transportation Fuels (see Appendix A) provides another estimate of the biomass resource base (NAS-NAE-NRC, 2009). It estimates that an annual supply of 400 million dry tons of cellulosic biomass could be produced sustainably, using technologies and management practices available in 2008, an amount that could likely be increased to about 550 million dry tons by 2020. The AEF alternative liquid fuels panel judges that those estimated quantities of biomass can be produced from dedicated energy crops, agricultural and forestry residues, and municipal solid waste with minimal impacts on U.S. food, feed, and fiber production, and with minimal adverse environmental impacts. The AEF alternative liquid fuels panel did not extend its estimate to 2035, as did the 2005 USDA/DOE report.

Electricity Generation from Biomass

Based on 2005 biomass production levels, full use of the 190 million dry tons of sustainable biomass produced in the United States, at 17 GJ (1 GJ = 1×10^9 J)/dry ton, and at 35 percent efficiency for conversion of the heat produced from bio-

---

[8]The perennial crops are crops dedicated primarily to energy and other products and will likely include a combination of grasses and woody crops.

mass combustion into electric energy, would provide energy of 1.1 EJ.[9] In other words, 100 percent of the sustainable biomass produced domestically in 2005, if used entirely for electricity generation, would produce 0.306 million GWh/yr of electricity, or 7.3 percent of the 2007 domestic electricity generation of 4.2 million GWh/yr. Using the AEF alternative liquid fuels panel's more recent resource average value of ~500 million tons of biomass (NAS-NAE-NRC, 2009), a total of 0.8 million GWh/yr of electricity could be produced, which is 19 percent of 2007 U.S. electricity generation.

Increasing the available biomass production to 1 billion tons and using it solely for electricity generation would produce 6 EJ, which is equal to 1.6 million GWh/yr of electricity, representing approximately 40 percent of the domestic electric generation in 2007. If, however, 75 percent of this biomass was used to produce cellulosic ethanol or other biofuels, then only 25 percent of the biomass would be available for electricity generation. Thus, 250 million tons of biomass, projected as potentially available in 35–40 years through the use of more than 60 percent of the land area of the continental United States, would be capable of producing 0.416 million GWh of electricity, or 10 percent of the 2007 U.S. electricity generation. This potential represents more than 7 times the actual electric generation from biomass in 2005 (0.054 million GWh, which accounted for just above 1 percent of the 2007 U.S. electricity generation).

FINDINGS

Shown below in bold text are the most critical elements of the findings of the AEF Panel on Electricity from Renewable Resources, based on its consideration of the U.S. resource base for generation of renewable electricity.

In summary, the United States has significant renewable energy resources, which, combined, have the potential, in principle, to provide more electric power than the total existing installed peak capacity and more electric energy annually than the total electricity consumed domestically in 2005. This resource base is spread widely across the United States. However, as described in the remainder of this report, many other factors will determine what portion of these resources will actually be incorporated into the electricity system; some of these factors include

---

[9] 1.9×10^8 tonnes $\times$ (1.7×10^{10} J/tonne) at 35 percent electric generation efficiency.

the costs of technologies needed to transform these resources into electricity; the expanded capacity and associated costs for transmission to bring this electricity into load centers; and the need to compensate for intermittency.

Solar and wind renewable resources offer significantly larger total energy and power potential than do other domestic renewable resources. Although solar intensity varies across the nation, the land-based solar resource provides a yearly average of more than 5×10^{22} J (13.9 million TWh) and thus exceeds, by several thousand-fold, present annual U.S. electrical energy demand, which totals 1.4×10^{19} J (~4,000 TWh). Hence, at even modest conversion efficiency, solar energy is capable, in principle, of providing enormous amounts of electricity without stress to the resource base. The land-based wind resource is capable of providing at least 10–20 percent, and in some regions potentially higher percentages, of current electrical energy demand. Other (non-hydroelectric) renewable resources can contribute significantly to the electrical energy mix in some regions of the country.

Renewable resources are not distributed uniformly in the United States. Resources such as solar, wind, geothermal, tidal, wave, and biomass vary widely in space and time. **Thus, the potential to derive a given percentage of electricity from renewable resources will vary from location to location. Awareness of such factors is important in developing effective policies at the state and federal levels to promote the use of renewable resources for generation of electricity.**

REFERENCES

ASES (American Solar Energy Society). 2007. Tackling Climate Change in the United States: Potential Carbon Emission Reductions from Energy Efficiency and Renewable Energy by 2030. Washington, D.C.

AWEA (American Wind Energy Association). 2007. 20 Percent Wind Energy Penetration in the United States: A Technical Analysis of the Energy Resource. Washington, D.C.

Chaudhari, M., L. Frantzis, and T.E. Holff. 2004. PV Grid Connected Market Potential Under a Cost Breakthrough Scenario. San Francisco: The Energy Foundation and Navigant Consulting.

DOE (U.S. Department of Energy). 2004. Water Energy Resources of the United States with Emphasis on Low Head/Low Power Resources. DOE/ID-11111. Washington, D.C.

DOE. 2006. Feasibility Assessment of the Water Energy Resources of the United States for New Low Power and Small Hydro Classes of Hydroelectric Plants. Washington, D.C.

DOE. 2008. 20% Wind Energy by 2030: Increasing Wind Energy's Contribution to U.S. Electricity Supply. Washington, D.C.

DOE/EERE (U.S. Department of Energy/Energy Efficiency and Renewable Energy). 2008. United States—Wind Resource Map. Washington, D.C. Available at http://www.windpoweringamerica.gov/pdfs/wind_maps/us_windmap.pdf.

EIA (Energy Information Administration). 2008. Annual Energy Review 2007. Washington, D.C.: U.S. Department of Energy, EIA.

EIA. 2009. Electric Power Annual. Washington, D.C. Available at http://www.eia.doe.gov/cneaf/electricity/epa/epa_sum.html; table available at http://www.eia.doe.gov/cneaf/electricity/epa/epat2p2.html.

Elliott, D.L., C.G. Holladay, W.R. Brachet, H.P. Foote, and W.R. Sandusky. 1986. Wind Energy Resource Atlas of the United States. Washington, D.C.: National Renewable Energy Laboratory.

Elliott, D.L., L.L. Wendell, and G.L. Gower. 1991. An Assessment of the Available Windy Land Area and Wind Energy Potential in the Contiguous United States. Richland, Wash.: Pacific Northwest Laboratory.

EPRI (Electric Power Research Institute). 2005. Final Summary Report, Project Definition Study, Offshore Wave Power Feasibility Demonstration Project. Washington, D.C.

EPRI. 2007. Assessment of Waterpower Potential and Development Needs. Washington, D.C.

Frandsen, S., R. Barthelmei, O. Rathmann, H.E. Jorgensen, J. Badger, K. Hansen, S. Ott, P.E. Rethore, S.E. Larsen, and L.E. Jensen. 2007. Summary Report: The Shadow Effect of Large Wind Farms: Measurements, Data Analysis and Modeling. Riso National Laboratory, Technical University of Denmark, Riskilde, Denmark.

Keith, D.W., J.F. DeCarolis, D.C. Denkenberger, D.H. Lenschow, S.L. Malyshev, S. Pacala, and P.J. Rasch. 2004. The influence of large scale wind power on global climate. Proceedings of the National Academy of Sciences USA 101:16115-16120.

Milbrandt, A. 2005. A Geographic Perspective on the Current Biomass Resource Availability in the United States. Golden, Colo.: National Renewable Energy Laboratory.

Miles, A.C. 2008. Hydropower at the Federal Energy Regulatory Commission. Presentation at the third meeting of the Panel on Electricity from Renewable Resources, January 16, 2008. Washington, D.C.

MIT (Massachusetts Institute of Technology). 2006. The Future of Geothermal Energy: Impact of Enhanced Geothermal Systems (EGS) on the United States in the 21st Century. Cambridge, Mass.

NAS-NAE-NRC (National Academy of Sciences-National Academy of Engineering-National Research Council). 2009. Liquid Transportation Fuels from Coal and Biomass: Technological Status, Costs, and Environmental Impacts. Washington, D.C.: The National Academies Press.

NREL (National Renewable Energy Laboratory). 2004a. Future for Offshore Wind Energy in the United States. Preprint. Washington, D.C.

NREL. 2004b. PV Solar Radiation: Annual. Washington, D.C. Available at http://www.nrel.gov/gis/images/map_pv_us_annual_may2004.jpg.

NREL. 2006. Geothermal—The Energy Under Our Feet. Geothermal Resource Estimates for the United States. Washington, D.C.

NREL. 2007. Very Large Scale Deployment of Grid Connected Solar PV in the United States. Washington, D.C.

Roy, B.S., S.W. Pacala, and R.L. Walko. 2004. Can large wind farms affect local meteorology? Journal of Geophysical Research 109:D19101.

USDA/DOE (U.S. Department of Agriculture/Department of Energy). 2005. Biomass as Feedstock for a Bioenergy and Bioproducts Industry: The Technical Feasibility of a Billion-Ton Annual Supply. Washington, D.C.

USGS (U.S. Geological Survey). 1979. Assessment of Geothermal Resources of the United States—1978. Geological Survey Circular 790. Arlington, Va.

WGA (Western Governors' Association). 2006a. Clean and Diversified Energy Initiative: Geothermal Task Force Report. Washington, D.C.

WGA. 2006b. Clean and Diversified Energy Initiative: Solar Task Force Report. Washington, D.C.

3

Renewable Electricity Generation Technologies

A renewable electricity generation technology harnesses a naturally existing energy flux, such as wind, sun, heat, or tides, and converts that flux to electricity. Natural phenomena have varying time constants, cycles, and energy densities. To tap these sources of energy, renewable electricity generation technologies must be located where the natural energy flux occurs, unlike conventional fossil-fuel and nuclear electricity-generating facilities, which can be located at some distance from their fuel sources. Renewable technologies also follow a paradigm somewhat different from conventional energy sources in that renewable energy can be thought of as manufactured energy, with the largest proportion of costs, external energy, and material inputs occurring during the manufacturing process. Although conventional sources such as nuclear- and coal-powered electricity generation have a high proportion of capital-to-fuel costs, all renewable technologies, except for biomass-generated electricity (biopower), have no fuel costs. The trade-off is the ongoing and future cost of fossil fuel against the present fixed capital costs of renewable energy technologies.

Scale economics likewise differs for renewables and conventional energy production. Larger coal-fired and nuclear-powered generating facilities exhibit lower average costs of generation than do smaller plants, realizing economies of scale based on the size of the facility. Renewable electricity achieves economies of scale prmarily at the equipment manufacturing stage rather than through construction of large facilities at the generating site. Large hydroelectric generating units are an exception and have on-site economies of scale, but not to the same extent as coal- and nuclear-powered electricity plants.

With the exception of hydropower, renewable technologies are often disruptive and do not bring incremental changes to long-established electricity industry sectors. As described by Bowen and Christensen (1995), disruptive technologies present a package of performance attributes that, at least at the outset, are not valued by a majority of existing customers. Christensen (1997) observes:

> Disruptive technologies can result in worse product performance, at least in the near term. Disruptive technologies bring to market very different value propositions than had been available previously. Generally, disruptive technologies underperform established products in mainstream markets. But they have other features that a few fringe customers value. Disruptive technologies that may underperform today, relative to what users in the market demand, may be fully performance-competitive in that same market tomorrow.

Traditional sources of electricity generation at least initially outperform non-hydropower renewables. The environmental attributes of renewables are the initial value proposition that have brought them into the electricity sector. However, with improvements in renewables technologies and increasing costs of generation from conventional sources (particularly as costs of greenhouse gas production are incorporated), renewables may offer the potential to match the performance of traditional generating sources.

This chapter examines several technologies for generation of renewable electricity. It discusses the technology associated with each renewable resource, the state of that technology, and research and development needs until 2020, between 2020 and 2035, and those beyond 2035.

WIND POWER

Wind power uses a wind turbine and related components to convert the kinetic energy of moving air into electricity and other forms of energy. Wind power has been harnessed for centuries—from the time of the ancient Greeks to the present. The modern era of wind-driven electrical generation began with the oil shocks of the 1970s and accelerated with the passage of the Public Utilities Regulatory Policies Act (PURPA). Both the development of wind technology and the installation of wind power plants have grown ever since.

Status of Technology

System Components

A typical wind turbine consists of a number of components: rotor, controls, drivetrain (gearbox, generator, and power converter), tower, and balance of system.[1] Each of these components has undergone significant development in the last 10 years, with improvements integrated into the latest turbine designs. In addition, improved understanding and better modeling capabilities have contributed to the rapid introduction of technical improvements. What were initially small clusters of 100 kW turbines in the early 1980s have grown to clusters of hundreds of machines, including machines of 1.5 MW or more.

In general, wind speed increases with height, and the energy capture capability depends on the rotor diameter. Figure 3.1 shows the change in rotor diameter and rated capacity over time. In 2006 the most common installed machine had hub heights of 275 ft (84 m) and a rotor diameter of 220 ft (67 m). Turbines as big as 5 MW have been installed in offshore locations; these have 505 ft (154 m) hub height and 420 ft (128 m) rotor diameter (IEEE, 2007a).[2] As noted in Chapter 1, the U.S. wind energy industry installed almost 14,000 MW of capacity during 2007 and 2008. The U.S. wind power capacity is now more than 25 GW and spans 34 states; the world's largest wind power plant, Horse Hollow Wind Energy center with a capacity of 750 MW, was recently commissioned in Texas (SECO, 2008). U.S. wind farms will generate an estimated 52,000 GWh of electricity in 2008, about 1.2 percent of the U.S. electricity supply. As discussed in Chapter 1, the installed wind power generating capacity worldwide at the end of 2006 was 75,000 MW.

---

[1]In general, the balance of system (BOS) is the system between the technologies that convert the renewable flux (wind or solar) into electricity and the electricity grid (for power production) or load (for direct use). The BOS might include the power-conditioning equipment that adjusts and converts the DC electricity to the proper form and magnitude required by an alternating-current (AC) load. For solar PV, the BOS consists of the structure for mounting the PV arrays and storage batteries. For wind turbines, it typically includes all the related electronics required to provide the connection to the grid.

[2]Background description and information on activities of the wind industry can be found on the American Wind Energy Association website at http://awea.org.

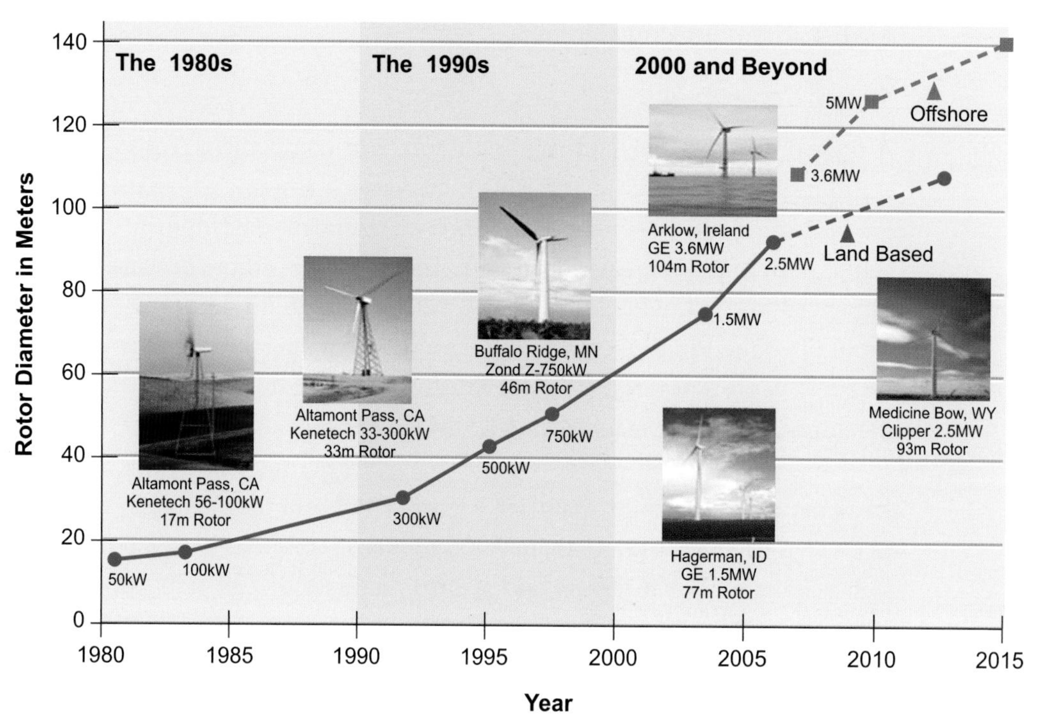

FIGURE 3.1 *Increase in rotor dimensions over recent past.*
Source: IEEE, 2005. Copyright 2005 IEEE. Reprinted by permission.

Electrical Output Controls

Besides the mechanical characteristics, the development of the turbine mechanical to electrical conversion characteristics have evolved from machines based primarily on fixed-speed induction generators (Type 1), to variable-speed machines with electronic control (Type 2), and then machines incorporating vastly different outputs and controls (Type 3). These Type 3 machines are able to control for low-voltage ride-through (LVRT),[3] voltage,[4] output[5] and ramp rate,[6] and volt-ampere-

[3]Under FERC order 661A, low-voltage ride-through is the capability to continue to operate down to 15 percent of rated line voltage for 0.626 s and continuously at 90 percent of rated line voltage. This capability keeps the plant from shutting down as a result of short-term voltage fluctuation.

[4]Voltage control ability provides control of wind turbine voltage output.

[5]Output control ability allows the power produced to be reduced by feathering the blades.

[6]Ramp rate management allows the power output to stay within the increase or decrease limits required by the system.

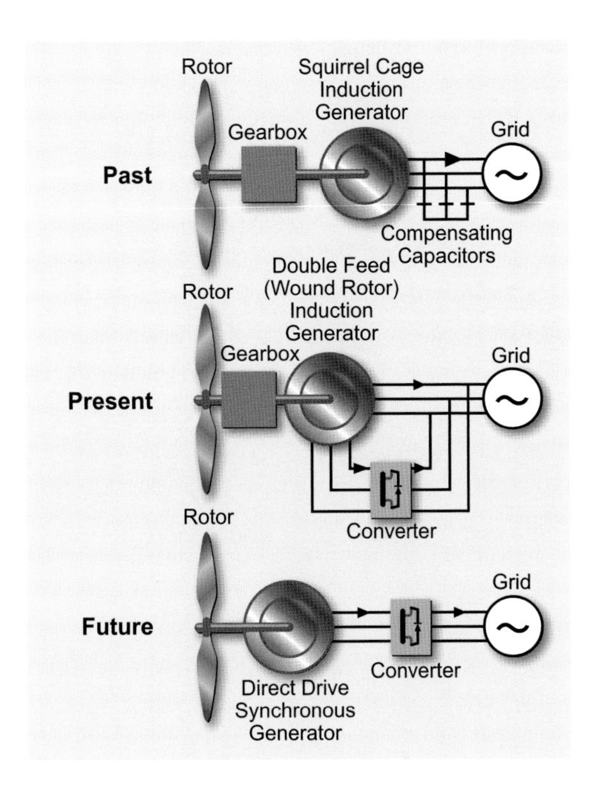

FIGURE 3.2 *Evolution of wind turbine technology.*
Source: IEEE, 2005. Copyright 2005 IEEE. Reprinted by permission.

reactive (VAR) support.[7] While wind generators have increased in height and rotor diameter, the major changes in internal operating characteristics are not as apparent. Figure 3.2 depicts the evolution of the internal operating characteristics. Many perceptions of wind technology's negative impact on the electrical system, such as the inability to remain connected to the electricity grid during voltage disturbances and the draw on the grid's reactive power resources, stem from Type 1 machines.

The evolution of control technologies has made wind generators and their electricity output easier to integrate into the utility system. With these new control technologies, wind power plants are better at mimicking traditional generating plants. This capability led to Federal Energy Regulatory Commission (FERC) Order 661-A, issued December 2005, which deals with machine design and system integration. It calls for wind facilities of 20 MW or larger to provide the ability

[7]VAR support provides reactive power compensation to aid in electricity grid stability.

to maintain operations, including LVRT, during disturbances on the electric grid; provide reactive power; and maintain continuous real-time communications and data exchange with the control area operator. These power integration capabilities have been incorporated into Type 3 machines. However, wind power generation takes place where and when the wind blows, and electricity must be used when it is generated. This intermittency has raised concerns about integrating wind power into the existing power system and requires wind turbines to provide LVRT, voltage control, output and ramp rate controls, and VAR support.

Integrating Type 3 machines into existing grids is not without its challenges. Circumstances such as wind fluctuations and overall grid stability are unique to each particular control area. Thus, even as technologies improve, it will be critical to carry out site-specific analyses of each control area, which will better aid grid operators in balancing the system within their control area.

Integration into Utility System Operation

A number of studies on the integration of wind power into a utility capacity and dispatch structure indicate that wind can be integrated at up to approximately 20 percent of the total electricity mix without requiring storage, although the exact level depends on the power system (Parsons et al., 2006; ETSO, 2007; DOE, 2008).[8] The specifics of these studies are discussed in this report in the chapters on economics (Chapter 4), deployment (Chapter 6), and scenarios (Chapter 7). As the studies point out, achieving such levels of renewables penetration will depend on upgrades to the grid (necessary regardless of the energy mix) and new transmission lines for more remote sources.

Modern electricity grid systems are designed to handle loss of the largest power plant without disruption; to have ramp up and ramp down capabilities: and to increase or decrease generation as demand increases or decreases. However, each system has its own generating capacity structure, transmission capabilities, and ability to purchase power outside its own boundaries, making wind power integration somewhat unique for each utility.

---

[8]A number of studies can be found on the Utility Wind Integration Group (UWIG) website at http://www.uwig.org.

Small Wind Systems

The vast majority of wind power is generated by large wind turbines feeding into the electricity grid, while small wind turbines generally provide electricity directly to customers. The United States is the leading world producer of small wind turbines. These residential turbines are erected and connected directly to the customer's facility or to the electricity distribution system at the customer's site. The manufacture and marketing of wind-powered electric systems sized for residential homes, farms, and small businesses have experienced major growth in the past decade. These small wind turbines (Figure 3.3), defined as 100 kW or less in capacity, have seen significant market growth, and the industry has set ambitious targets: growth at 18–20 percent through 2010.

FIGURE 3.3 *Small wind turbine, shown near home with rooftop photovoltaic panels installed.*
Source: Courtesy of National Renewable Energy Laboratory.

Key Technology Opportunities

Short Term: Present to 2020

The key technological issues for wind power focus on continuing to develop better turbine components and to improve the integration of wind power into the electricity system, including operations and maintenance, evaluation, and forecasting. Goals appear relatively straightforward: taller towers; larger rotors; power electronics; reducing the weight of equipment at the top and cables coming from top to bottom; and ongoing progress through the design and manufacturing learning curve (Thresher et al., 2007; DOE, 2008). Table 3.1 summarizes the incremental improvements under consideration.

Although no big breakthroughs are anticipated, continuous improvement of existing components is anticipated, and many are already being actively developed. For example, there are advanced rotors that use new airfoil shapes specifically designed for wind turbines instead of those based on the design of helicopter blades. These rotors are thicker at points of highest stress and reduce loads during turbulent winds by flying the blades using turbine control systems. Other improvements include the use of composite materials and advanced drivetrains. In particular, gearboxes are a major area of concern for reliability. Approaches for improving this component include direct-drive generators; greater use of rare-earth permanent magnets in generator design; possibility of single-stage drives using low-speed generators; and distributed drivetrains using the rotor to drive several parallel generators. Advanced towers are a major focus for innovation, given the current need for large cranes and transport of large tower and blade sections. Concepts under investigation include self-erecting towers, blade manufacturing on site, vibration damping, and tower–drivetrain interactions.

There is certain to be some development of offshore wind in the United States in the near term, but it is not expected that this will have a significant impact before 2020. Nonetheless, there is a near-term opportunity to learn from offshore projects in Europe and the United States, if offshore wind is going to have an impact in the medium term.

Other near-term opportunities will lie in improving the integration of existing wind power plants into the transmission and distribution system, which includes using improved computational models for simulating and optimizing system integration (Ernst et al., 2007). Chapters 6 and 7 discuss the deployment and integration of wind-generated electricity.

TABLE 3.1 Areas of Potential Wind Power Technology Improvements

Technical Area	Potential Advances	Performance and Cost Increments Best/Expected/Least (%) Annual Energy Production	Performance and Cost Increments Best/Expected/Least (%) Turbine Capital Cost
Advanced tower concepts	• Taller towers in difficult locations • New materials and/or processes • Advanced structures/foundations • Self-erecting, initial, or for service	+11/+11/+11	+8/+12/+20
Advanced (enlarged) rotors	• Advanced materials • Improved structural-aero design • Active controls • Passive controls • Higher tip speed/lower acoustics	+35/+25/+10	–6/–3/+3
Reduced energy losses and improved availability	• Reduced blade soiling losses • Damage-tolerant sensors • Robust control systems • Prognostic maintenance	+7/+5/0	0/0/0
Drivetrains (gearboxes and generators and power electronics)	• Fewer gear stages or direct-drive • Medium- to low-speed generators • Distributed gearbox topologies • Permanent-magnet generators • Medium-voltage equipment • Advanced gear tooth profiles • New circuit topologies • New semiconductor devices • New materials (gallium arsenide [GaAs], SiC)	+8/+4/0	–11/–6/+1
Manufacturing and learning curve[a]	• Sustained, incremental design and process improvements • Large-scale manufacturing • Reduced design loads	0/0/0	–27/–13/–3
Totals		+61/+45/+21	–36/–10/+21

[a]The learning curve results from NREL (2008) (Cohen and Schweizer et al., 2008) are adjusted from 3.0 doubling in the reference to the 4.6 doubling in the 20 percent wind scenario.
Source: DOE, 2008.

Medium Term: 2020 to 2035

Mid-term wind technology development will have two thrusts: the movement toward offshore, and its implications for turbine design; and the development of efficient low-wind speed turbines. Development of offshore wind power plants has already begun in Europe (approximately 1200 MW of installed capacity), but progress has been slower in the United States. Nine projects are in various stages of development in state and federal waters. In addition to technical risks and higher costs, these projects have been slowed by social and regulatory challenges (DOE, 2008).

In the mid-term, offshore turbines will have a larger size and generating capacity than onshore turbines, but, owing primarily to technical and cost concerns, development will likely lag behind onshore machines. Transmission siting issues with offshore wind power plants will be simplified because of fewer siting impediments. However, underwater cables must be carefully constructed, and there will likely be a move to develop microgrids with high-voltage direct current to integrate the offshore resources. Offshore wind technologies face several transition problems as they move from near-shore, land-based sites to offshore sites of various depths and, finally, floating designs. Assessment tools for sensitive marine areas, wind loads, and system design are not now ready for offshore development. Offshore projects must be built to handle both wind and wave loads, and components must be able to endure marine moisture and extreme weather. Offshore wind projects have a higher balance of station cost (approximately two-thirds of total costs) than do onshore projects, and thus will rely on cost reductions across the system in order to become more competitive. All of these developments pose both technological and organizational problems and will require continuous research and development in order to be feasible. It should be noted that challenges posed by the greater technical difficulties of offshore wind power development are being addressed by other countries. However, political, organizational, social, and economic obstacles may continue to inhibit investment in offshore wind power development, given the higher risk compared to onshore wind energy development (Williams and Zhang, 2008).

In terms of onshore development, as the higher wind speed sites are used, wind power development will move to lower wind speed sites, which will require turbines that are relatively efficient at lower wind speeds, necessitating larger rotors with lighter, stronger materials, as well as increased tower height.

Long Term: After 2035

At present, no revolutionary technology to extract energy from wind has been proposed, but several designs, e.g., vertical wind turbines or eggbeaters, are again under consideration. There have been conceptual proposals to access high-altitude winds using balloons or kites. Component improvements will continue, with additional emphasis on offshore turbine installation. Floating offshore platforms may gain interest, but first must come experience from anchored offshore wind facilities.

Summary of Wind Power Potential

Wind-power technologies are actively deployed today, and there are no technological barriers to continued deployment. Cost reductions will be possible as a result of wider deployment and incremental improvements in components. No other enhancing technologies are required for wind power to meet 20 percent and higher of U.S. electricity demand.

SOLAR PHOTOVOLTAIC POWER

Solar power involves the conversion of the radiant energy from the sun into electricity by using photovoltaics (PV) or concentrating devices. When sunlight strikes the surface of the PV cell, some of the photons are absorbed and release electrons from the solar cell that are used to produce an electric current flow, i.e., electricity. A solar cell consists of two layers of materials, one that absorbs the light and the other that controls the direction of current flow through an external circuit (Figure 3.4). The absorbing materials can be silicon (Si), which is also used in integrated circuits and computer hardware; thin films of light-absorbing inorganic materials, such as cadmium telluride (CdTe) or gallium arsenide (GaAs), that have absorption properties well matched to capture the solar spectrum; or a variety of organic (plastic) materials, nanostructures, or combinations.

Status of Technology

The PV industry has grown at a rate greater than 40 percent per year from 2000 through 2008. Much of this growth is the result of national and local programs targeted toward growing the PV industry and improving the competitiveness of PV in the marketplace. In 2007, PV modules supplying 3.4 GW were produced worldwide, and approximately 220 MW were installed in the United States.[9] Table 3.2 provides a breakdown of PV module shipments by technology type.

---

[9]See http://www.solarbuzz.com/Marketbuzz2008-intro.htm.

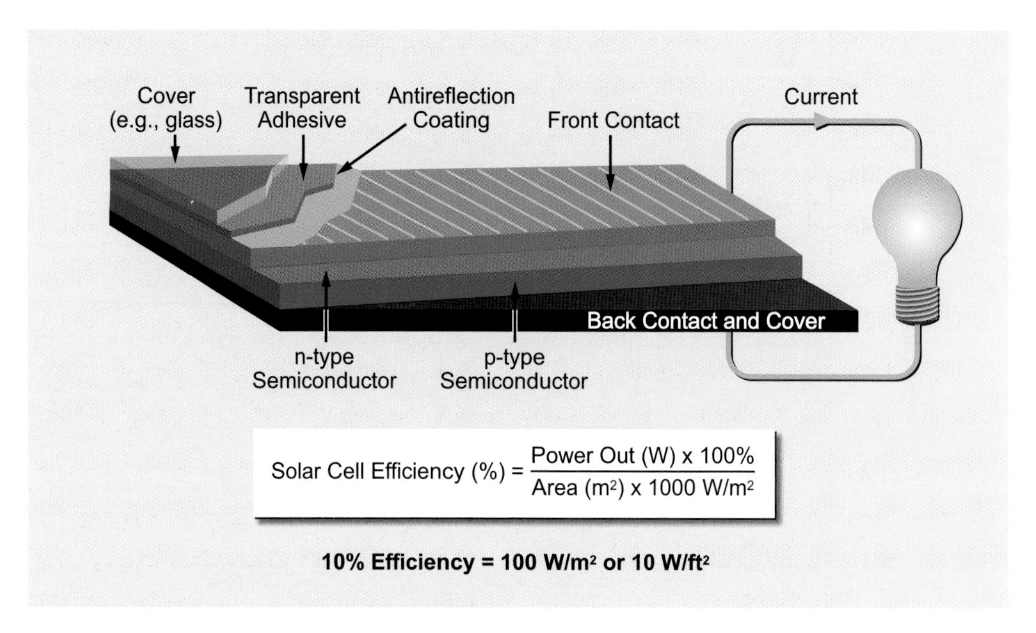

FIGURE 3.4 *Schematic of a typical solar cell.*
Source: DOE, 2005.

TABLE 3.2 PV Cell and Module Shipments by Type, 2005–2007

	Shipments (Peak Kilowatts)			Percent of Total		
Type	2005	2006	2007	2005	2006	2007
Crystalline silicon						
Single-crystal	71,901	85,627	128,542	32	25	25
Cast and ribbon	101,065	147,892	181,788	45	44	35
Subtotal	172,965	233,518	310,330	76	69	60
Thin-film	53,826	101,766	202,519	24	30	39
Concentrator	125	1,984	4,835	[a]	1	1
Other[b]	—	—	—	—	—	—
U.S. Total	226,916	337,268	517,684	100	100	100

Note: Totals may not equal sum of components due to independent rounding. —, no data reported.
[a]Less than 0.5 percent.
[b]"Other" includes categories not identified by reporting companies.
Source: EIA, 2008, Table 3.5.

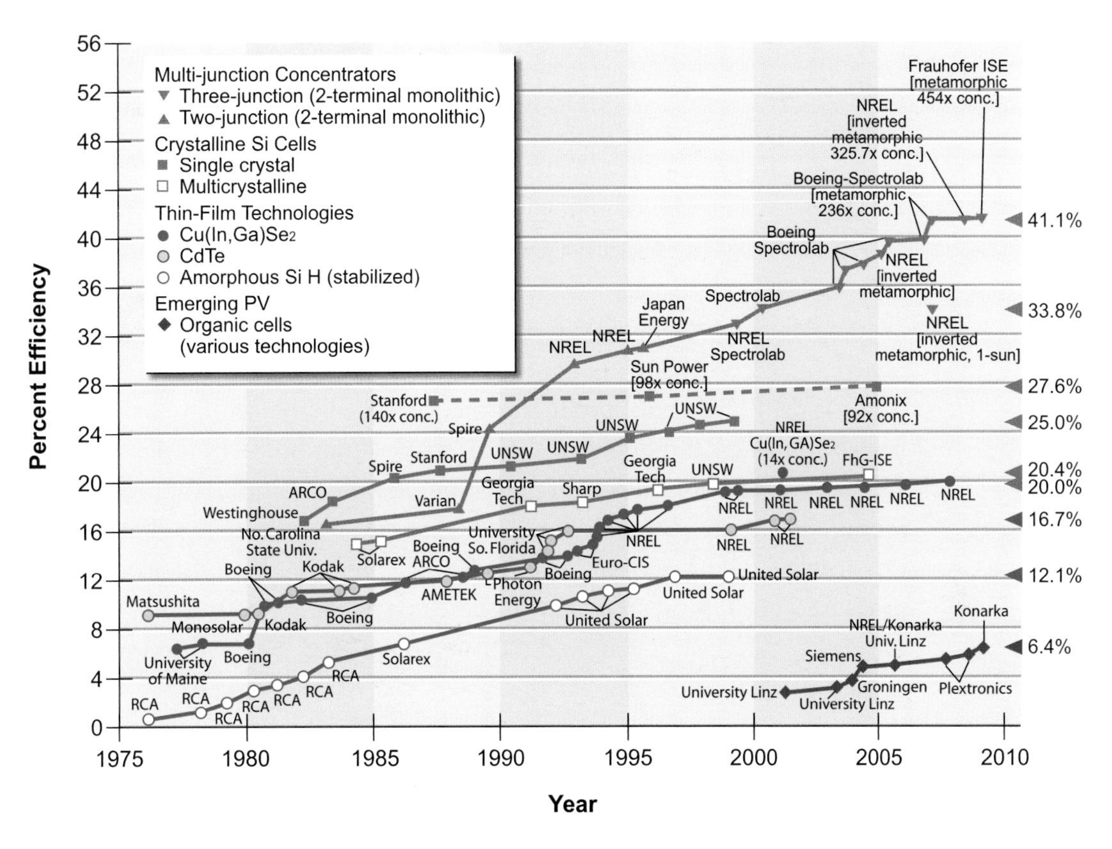

FIGURE 3.5 *Historical progress of solar cell efficiencies.*
Source: NREL, 2009.

Over the past 30 years, the efficiency of PV technologies has steadily improved. Figure 3.5 presents the historical progress of the best reported solar cell efficiencies through 2008 (NREL, 2009). Commercial (or even the best prototype) modules achieve, on average, only about 10–15 percent efficiency, which is 50–60 percent of the efficiency of the best research cells. Figure 3.5 includes several PV technologies: single-crystalline Si, thin films, multi-junction cells, and emerging technologies, such as dye-sensitized nanocrystalline TiO_2 cells, cells based on organic compounds, and plastic solar cells.

Flat-Plate PV Technologies

Photovoltaic technologies can be divided into two main types: flat plates and concentrators. Flat-plate technologies include crystalline silicon (from both ingot and ribbon- or sheet-growth techniques) and thin films of various semiconductor materials, usually deposited on a low-cost substrate, such as glass, plastic, or stainless steel, using some type of vapor deposition, or wet chemical process. Thin film cells typically are 1–20 μm in thickness and require one-tenth to one-hundredth of the expensive semiconductor material required by crystalline silicon (DOE, 2007f). Additionally, thin film deposition technology allows production of large-area solar cells, and though they exhibit lower efficiencies (upward of 10 percent) than crystalline silicon PV panels, their lower production costs can make them an attractive alternative. Even thinner layers are involved in some of the future generation technologies, such as organic polymers and nanomaterials (DOE, 2007j).

Of the PV modules produced today, nearly 88 percent are based on crystalline silicon wafer technologies. Of this total, about 30 percent are based on conventional, single-crystal silicon grown by the Czochralski ingot process,[10] 60 percent are based on polycrystalline (also referred to as multicrystalline) ingots cast in a crucible, and 3 percent are from silicon ribbons/sheet produced by various processes. The typical efficiency of these crystalline PV cells is 12–18 percent, and further development is required to increase the efficiency and to lower the production cost (DOE, 2007e).

Concentrator PV Technology

The key elements of a concentrator PV system are low-cost concentrating (reflective or refractive) optics, low-cost mounting and tracking systems (to track the movement of the sun), and high-efficiency III-V[11] or silicon solar cells (DOE, 2007g). The large-scale manufacturing capability for all components has already been demonstrated, including 27 percent efficient silicon cells and 28 percent efficient GaAs cells (DOE, 2007g; Surek, 2001). Concentrator systems using point-focus Fresnel lenses have been routinely fabricated. Module efficiencies of up to 20 percent have been demonstrated by commercially made 25 percent efficient silicon solar cells (DOE, 2005). Progress in multi-junction, III-V based solar cells for space applications has led to evaluating their terrestrial potential in concentrating applications (Bett et al., 1999; DOE, 2007g). An efficiency of 37.3 percent (at up to 600 times the sun's normal intensity) has been achieved for a $GaInP_2$/GaInAs/Ge triple-junction structure (King et al., 2004), and NREL has recently announced

[10]A method of crystal growth commonly used to obtain single crystals of semiconductors.

[11]III-V compounds (the III and V indicate the column location on the Periodic Table) are the basic materials for modern optoelectronic devices typically used in high-speed transistors (Bett et al., 1999).

an efficiency of almost 41 percent (at 380 times the sun's intensity) (NREL Press Release, August 13, 2008).

Concentrated photovoltaic (CPV) plants are composed of many aggregated photovoltaic modules, as are non-concentrator plants, but the required cell area is reduced by the concentration factor (DOE, 2005). The concentration ratio[12] of one-axis CPV systems is commonly 10:50. High-concentration PV (HCPV) systems use two-axis trackers with concentration ratios of 200:500. Concentration makes the use of the most efficient and expensive PV cells more practical. Mature HCPV systems are projected to cost about 40–60 percent of standard PV systems and to provide 10–20 percent more energy with the same power rating. Projections put the installed costs of CPV with multi-junction PV cells now under development at about $2/W (DOE, 2007g). The present cost of single-junction systems from Amonix and Solar Systems Pty Ltd. is, for example, about $4/W.

Potential Technology Development

Future directions for thin-film technologies include multi-junction thin films aimed at significantly higher conversion efficiencies, better transparent conducting oxide electrodes, thin polycrystalline silicon films, and organic inks.

Concentrator systems use only direct, rather than diffuse or global, solar radiation; therefore, their areas of best application (e.g., in the southwestern United States) are more limited than those for flat plates. There is also ongoing research to improve the long-term reliability of concentrator systems and to develop standard tests for concentrator cells and systems. Thus, most of today's remote and distributed markets for PV systems are not suitable for concentrator systems.

By far the fastest-growing segment of the PV industry is that based on casting large, multicrystalline ingots in some crucible that is usually consumed in the process. Manufacturers routinely fabricate large multicrystalline silicon solar cells with efficiencies in the 13–15 percent range; small-area research cells are 20 percent efficient. Silicon ribbon or sheet technologies avoid the costs and material losses associated with slicing ingots. The present commercial approaches in the field are the edge-defined, film-fed growth of silicon ribbons and the string ribbon process. Full-scale production of silicon modules based on micron-sized

---

[12]Defined as the average solar flux through the receiver aperture divided by the ambient direct normal solar insolation.

silicon spheres was recently announced. In this process, submillimeter-size silicon spheres are bonded between two thin aluminum sheets, processed into solar cells, and packaged into flexible, lightweight modules. Another approach uses a micromachining technique to form deep narrow grooves perpendicular to the surface of a 1- to 2-mm-thick single-crystal silicon wafer. This technique results in large numbers of thin (50 μm), long (100 mm), and narrow (nearly the original wafer thickness) silicon strips that are processed into solar cells just prior to separation from the wafer. In another technique, a carbon foil is pulled through a silicon melt, resulting in the growth of two thin silicon layers on either side of the foil. After the edges are scribed and the sheet is cut into wafers, the carbon foil is burned off, resulting in two silicon wafers (150 μm thick) for processing into solar cells.

Thin-film technologies have the potential for substantial cost advantages over wafer-based crystalline silicon because of factors such as lesser material use due to direct band gaps, fewer processing steps, and simpler manufacturing technology for large-area modules. Thin-film technologies commonly require less or no high-cost crystalline Si. Many of the processes are high throughput and continuous (e.g., roll-to-roll); they usually do not involve high temperatures and, in some cases, do not require high-vacuum deposition equipment. Module fabrication, involving the interconnection of individual solar cells, is usually carried out as part of the film-deposition processes. The major systems are amorphous silicon, cadmium telluride,[13] and copper indium diselenide[14] (CIS) and related alloys (DOE, 2007h). Future directions include multi-junction thin films aimed at significantly higher conversion efficiencies, better transparent conducting oxide electrodes, and thin polycrystalline silicon films.

Dye-sensitized Solar Cells

The dye-sensitized solar cell (O'Regan and Grätzel, 1991) has its foundation in photochemistry rather than in solid-state physics. In this device, also called the "Grätzel cell" after its Swiss inventor, organic dye molecules are adsorbed on a nanocrystalline titanium dioxide (TiO_2) film, and the nanopores of the film are filled with a redox electrolyte. The dyes absorb solar photons to create an excited

[13]CdTe PV cells require a small amount of semiconductor, and the production can be automated, which can increase its yield.

[14]CIS has higher efficiency and has the capability to be made on a flexible substrate, but large-scale production might be limited to the availability of indium.

molecular state that can inject electrons into the TiO_2. The electrons percolate through the nanoporous TiO_2 film and are collected at a transparent electrode. The oxidized dye is reduced back to its initial state by accepting electrons from the redox relay via ionic transport from a metal counter-electrode; this completes the circuit, and electrical power is delivered in the external circuit. Dye-sensitized solar cells are very attractive because of the very low cost of the constituent materials (TiO_2 is a common material used in paints and toothpaste) and the potential simplicity of their manufacturing process. Additionally, sensitized solar cells are tolerant to impurities, which allow ease in scaling up the production. Laboratory-scale devices of 11 percent efficiency have been demonstrated, but larger modules are typically less than half that efficient. Stability of the devices (e.g., dye materials and electrolyte) while maintaining high efficiency is an ongoing research issue (DOE, 2007k).

Organic and Nanotechnology Solar Cells

Organic semiconductors hold promise as building blocks for organic electronics, displays, and very low-cost solar cells. In an organic solar cell, light creates a bound electron-hole pair, called an exciton, which separates into an electron on one side and a hole on the other side of a material interface within the device. Polymers, dendrimers, small molecules and dyes, and inorganic nanostructures are materials that can be used (DOE, 2007j). Organic solar cells can be about 10 times thinner than thin-film solar cells. Consequently, organic solar cells could lower costs in four ways: low-cost constituent elements (e.g., carbon, hydrogen oxygen, and nitrogen sulfur); reduced material use; high conversion efficiency; and high-volume production techniques (e.g., high-rate deposition on roll-to-roll plastic substrates). Organic solar cells are the focus of DOE's research goals for 2020 (DOE, 2007j). Research examples in organic solar cells include quantum dots embedded in an organic polymer, liquid-crystal (small-molecule) cells, and small-molecule chromophore cells. Solar cell efficiencies to date are modest (less than 3–5 percent). Unresolved problems associated with this technology include large optical bandgap, unoptimized band offset, and fast degradation rate due to photoxidation, interfacial instability delamination, interdiffusion, and morphological changes (DOE, 2007j).

The use of nanotechnology for PV is especially promising, because the optical and electronic properties of the materials could be tuned by controlling particle size and shape (DOE, 2007m).[15] They may be easy to manufacture when the nanoparticles are produced by means of chemical solution. Some of these concepts are already being pursued commercially. Long-term stability of these devices is another major issue to resolve, along with increasing the efficiency.

Key Technology Opportunities

Short Term: Present to 2020

Currently, polycrystalline silicon PV technologies are well developed and commercially available. Given its higher cost compared to fossil-based electricity now and for the foreseeable future, deployment of the existing PV technology will be constrained only by the extent of financial incentives and the absence of policies that encourage use of solar electricity technology in the nation's electricity mix. Improvement in thin-film technology efficiencies, which cost less but are less efficient than Si-base cells, is important for the development of this technology.

Balance-of-systems costs must be brought down significantly to reduce the whole cost of a solar electricity system. For example, in California at present, approximately 50 percent or more of the total installed cost of a rooftop PV system is not in the module cost but in the costs of installation and of the inverter, cables, support structures, grid hookups, and other components. These costs must come down through innovative system-integration approaches, or this aspect of a PV system will set a floor on the price of a fully installed PV system, either freestanding or in a rooftop installation. In addition, PV interface devices must improve, including integrated PV inverters; disconnect, metering, and communications interfaces; direct PV-DC devices such as DC-driven end-use devices; and master controllers for use in buildings with PV, storage, and end users.

Medium Term: 2020 to 2035

Cost reductions are needed through new technology development and in the manufacturing that will accompany the scale-up of existing PV technologies. For example, new technologies are being developed to make conventional solar cells by using nanocrystalline inks of precursor as well as semiconducting materials.

---

[15]Includes nanowires, nanotubes, and nanocrystals, including single-component, core-shell, embedded nanowires or nanocrystals, as either absorbers or transporters.

New cell structures are being investigated to produce higher efficiency at lower cost.

Thin-film technologies have the potential for substantial cost reduction over current wafer-based crystalline silicon methods because of factors such as lower material use (due to direct band gaps), fewer processing steps, and simpler manufacturing technology for large-area modules. Thin-film technologies have many advantages, such as high throughput and continuous production rate, lower-temperature and non-vacuum processes, and ease of film deposition. Even lower costs are possible with plastic organic solar cells, dye-sensitized solar cells, nanotechnology-based solar cells, and other new PV technologies.

Long Term: After 2035

Widespread deployment of PV technology will depend on the ability to reach scale in manufacturing capacity and achieve cost reductions using technologies for ultralow-cost module production at acceptable efficiency. Reaching ultralow costs will probably require learning-curve-based cost reduction, along with development of future generations of PV materials and systems to increase efficiency. Next-generation PV cells will most likely have structures that will make optimal use of the total solar spectrum to maximize light-to-electricity conversion efficiency.

Summary of Solar PV Potential

A wide range of solar PV technologies are now at various levels of development. Silicon flat-plate PV technologies are mature and actively deployed today. Reduction in the production cost of the cell and an increase in efficiency and reliability will make silicon PV cells even more attractive to customers. New technologies such as thin film, which has great potential to reduce the module cost, are in a relatively mature development stage, with further research and testing required. Other competing technologies, such as dye-sensitized PV and nanoparticle PV, are at an early stage of development, and commercialization will require much more technology development.

The PV industry has a roadmap that sets a deployment goal of 200 GW peak (GW_p) in the United States by 2030 (SEIA, 2004). Chapter 7 describes the PV roadmap and other future scenarios for PV. Actual deployment rates will depend on national commitment and policy incentives. This 200 GW potential represents about a 500-fold increase over currently installed capacity in the United States, a much larger expansion than for the other renewable technologies examined in this report. There is no resource base limitation that would preclude reaching this level of PV deployment; rather, cost, technology, and policy issues are the main variables.

CONCENTRATING SOLAR POWER

Concentrating solar power (CSP) systems use optics to concentrate beam radiation, which is the portion of the solar radiation not scattered by the atmosphere. The concentrated solar energy converts the sun's energy into high-temperature heat that can be used to generate electricity or drive chemical reactions to produce fuels (syngas or hydrogen). CSP, similar to CPV, requires high-quality solar resources (6.75 kWh/m^2 per day or greater), and this restricts its application in the United States to the southwest part of the country (see Figure 2.3).

Status of Technology

Solar thermal electric generation comprises three technologies: parabolic troughs, power towers (also known as central receiver concentrator), and dish-Stirling engine systems (also known as parabolic dishes). Figure 3.6 shows the basic design for CSP technologies. The difference in these technologies is the optical system and the receiver where the concentrated solar radiation is absorbed and converted to heat or chemical potential.[16] These differences also define the potential plant size from the smallest (dish-Stirling concentrator) to the largest (parabolic troughs and power towers).

Parabolic Trough

The most mature technology is the parabolic trough combined with a conventional Rankine cycle steam power plant. The concentrator uses concave, parabolic-shaped mirrors to focus the direct beam radiation on a linear receiver. The mirrors track the sun from east to west during the day. The linear receiver is typically a stainless steel tube with a solar selective surface surrounded by an evacuated glass tube. The ratio of the collector area to the absorber area (the concentration ratio) is on the order of 100 or less. Recently, compact linear Fresnel reflec-

[16] See http://www.energylan.sandia.gov/sunlab/overview.htm#tower.

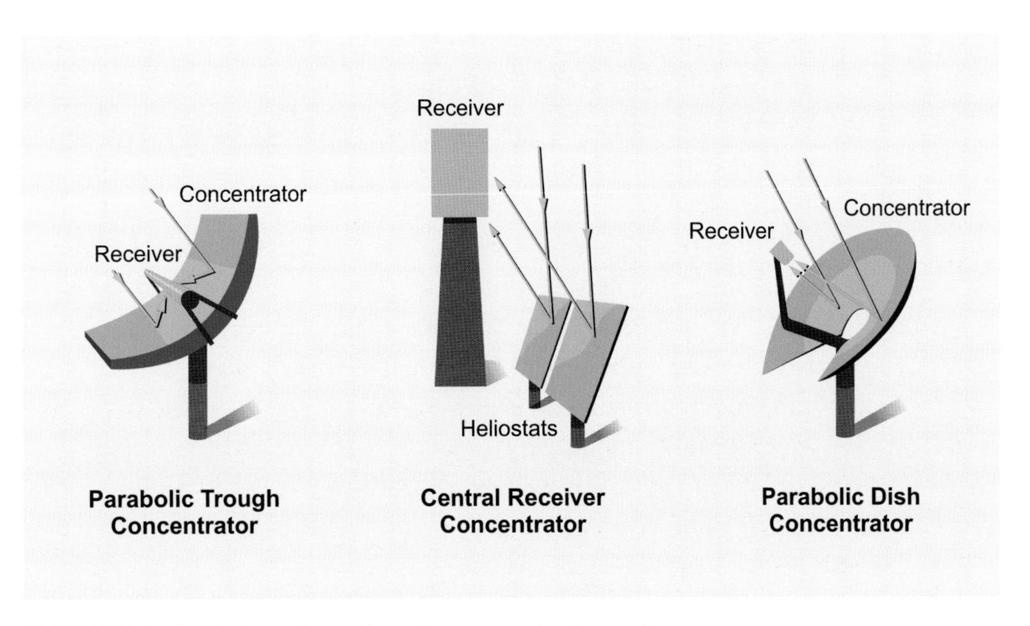

FIGURE 3.6 *Optical configurations for concentrating solar power.*

tors have been commercialized for use with stationary tubular receivers (Mills et al., 2004). These Fresnel reflectors may reduce the cost, but they have lower efficiencies and a shorter track record than the parabolic trough design. In the commercial parabolic trough systems, synthetic oil is circulated in the tubes. Oil can reach temperatures of about 370°C. The heated oil is used to superheat steam, which in turn drives a conventional turbine/generator to produce electricity. Individual trough systems can generate about 80 MW of electricity.[17] A collector field comprises many troughs in parallel rows aligned on a north-south axis.

The first parabolic trough plant—Solar Energy Generating Systems (SEGS)—was constructed in 1984 by Luz International in the California Mojave Desert near Barstow. In 1990, the installed capacity of the SEGS facility reached 354 MW. The plant has operated continuously since installation, and Southern California Edison purchases the electricity.

Parabolic trough plants can include solar energy storage capabilities, e.g., concrete, molten salt, and thermocline storage, that can extend generation for several hours. At present, many plants have a backup fossil-fired capability that

[17]For more information on the California SEGS design, see http://solar-thermal.anu.edu.au/high_temp/concentrators/basics.php.

can be used to supplement the solar output during periods of low solar radiation and at night. The SEGS facility includes natural gas generation. Annual solar-to-electric conversion efficiency is 12–25 percent, with capacity factors of 26–28 percent without storage. More recent plants in the United States are the 64 MW Nevada One plant, developed by Solargenix and operational since 2007, and the 1 MW Saguaro plant in Arizona. The Nevada One plant includes a natural gas component that may supply about 2 percent of the plant's total output.

The Integrated Solar Combined Cycle System (ISCCS) integrates a parabolic trough plant with a gas turbine combined-cycle plant. The ISCCS uses solar heat to supplement the waste heat from the gas turbine to augment power generation in the steam Rankine cycle.

Power Towers

Power towers consist of many two-axis mirrors (heliostats) that track the sun and direct the incoming beam radiation to a receiver located at the top of a tower. The first commercial plant is an 11 MW steam receiver plant developed by Abengoa and inaugurated in March 2007 near Sevilla, Spain. Known as PS10, the plant has a 114-meter tower and 624 heliostats, each 120 square meters. The plant uses a saturated steam receiver and includes a 20 MW_p water storage component. The developer reports a solar- to-electric conversion efficiency of 17 percent. Spain's electric feed-in law, set at 18 euro ¢/kWh at all times, and European Union (EU) and government subsidies for the plant totaling 6.2 million euros were the main drivers for the plant. A 20 MW power tower plant is under construction adjacent to PS10 at the Solúcar Solar Park. The solar field will consist of 1255 heliostats, each 120 square meters, and a 160-meter-high tower. Like PS10, the PS20 receiver will use steam technology.

Dish-Stirling Technology

Dish technology uses a two-axis parabolic dish to concentrate solar energy into a cavity receiver where it is absorbed and transferred to a heat engine/generator (Mancini et al., 2003). The concentration ratio is typically over 2000, and can be as high as 3000 with operation at temperatures of 750°C. Stirling engines are preferred over Brayton engines because of their high efficiencies (thermal-to-electric efficiencies are about 40 percent) and high power density (40–70 kW/liter). These systems are modular and as large as 25 kW, corresponding to a dish diameter of approximately 10 meters. The ideal concentrator shape is paraboloid, approximated with multiple spherically shaped mirrors or reflective membranes. The near-term markets identified by the developers of these systems include remote power, grid-connected power, and end-of-line power-conditioning applications. There is no large-scale solar dish Stirling plant to provide operational experience, but annual solar-to-electric efficiencies of 22–25 percent are predicted.

Potential Technology Development

A number of new CSP plants are under development or planned. In Spain, Abengoa is constructing a 20 MW power tower plant next to the PS10 plant. Recent developments include the AndaSol trough project, which is the first large-scale trough plant in Europe and the first anywhere with molten salt storage. The salt is a mixture of 60 percent sodium nitrate and 40 percent potassium nitrate. The Spanish government plans to have 10 GW of CSP within the next 5–7 years.[18]

There are a number of upcoming projects for CSP in the United States, particularly in California, which has an aggressive renewables portfolio standard (20 percent of investor-owned-utilities' loads to be served by renewables in 2010, with the same target intended for public utilities).[19] A number of utilities in the Southwest have formed a consortium to pursue 250 MW of new CSP plants.[20] The CSP industry estimates that 13.4 GW could be deployed for service by 2015 (WGA, 2006a). Purchase agreements for CSP of about 4 GW in the United States had been signed as of February 2009, but there is probably twice that capacity in planned projects.[21]

An evolving technology that relies on solar concentration is high-temperature chemical processing (Fletcher, 2001; Steinfeld, 2005; Perkins and Weimer, 2004). The concentrating component of these systems is identical to that of concentrated solar thermal processes for power generation, but the receiver placed at the focus of the concentrating reactor is designed to include a chemical reactor. These systems can provide long-term storage of intermittent solar energy, such as storage in the form of fuel or a commodity chemical. The global research community is

---

[18]Thomas Mancini, Sandia National Laboratories, personal communication, February 2, 2009.

[19]The lack of other strong renewable energy opportunities in the transmission-constrained state of California has pushed solar project bids ahead of wind power projects.

[20]See http://www.eere.energy.gov/news/news_detail.cfm/news_id=11474.

[21]Thomas Mancini, Sandia National Laboratories, personal communication, February 2, 2009.

pursuing a number of multi-step cycles, including production of hydrogen using water as the feedstock; decarbonization of fossil fuels; gasification of biomass; production of metals including aluminum; and processing and detoxification of waste. These systems are most likely to become cost-competitive when a cost is associated directly with a reduction in carbon emissions.

Key Technology Opportunities

Short Term: Present to 2020

CSP technologies are commercially available, and in the past few years new plants have been deployed in the United States and abroad, with trough systems dominating the U.S. CSP market. With nearly 4 GW of signed purchase agreements and additional planned projects, along with favorable financial policies, it is reasonable to expect significant growth by 2020. Most of the new plants are solar-only plants and do not include fossil-fuel backup on-site. During this timeframe, with the anticipated growth rate, CSP plants will continue to provide peaking power. With even more expanded growth, CSP technologies will probably be hybridized with fossil-fuel-fired components to share the generation portion of a fossil-fuel facility, as well as continue to serve as peaking plants.

In the short term, incremental design improvements will drive down costs and reduce uncertainty in performance predictions. With more systems installed, there will be increased economies of scale, both for plant sites and for manufacturing. Increasing the reflector size and working with low-cost structures, better optics, and high-accuracy tracking may reduce the cost of the heliostat or dish concentrators. There may also be design improvements in receiver technology.

Until 2020, long-term thermal storage, extending over days rather than hours, will not be a major roadblock. However, new storage technologies will be needed in the longer term to make solar dispatchable. Storage technologies, such as concrete, graphite, phase-change materials, molten salt, and thermocline storage, show promise. The number of molten salt tanks providing thermal storage on the order of hours will likely increase, as ancillary equipment such as pumps and valves are improved for greater reliability. Molten salt receivers, which provide storage at about 550°C to power a turbine, can extend storage up to 12 hours, but there are no molten salt receiver plants at this time.

Availability of water may not be a major deterrent, as water withdrawals are not large with CSP. However, as noted in Chapter 5, CSP consumes at least as much water as some conventional generation technologies. The primary water uses at a Rankine steam solar power plant are for condensate makeup, cooling for the condenser, and washing of mirrors. Historically, parabolic trough plants have used wet-cooling towers for cooling. With wet-cooling, the cooling tower makeup represents approximately 90 percent of the raw water consumption. Steam cycle makeup represents approximately 8 percent of raw water consumption, and mirror washing represents the remaining 2 percent. Dust-resistant glass is being explored as a possible means to reduce the mirror washing requirement. Chapter 5 includes additional discussion of water-use impacts.

Medium Term: 2020 to 2035

New demands on existing transmission systems may require new or upgraded lines. Longer-term storage on the order of days will be needed if CSP is to be a major source of electricity. Research and development will continue to accelerate design improvement and drive down manufacturing costs. Development of less expensive yet durable optical materials will help control cost and water use, including selective surfaces for receivers in towers and dishes, transparent polymeric materials that are cheaper than glass, and reflective surfaces that prevent dust deposition.

Long Term: After 2035

In the longer term, the use of concentrated solar energy to produce fuels and thus provide storage via a number of reversible chemical reactions is promising. Fuels produced from concentrated solar energy may provide a means of generating electricity during periods of low insolation or at night. Much of the scientific work to date has focused on the production of hydrogen and synthesis gas through various processes, including direct thermolysis of water and a number of metal oxide reduction/oxidation cycles. Direct water splitting is not feasible, because the required temperatures exceed the capability and material limits of modern concentrating systems, and separation of the products at such temperatures is impractical. Multi-step metal oxide reactions are more promising. A two-step process involves endothermic dissociation of a metal oxide (M_xO_y) to the metal (M) and oxygen in a solar reactor, followed by hydrolysis of the metal to produce hydrogen and the corresponding metal oxide. Carbothermal reduction in a solar reactor reduces the required operating temperature and yields syngas. The process is technically feasible but has not been demonstrated at production scale. Gasification of cellulosic biomass is another promising route to produce synthesis gas (Perkins and Weimer, 2009).

Summary of Concentrating Solar Potential

Concentrating trough and power tower systems are potentially the lowest-cost utility-scale solar electricity for the southwestern United States and other areas of the world with sufficient direct normal solar radiation. In the short term, incremental design improvements will drive down costs and eliminate uncertainties in performance predictions as more systems are installed (the learning curve), increasing economies of scale both for plant sites and for manufacturing. In the medium term, advances in high-temperature and optical materials are needed to reduce costs and improve performance further. An evolving long-term technology that relies on solar concentration is high-temperature chemical processing. Solar thermochemical production of fuels is a promising mechanism for storage of solar energy.

GEOTHERMAL POWER

Today, geothermal electricity is produced by conventional power-generating technologies using hydrothermal resources, hot water or steam, accessible within 3 km of Earth's surface. Existing plants operate 90–98 percent of the time and thus can provide baseload electricity. Growth of conventional hydrothermal energy is expected to be modest and regional in nature, occurring primarily in the western United States. More aggressive growth would be possible if the heat stored deeper below Earth's surface could be successfully mined. Enhanced geothermal systems[22] (EGSs) would use hydraulic stimulation to mine the heat stored in natural rock reservoirs. In the case of deep, low-permeability rock, hydraulic stimulation would create a porous or fractured reservoir through which fluid could be circulated and heated for use in a conventional generation plant. Figure 3.7 shows a schematic of the EGS system known as "hot dry rock geothermal." In sites with sufficient natural liquids, stimulation would open up flow paths for dry steam or superheated liquid water. For example, the Iceland Deep Drilling Project plans to access a high-temperature (400–650°C) hydrothermal resource 4–5 km deep at the Krafla, the Hengill, and the Reykjanes geothermal fields.[23]

---

[22]EGS is the term commonly used by DOE. It is synonymous with the earlier term "hot dry rock," which is still widely used.

[23] Bjorn Stefansson, Bjarni Palsson, and Guomundur Omar Frioleifsson, 2008, "Iceland Deep Drilling Project, exploration of supercritical geothermal resources," in IEEE Power and Energy

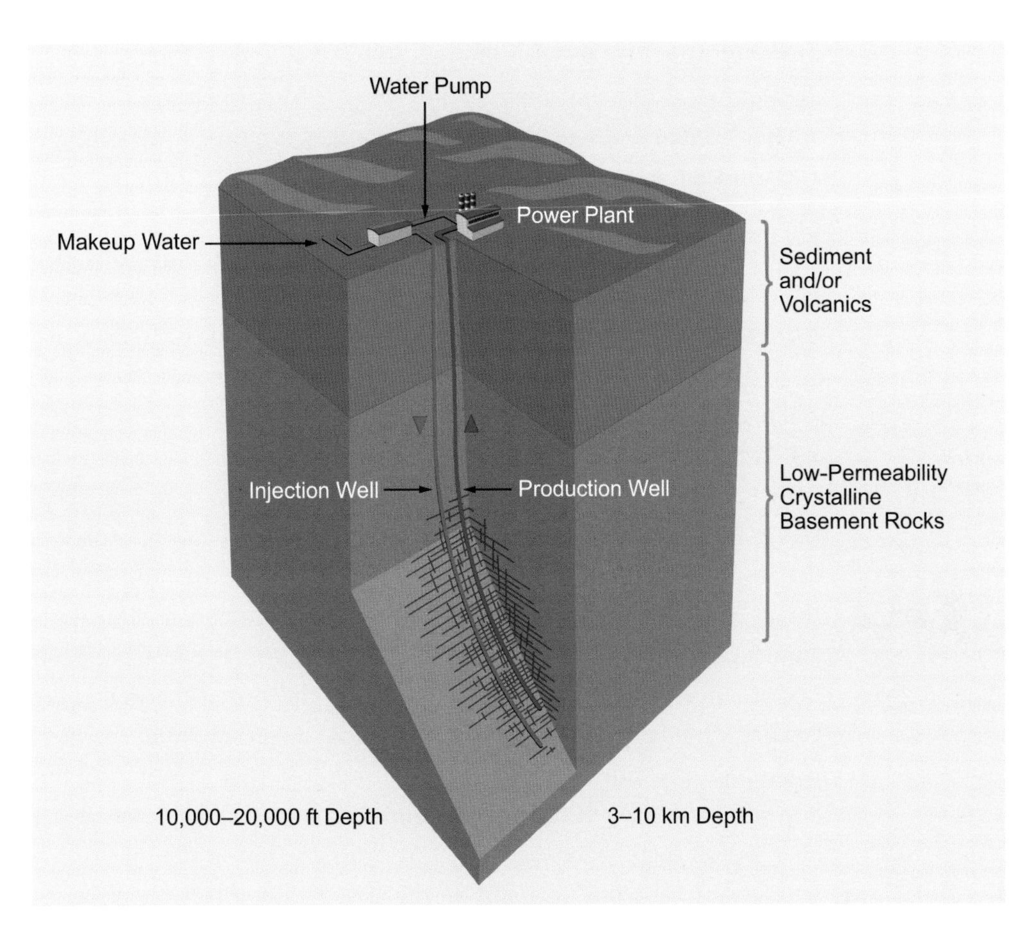

FIGURE 3.7 *Schematic of enhanced geothermal systems using an injection and production well.*
Source: MIT, 2006. Copyright 2006 MIT.

Status of Technology

Hydrothermal reservoirs are generally classified as either low temperature (less than 150°C) or high temperature (greater than 150°C), with high-temperature reservoirs more suitable for electricity production. Hydrothermal power plants are binary or steam. Binary plants are more prevalent than steam plants, because lower-temperature reservoirs suitable for binary plants are far more common than steam reservoirs. In addition, binary plants may be the best option at any temper-

Society 2008 General Meeting: Conversion and Delivery of Electrical Energy in the 21st Century, July 20–24, 2008, Pittsburgh, Pa.

ature for applications with limited water availability. Binary cycle plants convert geothermal waters, normally from 90 to 175°C, to electricity by routing the hot water through a closed-loop heat exchanger, where a low-boiling-point hydrocarbon, such as isobutane or isopentane, is evaporated to drive a Rankine power cycle. The cooled or "spent" geothermal fluid is returned to the reservoir. Because binary plants use a self-contained cycle, there are no emissions other than water vapor. Current electrical generation costs are 5–8¢/kWh (NREL, 2008). Steam plants either use steam directly from the source to directly drive a turbine, or use flash plants to depressurize hot water from the source (175–300°C) to produce steam. Energy produced via steam generation costs between 4–6¢/kWh (NREL, 2008).

Enhanced (or engineered) geothermal systems are not in operation yet, but if successfully developed, they would recover thermal energy stored at depths ranging from 3 to 10 km. This resource is vast, but it exists at great depths and low fluxes (see Chapter 2). Broad implementation presents technical and economic challenges because of the required drilling depths, the low permeability, and the need for reservoir enhancement. Accessing the stored thermal energy would first require stimulating the hot rock by drilling a well to reach the hot rock, and then using high-pressure water to create a fractured rock region. Drilling injection and production wells into the fractured region would follow next, and the stored heat would then be extracted by circulating water in the injection well. The heat extraction rate would depend on the site. Technologies for electricity generation from the hot fluid would be similar to those for hydrothermal power plants.

Potential Technology Development

Growth of conventional hydrothermal electricity is expected to be modest and to occur primarily in the western United States. As described in Chapter 2, the Western Governors' Association (WGA) assessed the potential for new development by 2015 of about 140 known and accessible geothermal sites. The WGA concluded that the western states share an untapped capacity of 5.6 GW that could be developed within the next 10 years, with levelized costs of energy (LCOE) of about 5.3–7.9¢/kWh, assuming that federal production tax credits (PTCs) remain in place (without the PTC, LCOE values would be 2.3¢/kWh higher) (WGA, 2006b). Table 3.3 provides a state-by-state list of potential capacity expansions. The Geothermal Energy Association has identified more than 100 geothermal projects under development in 13 states, which represents more than a doubling of con-

TABLE 3.3 Summary of Western States' Near-Term New Geothermal Power Capacity

	Capacity (in Megawatts)	Number of Sites
Alaska	20	3
Arizona	20	2
Colorado	20	9
California	2400	25
Hawaii	70	3
Idaho	860	6
Nevada	1500	63
New Mexico	80	6
Oregon	380	11
Utah	230	5
Washington	50	5
Total	5630	138

Note: Summary does not include the capacity of Wyoming, Montana, Texas, Kansas, Nebraska, South Dakota, and North Dakota.
Source: Based on data analyzed during the July 25 Geothermal Task Force subgroup meeting on supply, as presented in WGA, 2006a.

ventional geothermal capacity in the coming decade. No additional technological developments are required to tap these resources, although advances in exploration and resource assessment could affect growth of new plants.

The studies cited previously do not include EGS. Extensive development of EGS is less certain because of the lack of experience in recovering the heat stored at 3- to 10-km depths in low-permeability rock. The primary technical challenges are accurate resource assessment and understanding how to achieve sufficient connectivity within the fractured rock so that the injection and production well system can yield commercially feasible and sustainable production rates. Other unresolved issues involve induced seismicity, land subsidence, and water requirements. Modeling analysis shows a large capability for these wells to yield significant heat (MIT, 2006). However, given the depths needed, there has been limited experience and success in developing EGS wells at sufficient flow rates in the field. Issues associated with EGS, including reservoir operation and management, are summarized in the MIT report (MIT, 2006) and in a series of reports summarizing workshops sponsored by the DOE (DOE, 2007a,b,c,d).

Key Technology Opportunities

Short Term: Present to 2020

In the near term, development of geothermal sites will continue to rely on conventional extraction methods and technologies. Technology is not a major barrier to developing conventional hydrothermal resources, but improvements in drilling and power conversion technologies could result in cost reductions and greater reliability. There is a need for continued and updated resource assessment. There will also be additional EGS field demonstrations.

Medium Term: 2020 to 2035

As indicated in Table 3.4, the largest source of geothermal energy resides in the thermal energy stored in rock formations that require EGS technology for extraction. Implementation of EGS has not been demonstrated at large scale, and there are unanswered questions about the extent of economical power available. Reaching depths of 3 to 5 km is feasible for conventional drilling methods used in the oil and natural gas industry. However, a significant uncertainty is the flow rate achievable in an enhanced reservoir and the heat flux associated with this flow rate. Drilling for geothermal resources is somewhat different from drilling for oil and natural gas, especially since geothermal systems typically occur in crystalline rocks as opposed to much softer sedimentary rocks targeted by oil and

TABLE 3.4 Estimates of U.S. Geothermal Resource Base to 10-km Depth by Category

Category of Resource	Thermal Energy (in Exajoules; 1 EJ = 10^{18} J)	Reference
Conduction-dominated EGS		
Sedimentary rock formations	>100,000	MIT (2006)
Crystalline basement rock formations	13,900,000	MIT (2006)
Supercritical volcanic EGS[a]	74,100	USGS Circular 790
Hydrothermal	2,400–9,600	USGS Circulars 726 and 790
Coproduced fluids	0.0944–0.4510	McKenna et al. (2005)
Geopressured systems	71,000–170,000[b]	USGS Circulars 726 and 790

Note: EGS, enhanced geothermal systems; USGS, U.S. Geological Survey.
[a]Excludes Yellowstone National Park and Hawaii.
[b]Includes methane content.
Source: Adapted from MIT, 2006.

natural gas exploration. With present EIA projections of the price of electricity, successful implementation of EGS would require sustained production at 80 kg/s (equivalent to the rate at a productive hydrothermal reservoir) at a temperature of 250°C, which would generate about 5 MW per well (DOE, 2007b). The EGS project at Soultz, France (5,000-m-deep wells through crystalline rock), which is the best-performing project to date, has achieved a well productivity of about 25 kg/s. Advances in stimulation and higher productivity are likely as more field demonstrations are conducted. Figure 2.5 shows that temperatures of 250°C exist primarily at depths of 5.5 km and deeper. On the other hand, the MIT study cited very-high-grade EGS on the margins of hydrothermal systems or in high-thermal-gradient regions that could work well at depths of 3 km. Clear Lake, California, and the Fenton Hill, New Mexico, sites are good examples of these.

Field demonstrations at different high-grade thermal areas would aid a realistic assessment of the risks and potential of EGS. For cost-effective commercial extraction, the studies should demonstrate that EGS technology that is successful at one site can be applied successfully to other sites with different geologic characteristics. The challenges are the technical and economic uncertainty of site-specific reservoir properties, such as fractured rock permeabilities, porosities, and in situ stresses, and the difficulties of stimulating sufficiently large productive reservoirs, and connecting them to a set of injection and production wells.

Long Term: After 2035

Initial field studies of EGS will most likely focus on moderate depths (up to ~5.0 km). If successful, exploration at greater depth may be warranted and bring improved prospects for private investment and commercial deployment.

Summary of Geothermal Power Potential

Geothermal energy is a renewable energy resource that can provide baseload power without storage. Existing geothermal power plants rely on well-understood power plant technology but are restricted to hydrothermal resources within 3 km of Earth's surface. Large expansion of the U.S. geothermal electricity-generating capacity will rely on resources that are much less accessible. It will be necessary to access the hot rock at depths as great as 10 km. The technical challenge is economically bringing the stored thermal energy to the surface where it can be used to generate electricity. Advances in stimulation and higher productivity are likely as more field demonstrations are conducted.

HYDROPOWER

Technologies for converting energy from water to electricity include conventional hydroelectric technologies and emerging hydrokinetic technologies that can convert ocean tidal currents, wave energy, and thermal gradients into electricity. Conventional hydroelectricity, or hydropower, the largest source of renewable electricity, comes from capturing the energy from freshwater rivers and converting it to electricity.

Status of Technology

Conventional hydroelectricity is one of the least expensive sources of electricity. Hydropower has played a long and important role in the history of electrification in the United States. Federal development of large-scale hydropower projects during the 1930s and 1940s, such as those constructed as part of the Tennessee Valley Authority system and the Grand Coulee, Bonneville, and Hoover dams, aided in rural electrification and the development of the country's industrial base. Most hydropower projects use a dam to back up and control the flow of water, a penstock to siphon water from the reservoir and direct it through a turbine, and a generator that converts the mechanical energy to electricity. The amount of electricity produced is a function of the capacity of the turbines and generators, the volume of water passing through the turbines, and the hydraulic head (the distance that the water drops in the penstock). Different categories of hydropower include large conventional hydropower with generating capacity greater than 30 MW, low-head hydropower with a hydraulic head of less than 65 feet and a generating capacity of less than 30 MW, and micro-hydropower with a generating capacity of less than 100 kW. All of these categories rely on the same basic technologies.

Potential Technology Development

Conventional Hydropower

Since this resource has been extensively exploited, many prime sites are no longer available. Furthermore, there is increasing recognition of negative ecosystem consequences from hydropower development. Future hydropower technological developments will relate to increasing the efficiency of existing facilities and mitigating the dams' negative consequences, especially on anadromous fish. Existing hydropower capacity could be expanded by increasing capacity at existing sites; installing electricity-generating capabilities at flood control, irrigation, or water supply reservoirs; and developing new hydropower sites (EPRI, 2007a). Turbines at existing sites also could be upgraded to increase generation. However, none of these require new technologies. The future of hydropower will play out in the public policy debate, where the benefits of the electric power are weighed against its effects on the ecosystem.

Hydrokinetic Power

New technologies to generate electricity from waterpower include those that can harness energy from currents, ocean waves, and salinity and thermal gradients. Many pilot-scale projects are demonstrating technologies that tap these sources, but only a few of them are commercial-scale power operations at particularly favorable locations. Tapping tidal, river, and ocean currents is done using a submerged turbine. An example of one design is shown in Figure 3.8. There is no

FIGURE 3.8 *Verdant Power's 35 kW turbine design for converting tidal currents into electricity.*
Source: Verdant Power; available at http://www.nature.com/news/2004/040809/full/news040809-17.html.

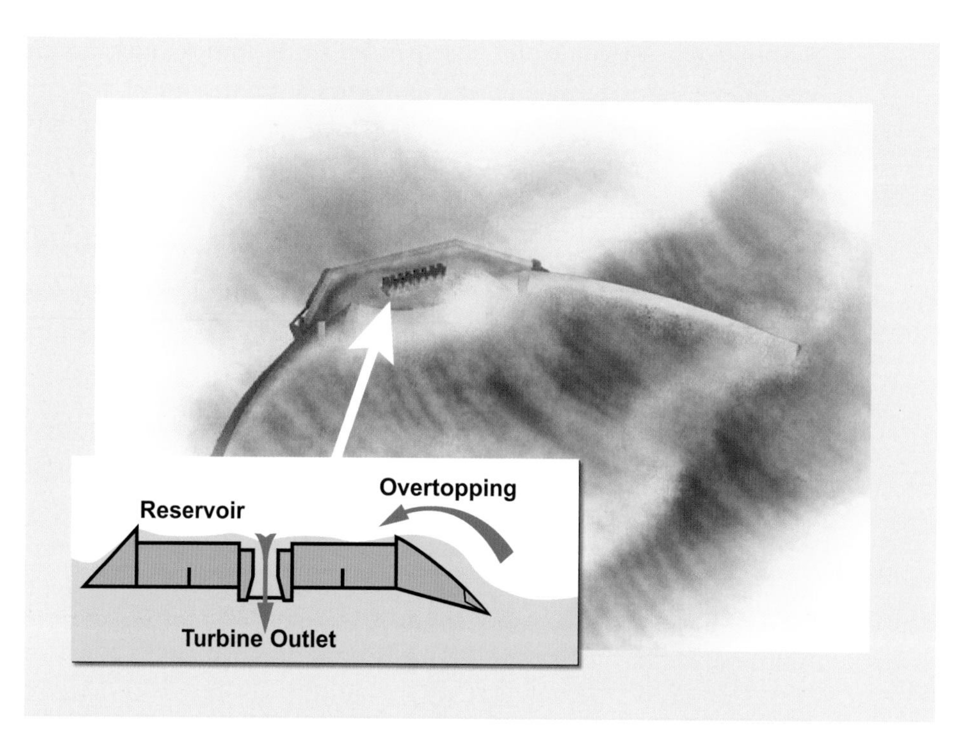

FIGURE 3.9 *Design of the Wave Dragon device.*
Source: Wave Dragon. Reprinted by permission.

single approach to converting the energy in waves into electricity. Approaches include floating and submerged designs that tap the energy in the impacting wave directly or that use the hydraulic gradient between the top and bottom of a wave (Minerals Management Service, 2006). Figure 3.9 shows the Wave Dragon device, which concentrates waves and allows them to overtop into a reservoir, generating electricity as the water in the reservoir drains out through a turbine.

Other approaches include long multi-segmented floating structures that use the differing heights to drive a hydraulic pump that runs a generator. Ocean thermal energy conversion converts solar radiation to electric power using the ocean's natural thermal gradient to drive a power-producing cycle. Designs using salinity gradient power would rely on the osmotic pressure difference between freshwater and salt water, although none of these have moved beyond the conceptual stage. In general, even though waves, currents, and gradients contain substantive amounts of energy resources, there are significant technological and cost issues to address before such sources can contribute significantly to electricity generation.

Storms and other metrological events also pose significant issues for hydrokinetic technologies.

Key Technology Opportunities

Short Term: Present to 2020

Key short-term technological developments to expand electricity from waterpower will occur in the area of conventional hydropower. The focus will be on developing and deploying technologies to improve fish passage and water quality, increase turbine efficiencies, and design enhanced tools for monitoring and managing water resources. Environmentally advanced hydropower turbine designs can improve fish survival and improve water quality.[24] The Grant County Public Utility District Advanced Hydropower Turbine System program is one example where the need for turbine replacement at a hydropower facility on the Columbia River is resulting in new turbines that have greater efficiencies and improved fish passage survival.[25] Other activities, such as those in the Oak Ridge National Laboratory's Environmental Sciences Division, are directed toward improving the balancing of hydropower production with other objectives for which dams are also operated, such as flood control, recreation, and ecosystem benefits, through mathematical modeling of complex hydrological systems.

Medium Term: 2020 to 2035

Over the next 10 years, many large-scale demonstration projects will be completed to help assess the capabilities of new waterpower sources, including wave and current technologies. It will take at least 10 to 25 years to know whether these technologies are viable for the production of significant electricity. Verdant Power, in a test of the technology shown in Figure 3.8, will install six turbines of this design, with a combined generating capacity of 200 kW, in the East River in New York City. Although an early test of this technology yielded successful generation of grid-connected electricity, the turbine blades failed within a short period of time. Another project in the United States is the Makah Bay, Washington, project, where four 250-kW floating buoys have recently been licensed by the Federal Energy

---

[24]See http://hydropower.inel.gov/turbines/index.shtml.
[25]See http://www.gcpud.org/aboutus/news/index.htm.

Regulatory Commission (Miles, 2008). The electricity from these buoys will be connected to a shore station via a 3.7-mile-long submarine transmission cable.

There are also many projects under way in Europe. The 4 MW Wave Dragon depicted in Figure 3.9 is scheduled to be installed in 2008 off the coast of Wales. The device will be tested for 3–5 years and then disassembled. A full-scale prototype of the Pelamis device, a four-segment device rated at 750 kW, was sea-tested for 1,000 hours in 2004.[26] An application of this technology began operation in 2008 with three devices rated at 2.25 MW located 5 km off the coast of Portugal. One of the leaders in the development of ocean wave and current electricity is the United Kingdom. The waters around that country are potentially an abundant source of clean renewable energy that could contribute up to 20 percent of its electricity needs (RAB, 2008). However, as noted in its recent assessment of the U.K. ocean energy program, while many prototypes demonstrating wave and tidal power have been deployed, progress has been slower than hoped (RAB, 2008). Particularly, there has not been the level of demonstration at a fully commercial scale as had been expected, although there have been some large demonstrations. One explanation is that the technical challenges, particularly of operating in the marine environment, are more difficult than originally expected.

The key technological challenge will be to develop designs that can withstand the deployment environment without causing harm to the ecosystem. Ultimately, if these new hydropower technologies are to scale up to levels that would contribute a significant amount to electricity generation, there would be deployment issues related to workforce, capital, and other industrial matters that are discussed in detail in Chapter 6.

Long Term: After 2035

Over the long term, deploying large-scale installations for wave and current technologies would depend on technological innovations now imagined but as yet undeveloped. There would likely be significant technological problems arising from moving from pilot plant and full-scale demonstration project operations at individual locations to utility-scale deployment. Future innovations would include standardization of generating technologies, technologies to integrate these new sources of power generation into the electricity system, and technologies to miti-

[26]See http://www.pelamiswave.com/index.php.

gate or reduce the potential impacts of use of such technologies on other uses of the ocean.

Other significant potential technologies that use ocean thermal and salinity gradients to generate electricity may also be investigated. These technologies exist in little more than conceptual designs, laboratory experimentation, and field trials. Ocean thermal energy conversion (OTEC) could convert the ocean's natural thermal gradient—that is, the varying of the ocean temperatures with depth—to drive electricity production (SERI, 1989). If temperatures between the warm surface water and the cold deep water differ by 20°C (36°F) or greater, an OTEC system could theoretically produce significant amounts of electricity, although there are major obstacles, including low temperature gradients, high costs, and potential for biofouling. According to NREL, OTEC research needs include improved turbine concepts and heat exchanger systems and actual experience with OTEC plant operation at demonstration plants.[27] Another concept for generating electricity in the open ocean is to use salinity gradients to generate electricity using the osmotic pressure differences between salt water and freshwater and waters of varying salinities. In reverse electrodialysis, a salt solution and freshwater are passed through a stack of alternating cathode and anode exchange membranes, and the chemical potential difference between salt water and freshwater generates a voltage over each membrane (Jones and Finley, 2003).

Summary of Hydropower Potential

The pressure to increase generation from traditional hydropower technologies due to their ability to provide low-cost, low-carbon electricity is countered with the understanding that damming freshwater rivers reduces their ecosystem benefits. There are significant pressures to return river systems back to free-running conditions. While removal of major generating facilities is unlikely, environmental and social forces will likely force the removal of some small dams and put a halt to any new hydroelectric dam development. At present, there is also great uncertainty about the future for new current, wave, and tidal generators. Scale demonstrations are under way, and some of these have been connected to the grid. However, there are no uniform designs or long-term experiences with the technologies. Tapping the oceans' huge reservoirs of energy on a large scale is clearly a distant prospect.

---

[27]See http://www.nrel.gov/otec/research.html.

BIOPOWER

Broadly defined, biomass is organic material produced on a short timescale by a biological process. Types of biomass for energy production fall into three broad categories: (1) wood/plant waste; (2) municipal solid waste/landfill gas (LFG); and (3) other biomass, including agricultural by-products, biofuels, and selected waste products such as tires (EIA, 2007). Dedicated energy crops are at present an insignificant portion of the U.S. biomass energy supply. However, there is increasing interest in biomass for alternative liquid transportation fuels (biofuels), which is already beginning to change the methodology of documenting biomass usage. A more complete discussion of biomass for alternative liquid fuels, including co-generation of biofuels and electricity, can be found in the forthcoming report of the America's Energy Future Panel on Alternative Liquid Transportation Fuels (NAS-NAE-NRC, 2009b) and the upcoming report from the Committee on America's Energy Future (NAS-NAE-NRC, 2009a).

Biomass is abundant, accounting for almost 50 percent of the national renewable energy resources in 2005, the largest single source of renewable energy (EIA, 2007). In 2005, biomass provided about 10 percent (9,848 MW) of the renewable electricity capacity in the United States, second only to hydroelectric power as a source of renewable electricity (EIA, 2007). From this installed capacity, 60,878 million net kWh of electricity was generated (17 percent of all renewable electricity generation, or 1.5 percent of total electricity generation). However, development of this renewable electricity source has not seen much recent growth. The nature of biomass use is such that electricity and heat are often co-generated. An attractive feature of biomass is that, as a chemical energy source, biomass energy is available when needed, which also makes it attractive for competing applications, such as transportation fuel.

Status of Technology

Because biomass includes a wide variety of resource types with a wide variety of characteristics (solid vs. liquid vs. gas; moisture content; energy content; ash content; emissions impact), a variety of electrical energy generation technologies are employed in biomass use. Despite differences, several commonalities exist. Production of electricity from biomass occurs in much the same manner as from fossil fuels. Similar to coal-fired power plants, the vast majority of biomass-fired power plants operate on a steam-Rankine cycle in which the fuel is directly combusted and the resulting heat is used to create high-pressure steam. The steam then serves as the working fluid to drive a generator for electricity production. With a gaseous fuel, electricity is produced with a more efficient turbine engine using the gas-Brayton cycle, in a manner similar to natural-gas-fired power plants. In addition to a gas turbine, a gas-reciprocating engine is also frequently used for <5 MW installations where a turbine would be too expensive.

A key difference between dedicated biomass power plants and coal-fired power plants is the size of the power plant, with wood-based biomass power plants (accounting for about 80 percent of biomass electricity) rarely reaching 50 MW, as compared to the 100–1500 MW range of conventional coal-fired power plants. Similarly, LFG power plants have capacities in the 0.5 MW to 5 MW range, whereas those operating on natural gas average about 100 times larger, in the 50 MW to 500 MW range. Because of their smaller sizes, dedicated biomass power plants are typically less efficient than their fossil-fuel-fired counterparts (in the low 20 percent range as opposed to the high 30 percent range for coal), since the cost of implementing high-efficiency technologies is not economically justified at the small scale.

The size difference of coal and biomass plants results, in part, from the high cost of shipping low-energy-content biomass. For example, typical wood has a moisture content of about 20 wt-percentage and an energy content, even after drying, of about 9,780 Btu/lb (18.6 MJ/kg), compared to about 14,000 Btu/lb (25 MJ/kg) for coal. In the case of LFG, shipping costs are eliminated by locating the power plant directly at the landfill site. The size of the power plant is determined by the rate of LFG production, which, in turn, is determined by the overall size of the landfill. Co-location and size matching are also characteristics of biomass power plants operated on black liquor, the lignin-rich by-product of fiber extraction from wood. The power plant, a key component of the paper mill, is sized to match the waste-product stream to meet the overall electrical and process steam needs of the pulping operation, often supplemented by purchases of grid electricity.

An increasing use of biomass is in co-fired power plants that burn coal as the primary fuel source and solid, typically woody, biomass as a secondary source. In co-fired plants, high efficiencies owing to large size are combined with the benefits of reduced CO_2 emissions from use of a renewable fuel input. With optimal design, co-fired plants can operate over a range of coal-to-biomass ratios, providing for attractive economics because the cheaper input fuel can be used when it is available. Co-fired plants tend to produce lower SO_x and particulate emissions and ash residue compared to purely coal-fired power plants, although NO_x emissions can be higher due to the presence of nitrogen in the biomass. The environmental tradeoffs depend on the specific characteristics of the biomass. An important unresolved issue is the impact of biomass co-firing on the effectiveness of selective catalytic-reduction technologies.

Although municipal solid waste (MSW) contains substantial energy content, designation of this fuel source as renewable is not justified, because much of the carbon in waste products derives from petroleum sources. Storage of that carbon in landfill sites can be viewed as a "carbon sequestration" solution. As a consequence, several states do not include MSW in their renewable portfolio standards. Nevertheless, the use of MSW for electricity production follows that of typical biomass power plants, relying on direct combustion to create steam that subsequently powers a generator. LFG is the gaseous product that results from the anaerobic decomposition of solid waste and contains about 50 percent CH_4, 50 percent CO_2, and trace components of other organic gases. In contrast to solid waste, LFG by definition cannot be sequestered in a landfill, and the released methane is about 20 times more potent than CO_2 as a greenhouse gas. As of December 2007, approximately 445 LFG energy projects operated in the United States, generating approximately 11 billion kWh of electricity per year and delivering 236 million cubic feet per day of LFG to direct-use applications, amounting to just under 20 percent of biomass electricity generation.

Potential Technology Development

Short Term: Present to 2020

The Energy Information Administration (EIA) (2001) estimated that biomass-fired electricity generation capacity could increase under the reference case (business-as-usual) from 6.65 GW in 2000 to 10.40 GW in 2020, thus adding 188 MW of capacity annually. In fact, according to the EIA (2007), the net summer capacity for biomass-derived electricity was essentially flat from 2001 to 2005, rising from 9.71 GW to 9.95 GW. Thus, the average annual growth rate, only 60 MW, was lower than anticipated, but the total capacity is already almost at the prediction for 2020. Existing technologies are sufficient for growing the biopower electrical capacity to 10.40 GW; any barriers are related to deployment. Factors affecting deployment are discussed in Chapter 6.

Technological advances in the short term would likely relate to power plant design to ensure fuel flexibility, particularly in co-fired plants, which in turn implies designing fuel feed and emissions control systems that can adjust to the variable characteristics of biomass fuel. Strategies include premixing coal and biomass in a single-feed system or providing separate coal and biomass inlets. With such advances, production of biomass electricity at competitive prices (depending on input fuel prices), high efficiency (about 30 percent), and high capacity factors (reaching 100 percent) could become widespread (Wiltsee, 2000).

Some fossil-fuel plants are being converted to 100 percent biomass combustion plants. These tend to be smaller-scale plants (e.g., the 24 MW Peepekeo plant near Hilo, Hawaii), but this trend may be accelerated in the United States, particularly if policy initiatives put a price on carbon. Progress here could also have ramifications in the medium term, if carbon capture and storage technologies are applied to biomass combustion plants. Capturing this carbon would result in net reductions of greenhouse gas emissions, and although no demonstration plant now exists, this potential is being reflected in modeling scenarios, notably in the European Union.

In parallel with improved use of woody biomass, the use of LFG for electricity production can be expected to increase in the near future, because it not only generates electricity in urban settings close to demand points, but also mitigates the release of methane, an extremely potent greenhouse gas. However, over the 2001–2005 time period, the portion of biomass capacity due to MSW/LFG has not changed to reflect these environmental benefits, suggesting the existence of other barriers. Furthermore, methane emissions from landfill sites have steadily decreased in the past decade, largely as a consequence of flaring the recovered methane (simply burning to convert the methane to carbon dioxide and water) rather than using the energy content. As of 2007 the EPA had identified approximately 560 candidate landfills with a total annual electric potential of 11 million MWh, amounting to just over one-quarter of 1 percent of the current U.S. electricity demand (EPA, 2008).

Medium Term: 2020 to 2035

In the medium term, it is likely that new biopower capacity, if pursued, will incorporate a pretreatment step in which the biomass is converted to a gaseous or liquid fuel more suitable for power generation, rather than direct-firing as is the norm today. As with all thermal power plants, higher operating temperatures generally result in higher efficiencies. Engines based on steam cycles (Rankine cycle) are inherently restricted to maximum temperatures of 580°C owing to the nature of the working fluid, water. In contrast, those based on open-air systems have a high exhaust-gas temperature because of the nature of the working fluid, air. These differences imply a maximum Rankine cycle efficiency of about 42 percent, whereas for the Brayton cycle (gas turbine engine), it is approximately 50 percent. A combined cycle, which uses the hot exhaust gas of the Brayton cycle to operate a lower-temperature Rankine cycle (steam engine), can potentially obtain a combined efficiency of ~65 percent. A solid fuel cannot be directly used for operation of a gas turbine engine and thus must be converted to a gas or liquid by a method commonly called gasification. Therefore, the efficiency of a biomass gasifier has a direct impact on the electricity production through this route. Biomass gasifiers would require improvement to be a viable option, as the present efficiency of biomass gasifiers is low (~30 percent) compared to the efficiency levels (~75 percent) of today's coal gasifiers, which are generally larger.

A power plant operated on a solid fuel but incorporating these three components (gasification, high-temperature Brayton cycle, low-temperature Rankine cycle) is known as an integrated gasifier combined cycle (IGCC) power plant (Figure 3.10) (Bain, 1993). Figure 3.11 presents a comparison of the efficiencies of the three generation technologies operated on biomass as a function of power plant size (Bridgewater, 1995).

While large-scale IGCC systems address the need to enhance system effi-

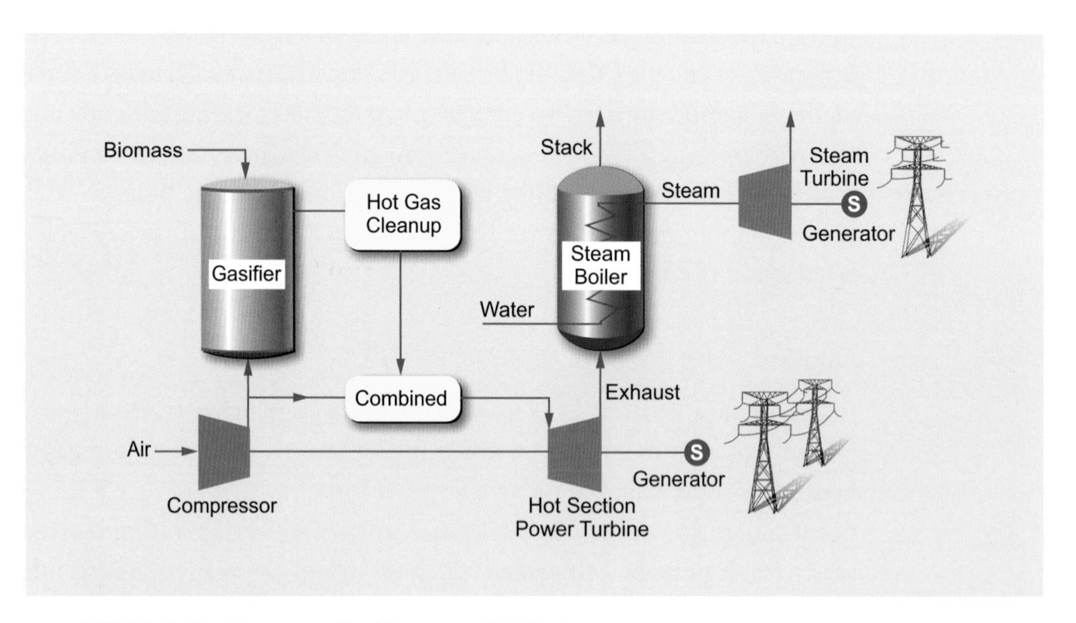

FIGURE 3.10 *Diagram of a biopower IGCC plant.*
Source: Bain, 1993.

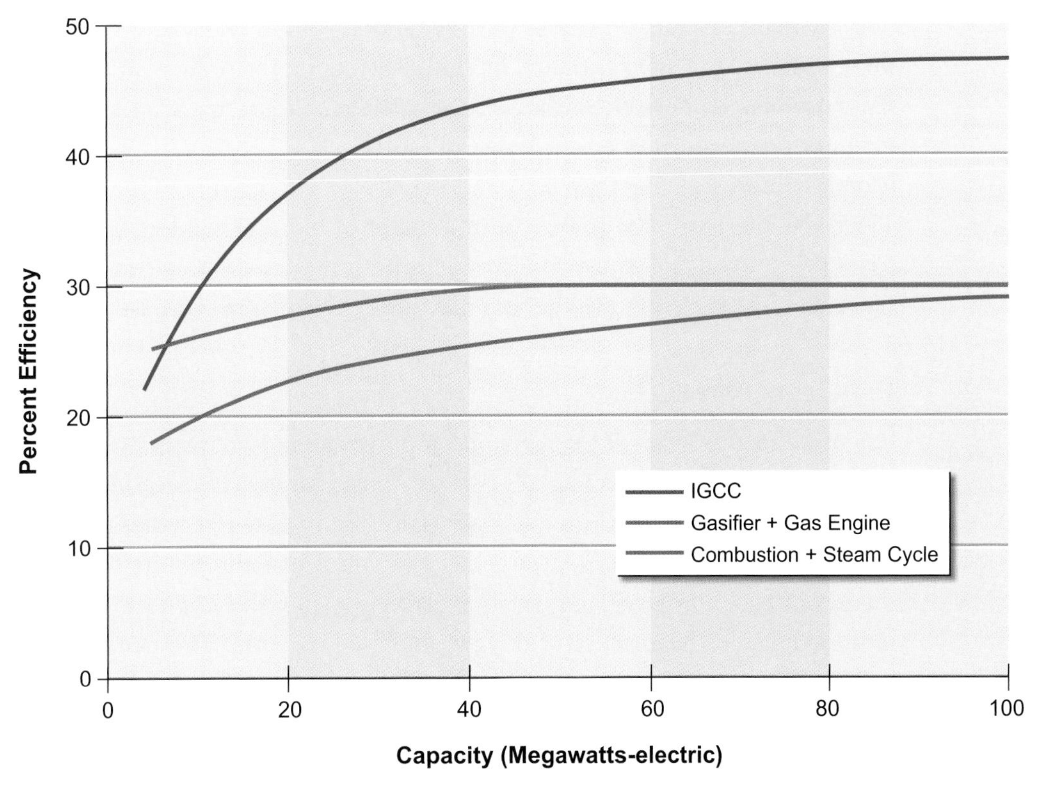

FIGURE 3.11 *Typical efficiencies of three classes of biomass power plants as a function of size.*
Source: Bridgewater, 1995.

ciency, at smaller scales (<25 MW_e) efficiency gains are lower. To obtain high efficiency at the scales typical of biomass power plants, one potential alternative is a fuel cell, in which chemical energy is directly, through electrochemical reactions, converted to electrical energy. Fuel cells are modular in nature, and their efficiency is largely independent of size. Consequently, they can be well matched to biomass power plants. High-temperature fuel cells have chemical-to-electrical conversion efficiencies of ~50–60 percent, and, as with the gas turbine, the high-temperature fuel cell exhaust can be supplied to a steam engine for even higher system efficiencies.

Mid-term developments of biopower can be anticipated in two primary directions: biomass gasification to enable widespread IGCC implementation; and improvements in lifetime and unit costs of fuel cells. In parallel, lower-cost high-temperature materials for both steam engines and gas turbines are potential development areas. In all cases, such advances would also benefit fossil-fuel-fired power plants, and substantial technology leveraging from those industries for biomass use may be possible, although some of the unique characteristics of biomass may not enable direct transfer between industries. It is noteworthy that biomass is generally more reactive than coal and hence easier to gasify (Williams and Larson, 1996). Furthermore, the lower sulfur content of biomass renders the produced gases more amenable to use in a fuel cell. Both molten carbonate and solid oxide fuel cells can efficiently use the fuel mixture derived from biomass gasification.

Long Term: After 2035

Potential long-term breakthroughs in biopower lie in two distinct areas. The first, and perhaps more tractable, is in advanced biological methods for converting raw biomass into clean fuels. Essentially, the high-temperature catalytic steps of gasification, or pyrolysis, are replaced by ambient-temperature steps through the use of bacteria. Here, natural consortia of bacteria decompose organic matter into methane in the absence of oxygen in closed reactors. This process, anaerobic digestion, is similar to the natural decomposition of waste in landfills, from which methane can also be harvested. Many farm- and community-based systems (particularly in Germany, Denmark, and several developing countries, but also in the United States) already use anaerobic digestion to produce biogas from wastes such as manure, food, and other organics. The biogas is then used in an internal combustion engine to produce electricity or is used directly for heating and cooking. Although much of the biomass resource might be dedicated to biofuel production (thus diminishing its role in electricity generation), biogas technologies could provide a small but nontrivial part of a renewable electricity portfolio, particularly given their flexibility and potential for distributed generation.

The second, more speculative potential breakthrough is in bioengineering new plants to radically enhance the efficiency of photosynthesis. As noted in Chapter 2, the solar-to-biomass conversion in typical plants is only ~0.25 percent; subsequent conversion from biomass to electricity proceeds with another efficiency penalty of at least 50 percent. Thus, solar-to-electric energy conversion efficiency is on the order of 0.1 percent, which is far below the 10–20 percent efficiency achievable with state-of-the-art photovoltaic and concentrating solar power systems. It is unclear, however, whether agricultural practices using bioengineered plants would be sustainable, even if photosynthesis could be enhanced through genetic modification. Even with today's candidate energy crops (e.g., willow, miscanthus, poplar, and switchgrass), it is unknown how much of the biomass must be left in the fields to ensure soil health. A complete evaluation of these uncertainties is beyond the scope of this analysis.

Summary of Biomass Potential

In the absence of a program to grow dedicated energy crops, biomass from waste streams (e.g., forestry, agricultural, and urban) is likely to grow but remain a relatively small contributor to the nation's electricity supply. As stated in Chapter 2, the long-term potential of biomass is limited by the low conversion efficiency of the photosynthesis process. Further, biomass's potential depends very much on its competing uses for fuel and electricity. In particular, conversion from raw biomass into syngas or other fuels renders biomass attractive for transportation applications, and competition between the two end uses must be considered. Indeed, the DOE has essentially stopped its biopower programs in favor of biofuels for transportation (Beaudry-Losique, 2007). However, this priority may once again shift if there is a move toward electrified transportation systems (e.g., plug-in hybrids or all-electric vehicles), which would again favor biomass for use in power systems.

ENHANCING TECHNOLOGIES FOR ELECTRICITY SYSTEM OPERATION

There are a host of technologies, operational modifications, and system upgrades that could enhance renewable energy resource use. These include storage, expansion of transmission capacity, and improvements in the intelligence of the electricity transmission and distribution (T&D) system. Because each local electricity system has its own generating capacity, transmission capability, and ability to purchase power outside its own territory, each system's needs for enhanced technologies for integrating renewables are unique.

New technologies and tools would be required to enable reliable transmission and integration of large-scale renewables, in addition to expanding transmission capacity to connect new renewables to the grid. These include technologies that support the transmission grid by adding reactive power and enabling low-voltage ride-through; advanced transmission planning for integrating intermittent generation; methods of determining supply capacity and reserve requirements for high wind power penetrations; and methods and tools for accommodating high penetrations of wind generation. Integrating high levels of distributed solar PV electricity would require improvements in PV interface devices and deployment of advanced metering technologies that focus on households or other end users. Integrating large amounts of PV also would require planning models that address PV deployment under two scenarios, existing distribution systems and possible future distribution systems. Modernization of the electricity system is discussed in some detail in the upcoming report of the Committee on America's Energy Future (NAS-NAE-NRC, 2009a).

Storage

Efficient and cost-effective storage of electrical energy would have a significant impact on the U.S. electrical power infrastructure, irrespective of the role of renewables. Storage requirements depend on where the storage occurs, the mix of renewables deployed, the temporal correlation of generation sources, and other features such as demand-management capabilities or vehicle-to-grid storage. Electricity consumption varies over the course of the day, whereas coal, nuclear, and hydropower electricity plants are generally designed to provide baseload electricity at some optimal level of generation.[28] Renewable resources such as solar and wind are intermittent by nature, and that intermittent supply can be mismatched with demand. Thus, neither baseload nor intermittent electricity generation technologies supply electricity in alignment with demand.

Despite this mismatch, electricity systems in the United States are managed today with little or no storage; pumped hydropower storage, the largest storage medium, provides a capacity that is less than 3 percent of the total electricity generation capacity. In the absence of storage, electricity-generating utilities are designed with a capacity sufficient to meet peak rather than average demand, which means each system's capacity is, on average, underused by roughly 40 percent or more.[29] Similarly, storage would be incorporated and designed to reflect not average scenarios but worst-case scenarios, to ensure reliability during low-probability/high-impact events. To date, the mismatch between electricity supply and demand has been handled largely by ramping power output up and down

---

[28]Baseload electricity plants are the generation facilities used to meet some or all of a given region's continuous electricity demand. These plants produce electricity at a constant or slowly varying rate and tend to be lower-cost generation plants relative to other capacity available to the system.

[29]For example, the New York Independent System Operator reported a New York state peak hourly demand of 33.5 GW in 2006 and an average hourly demand of 18.5 GW (NYISO, 2008).

from natural-gas-fired peaker plants and other peak-power plants. Large penetrations of renewable electricity from wind and solar, which are inherently intermittent, would exacerbate the challenges of load management. However, at moderate penetrations, up to at least 20 percent in the case of wind power, studies indicate that the existing management approaches suffice, and storage is not an immediate necessity for successful integration of renewable resources. These studies are discussed in Chapters 6 and 7.

Storage technologies are differentiated in terms of the time and scale at which they are useful (Figure 3.12). Rapid energy discharge, a feature that would be useful to maintain the quality of the electrical power supply, could some day be achieved with devices such as supercapacitors and high-power flywheels. More relevant to the integration of intermittent renewable technologies into the electrical grid are high-power systems that store energy for at least several hours. These

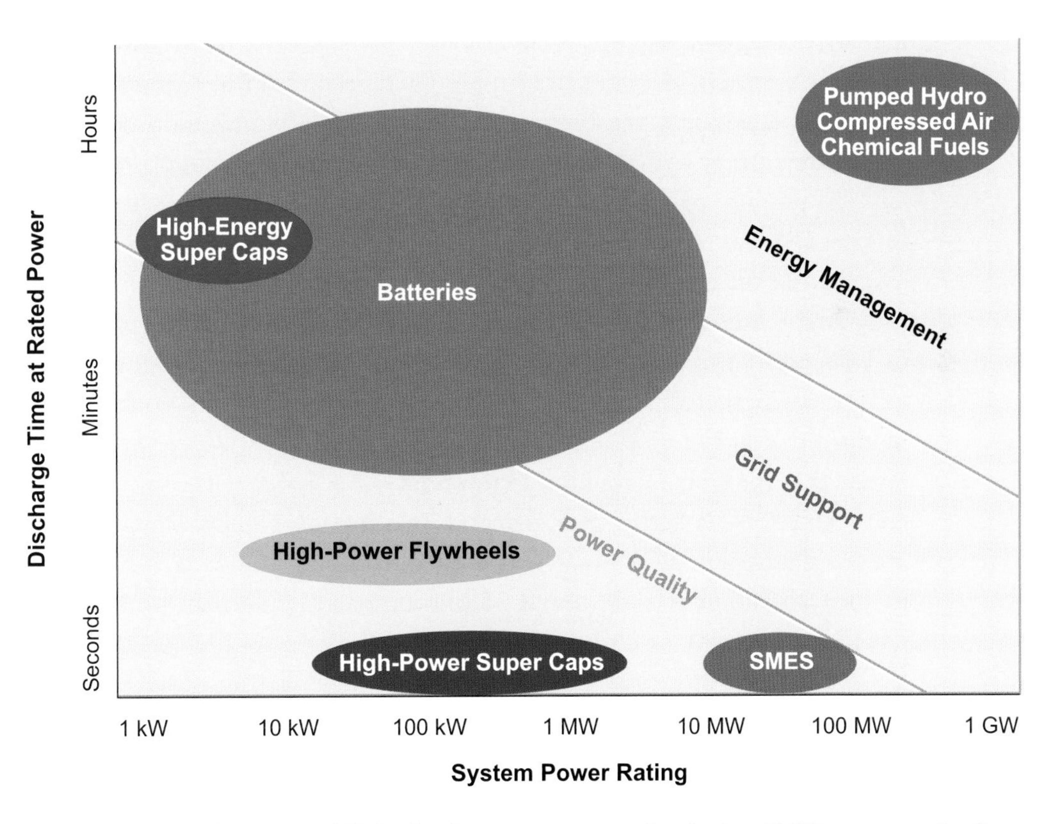

FIGURE 3.12 *Capabilities for future storage technologies. SMES, superconducting magnetic energy storage.*
Source: Developed from information in Gyuk (2008) and Rastler (2008).

include pumped hydropower, compressed air, some types of batteries, and systems for converting electricity into a chemical fuel such as hydrogen. In addition, some renewable electricity generation technologies, solar thermal and biomass in particular, naturally provide storage solutions. Energy storage in the form of chemical fuels, including biomass and batteries, has direct implications for transportation and underscores the likelihood of increasing overlap between the electricity and transportation sectors in future years.

Pumped Hydropower

Energy storage via pumped hydropower involves the use of electrical energy to move water into an elevated hydropower reservoir by operating the generator as a motor and running the hydroturbine in reverse. When electricity is needed, the water in the upper reservoir is released through the turbine, which operates the motor as a generator to produce electricity. Pumped hydropower is a mature and effective technology that provides the only source of electricity storage today to buffer electricity demand and supply fluctuations. Further growth of pumped hydro is limited, however, because of the lack of environmentally acceptable sites, just as the further growth of hydroelectric power itself is limited.

To put into perspective the scale of possible energy storage requirements to meet U.S. electricity demand and its implications for pumped hydropower, assume that the U.S. peak electricity consumption rate is 7.8×10^{11} J/s = 0.78 TW. Providing 6 hours of electricity at that level of demand would require storage of 1.68×10^{16} J of energy.[30] If storage met only 25 percent of that amount (the rest met by baseload power), it would require a mass of 4.3×10^{12} kg of water pumped to a height of 100 meters.[31] Using the density of water (1×10^3 kg/m^3) means that the system would need to pump 4.3×10^9 m^3 of water up 100 meters and then release it during peak demand. In a low-probability scenario (assuming for this discussion that storage was needed to supply 100 percent of peak electricity over a 12-hour period), nearly 35 km^3 of water (equivalent to the volume of Lake Mead) would have to be pumped up 100 meters and released.

Pumped hydropower is a relatively low-energy-density storage solution, as demonstrated from another perspective. The energy density of petroleum is 45 MJ/kg, whereas the potential energy of 1 kilogram of water at a height of

---

[30] 7.8×10^{11} J/s · 3.6×10^3 s/hr · 6 hr = 168×10^{14} J = 1.68×10^{16} J.
[31] Mass = E/gh = 4.2×10^{15} J/(9.8 · 100) = 4.3×10^{12} kg.

100 meters is 1,000 J/kg. Hence, accounting for the density difference between gasoline and water, storing the energy contained in 1 gallon of gasoline would require pumping more than 50,000 gallons of water up the height of Hoover Dam.

Compressed Air Energy Storage

Compressed air energy storage (CAES) refers to the storage of energy as compressed air, usually in an underground air-tight cavern. Other options include storing the compressed air in depleted natural gas fields and aboveground storage tanks. Demonstrated CAES systems (two exist in the world today; one is located in McIntosh, Alabama) use a diabatic[32] storage process in which air is cooled before it enters the cavern and, upon increased electricity demand, is expanded using external heating in a modified gas turbine that, in turn, operates an electric generator. CAES allows less expensive nighttime electric energy to be stored and used to replace relatively more expensive, peaking daytime energy (EPRI, 2007b). CAES may reduce the need to build fossil-fired power plants that meet peak rather than average capacity, yet CAES storage must be operated in conjunction with combustion. Because diabatic CAES power plants share similarities with conventional, natural-gas-fired power plants, the two existing systems have operated together reliably since their commissioning, and the technology is considered mature. Overall, the storage capacity provided by these plants is small relative to total U.S. electricity consumption. For example, the McIntosh plant in Alabama has a 110 MW capacity, and the storage cavern allows for 26 hours of continuous operation at the rated power before significant drawdown occurs. The second CAES system, the Huntorf plant in Germany, operates jointly with a nuclear power plant, with the goal of managing the mismatch between the baseload power generation and the variable consumer demand. The storage capacity is smaller, but the discharge rate is higher. New approaches to diabatic compressed air storage are directed toward microscale systems that use smaller volumes and capitalize on underground natural gas storage or storage in depleted gas fields.

Adiabatic CAES systems eliminate the need for combustion fuels by storing not only the mechanical energy of compression, but also the thermal energy

---

[32]In diabatic storage the heat produced during the compression of air escapes to the atmosphere and is wasted, whereas in adiabatic storage the heat produced during compression is also stored.

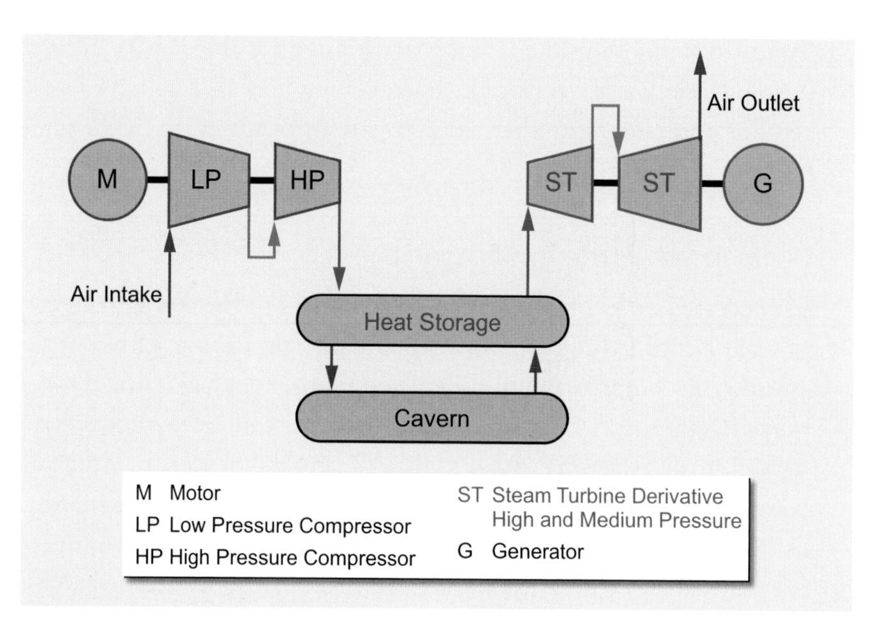

FIGURE 3.13 *Function diagram of an adiabatic CAES power plant with a single-stage configuration.*
Source: Bullough et al., 2004.

produced when air is compressed. Electric power generation from such a system (Figure 3.13) uses the hot air to operate a turbine (in the absence of combustion), which, in turn, operates an electric generator. Adiabatic compressed air energy storage has not yet been demonstrated, but the majority of the components indicated in Figure 3.14 are known technologies. A concept study supported by the European Union outlines some of the technical challenges and concludes that they are linked largely to system integration and optimization, rather than to individual component development (Bullough et al., 2004). However, the "thermal energy store" unique to adiabatic CAES will require particular attention.

Beyond the technical challenges of constructing and operating CAES power plants, it is of value to consider the storage volume (geologic) requirements for maintaining compressed air energy storage at a scale that would be significant compared to present-day electricity consumption. Operation of the 110 MW McIntosh plant, for example, requires 155 kg/s of compressed air supplied to its turbines, implying a required flow rate of 1.4 kg/s per MW. Given the density of air (1.2 kg m^3), this equates to a volumetric flow rate of 1.2 m^3 s^{-1} MW^{-1}. The total deliverability from all of the known natural gas reservoirs in the United States is ~1 × 10^{11} ft^3/day, equal to 3 × 10^4 m^3 s^{-1}. Dividing this total by the

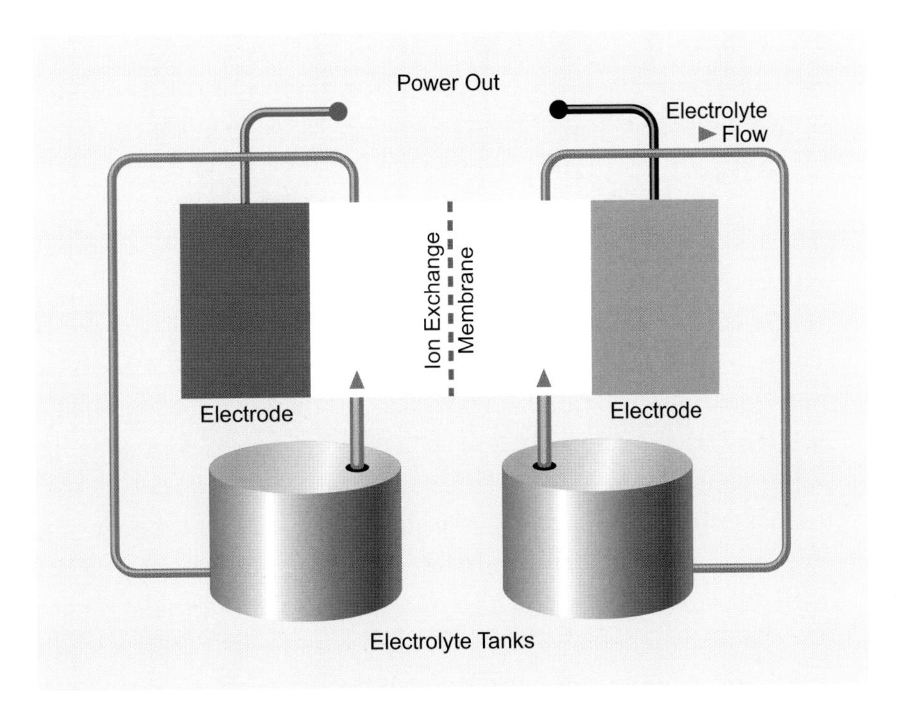

FIGURE 3.14 *Schematic of a flow battery.*

per-megawatt flow rate required for electricity generation from a compressed air cavern would result in a total generation capacity of 26 GW, which amounts to ~5.5 percent of the U.S. average 2005 load of 460 GW. Though some CAES would be available in aboveground storage tanks, using CAES on a large scale would require extensive, if not immense, amounts of geologic storage.

Batteries

Battery technologies cover an enormous range of chemistries, including lead-acid, lithium ion, and sodium sulfur, and storage efficiencies range from 65 to 90 percent. These values depend not only on the particular chemistry but also on the details of the charge and discharge profile. Furthermore, as in the case of chemical fuel production, present-day activities in battery development and demonstration focus largely on the transportation sector, but with a growing recognition of the importance of utility-scale electricity storage.

A battery is generally constructed with two reactive electrode materials separated by an electrolyte membrane that allows only selected ions to pass through it. During discharge, because of the presence of this separator membrane, the reaction between the two electrode materials must occur via a multi-step process in which a species from one electrode either accepts or rejects electrons to become an ionic species that can pass through the electrolyte. On reaching the second electrode, the ionic species reacts with the material of the second electrode and simultaneously either rejects or accepts electrons to regain its initial charge state. The ion current through the electrolyte is balanced by the electron current through an exterior circuit that draws the power. Depending on the nature of the reaction products that form at the electrodes, the battery may or may not be rechargeable. For rechargeable systems, application of a voltage induces the reserve reactions and regenerates the electrode materials. Rechargeable systems include lithium-ion, lead-acid, nickel-cadmium, and sodium-sulfur batteries. Among these, the sodium-sulfur batteries, because of the favorable balance between system complexity and overall efficiency, are usually considered for utility-scale applications. Lead-acid and nickel-cadmium batteries require the use of rather toxic metals, and lithium-ion batteries are costly and have shown significant degradation on deep discharge.

Flow batteries are alternatives to conventional batteries in which the electrode materials are consumed through the electrochemical reaction. In flow systems, the electrodes are inert, serving simply as current collectors, and the overall reaction takes place between two chemical solutions separated again by an electrolyte membrane (see Figure 3.14). Flow batteries are similar to fuel cells. The key difference is the nature of the reactant species. Fuel cells use gases, supplying hydrogen to the anode and oxygen to the cathode (Figure 3.15), whereas in the flow battery liquid electrolyte solutions are supplied to each electrode chamber. As in either conventional batteries or fuel cells, the direct reaction between the chemical species in the anode and cathode chamber is prevented by the presence of the electrolyte. The flow of ions across the membrane is balanced by a flow of electrons through an exterior circuit, in turn providing power generation. Much like a fuel cell, the energy capacity of a flow battery is fixed by the storage volume of the reactant solution, and not by the dimensions of the electrodes, as is the case in a conventional battery. Like fuel cells, however, flow batteries are complex systems involving pumps, valves, the flow of corrosive fluids, and the requirement to regenerate the spent solution in a subsequent step. The separation between the energy storage and energy delivery functions in a flow battery makes a flow battery more useful to utility-scale storage than a conventional battery, but the system complexity renders flow batteries difficult for portable applications. It is unclear where and how fundamental breakthroughs can bring revolutionary advances in battery technologies. For energy storage, the energy density stored in gasoline is

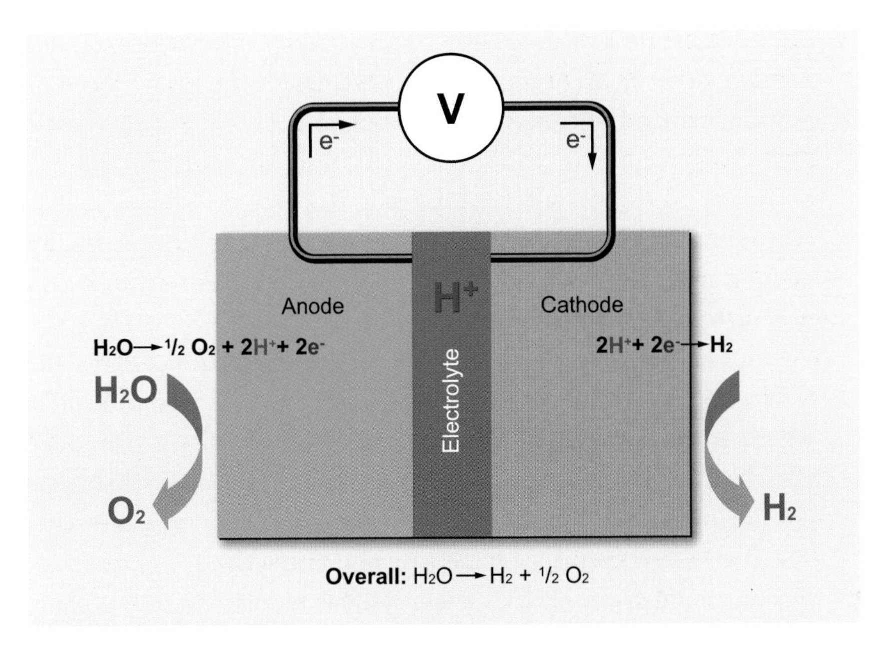

FIGURE 3.15 *Operation of an electrolysis cell. A fuel cell is an electrolysis cell operated in reverse, and accordingly the anode and cathode functions are also reversed.*

much greater than that storable in existing technologies for lithium-ion or flow batteries.

Chemical Energy Storage

Chemical energy storage refers to synthetic routes to producing fuels from energy resources. Depending on its nature, a fuel can subsequently be used for electricity production via fuel cells or used in conventional combustion systems. By far the simplest fuel to consider in this scenario is hydrogen, created according to the following reaction:

$$H_2O + \text{(renewable) energy} \rightarrow H_2 + \tfrac{1}{2}\, O_2.$$

Regardless of how hydrogen is produced, the fuel must be stored, which is a daunting challenge. For example, compressing hydrogen to a pressure of 800 bar incurs an energy penalty of ~13 percent. At any pressure, the volumetric energy density of methane, a fuel more familiar to the electricity industry, is more than three times greater than that of hydrogen stored at an equal pressure (Bossel and Eliasson, 2003). Furthermore, after storage, hydrogen would be used either in a combustion process or in a fuel cell to provide electricity, both incurring additional efficiency penalties (~60 percent loss for combustion and ~30 percent loss for the fuel cell), resulting in a maximum "round-trip" efficiency of ~60 percent, assuming a 70 percent efficient fuel cell and 87 percent efficient compression, excluding energy penalties for the hydrogen production itself. With these caveats, it is nevertheless useful to consider methods of renewable hydrogen generation.

If the energy input for splitting water is electricity, the reaction occurs simply by electrolysis. In the context of renewable electricity, generation is from solar, wind, or other renewable resources, and the electricity is then directed to a separate electrolysis cell. Small-scale electrolyzers are commercially available for the production of hydrogen for technical purposes. However, these systems' overall efficiency, 65–70 percent, renders them unattractive for large-scale energy storage (Bossel and Eliasson, 2003). These systems require the use of platinum (Pt) at a quantity that can be estimated from the platinum used in state-of-the-art polymer electrolyte membrane fuel cells, which essentially operate in reverse relative to electrolyzers. A DOE target for platinum use is 1 g/kW. Storage for 46 GW average capacity (amounting to 10 percent of the U.S. average) would require 46×10^3 kg of platinum, which is a relatively small amount compared to both the known platinum reserves, ~ 7×10^6 kg, and the present rate of platinum consumption, ~ 250×10^3 kg per year (Wilburn and Bleiwas, 2004). Because of the inverse relationship between electrolyzers and fuel cells, there has been some research on electrochemical cells that could operate in either mode, particularly in the case of high-temperature ceramic electrolyte systems. These dual attributes would be attractive, because costs would be reduced as a result of the multi-functionality of the electrochemical cell, and the high-temperature operation would obviate the need for precious metal catalysts.

In the case of solar energy, direct photo-electrochemical production of hydrogen is an attractive alternative to the two-step process (renewable energy → electricity → fuel). In direct photo-electrochemical production, a semiconductor material, immersed in water, absorbs light, exciting electron-hole pairs across the band gap of the semiconductor. These electronic species are then available to perform reduction and oxidation reactions at the electrodes of the cell. As with the ambient-temperature electrolysis cell, developing robust and efficient, non-precious-metal catalysts remains a daunting challenge for this approach. However, the recognition that biological systems carry out such reactions (i.e., photosynthesis) using base-metal compounds as catalysts suggests that success could ultimately be achieved. The DOE is attempting to increase investment in this area, reflecting the potential offered by recent advances in this approach (e.g., Muckerman et al., 2008).

Yet another alternative for hydrogen production is the thermochemical cycle. In this approach, thermal energy, ideally solar-thermal energy, is the renewable input applied to a material that occurs in oxidized form at low temperatures and undergoes dissociation/reduction at high temperatures. The process of cycling between these two states under appropriate gaseous atmospheres releases the desired reduced chemical fuel. For example, if one considers the FeO/Fe_3O_4 system, the hydrogen production cycle can be described as

$$Fe_3O_4 \rightarrow 3FeO + \tfrac{1}{2}\,O_2\,(g) \qquad \text{high temperature and}$$
$$3FeO + H_2O \rightarrow Fe_3O_4 + H_2\,(g) \qquad \text{low temperature.}$$

The success of the thermochemical approach relies fundamentally on the chemical thermodynamics of oxide stability. Rapid reaction kinetics and strong coupling of the solar radiation to the material for effective heating are also essential. There are no commercial activities in thermochemical fuel production, but there are ongoing large-scale demonstration plants at Sandia National Laboratories and at ETH Zurich.

Alternatives to hydrogen fuel production are under consideration, because converting renewable energy to hydrogen fuel merely transfers the energy storage problem to a different part of the energy delivery infrastructure. Alternatives typically employ biological processes to produce alcohols, alkanes, or other carbon-containing fuels, and can be considered advanced biomass approaches, such as production of biodiesel from algae. The few synthetic chemistry approaches that are being investigated center largely on electrochemical reduction of CO_2 to CO, whereby the combined carbon monoxide and hydrogen, or syngas, becomes the input in known industrial processes for the creation of a more suitable fuel. These approaches, still in the laboratory research stage, focus on chemical reaction pathways rather than potential scale-up to provide an energy solution. Because CO and H_2 are produced electrochemically, it is theoretically possible to react them further to generate methane, a fuel familiar to the electricity industry and thus likely to have more immediate impact than penetration of renewable electricity. Because natural-gas peaking plants are often co-sited with solar and wind farms, direct production of methane using the output of the combustion power plant could provide a closed-loop system in which methane would not have to be transported.

Summary of Storage Potential

Analysis of the future for the various storage technologies is beyond the scope of this panel, but some summary statements are in order. In the near term, diabatic CAES and various battery technologies, especially sodium sulfur batteries, have found initial applications in the electricity sector. In the longer term, when penetrations of renewables in the electricity sector might reach levels requiring energy storage, there may be a variety of approaches, including adiabatic CAES or the use of renewable energy in the production of chemical fuels. Advances in ultracapacitors and other short-term storage solutions may provide additional mechanisms to effectively integrate and stabilize intermittent resources.

Energy storage is a system resource that should be operated for the overall benefit of the system. The greatest value of energy storage is realized when it is operated for the benefit of the entire system, and not dedicated to balancing any particular resource on the system. Storage tied to smart transmission and distribution grids would become a valuable component of any power system, and could provide numerous benefits to the system. Storage benefits the system without renewables, and renewables benefit the system without storage. The task is to manage variability with flexibility.

Improved Grid Intelligence—the Smart Grid

The architecture needed to improve integration of renewables into the electricity grid would incorporate a variety of technologies, such as advanced sensors; smart meters (net metering, turn-on/turn-off capability, and the capability to enable time-of-day pricing); power converters, conditioners, and other power-quality technologies; source and load controls; improved software, including forecasting and operations models; and storage technologies (Kroposki, 2007). Most of these technologies are part of the broad initiative to improve the intelligence of the modern grid.[33] The objectives to meet in modernizing the electricity grid go beyond

---

[33]The term "Smart Grid" has often been used to describe this initiative. The Smart Grid may be described as the overlaying of a unified electronic control system and two-way communication over the entire power delivery infrastructure. Smart Grid capabilities optimize power supply and delivery, minimize loss, and enable maximum use of electricity generation resources, energy efficiency, and demand responses. However, this term suffers from overuse and multiple interpre-

increasing intermittent renewables, and include improving security and power quality and creating a more efficient, adaptive electricity system. Demonstrations are under way in several U.S. cities (e.g., Boulder, Colorado), but widespread deployment is expected to take decades.[34] More details on the objectives and technologies involved in creating a future electricity grid with increased capacity and intelligence are presented in the upcoming report of the Committee on America's Energy Future (NAS-NAE-NRC, 2009a).

A truly intelligent modern grid would anticipate the fluctuations in the power output from intermittent renewable energy sources and maintain absolute supply/demand equivalency on a given transmission or distribution circuit, while requiring less compensating backup power and storage capacity. Instantaneous electronic control of the grid would allow each transmission line to operate at a higher load factor without risking thermal overload than is now feasible on the electromechanically controlled transmission system. This level of coordinated control would require improved communications and seamless connectivity, or interoperability,[35] which would make the grid a dynamic, interactive infrastructure for the real-time exchange of power and information. Open connectivity architecture would create a plug-and-play environment that would securely network grid components and operators. The current lack of uniform interconnection and operations codes and standards, as well as the acceptance of standardized open communications architecture, is restricting the timely implementation of the modern grid. A system-wide integrated cyber security capability is also an important dimension of this communications architecture.

The Smart Grid's emphasis today is primarily on creating interstate high-voltage transmission capabilities to facilitate bulk wind power access. While important, transmission is only one element of the nationwide grid modernization effort needed to realize the potential benefits of renewable energy. The electronic modernization of the local electricity distribution network is equally essential to incorporating distributed renewable energy technologies such as photovoltaics and wind power. One critical objective of smart distribution grids is to enable the

---

tations. The panel instead uses the improved term "grid intelligence" to refer to the collection of technologies needed to improve the integration of renewables into.

[34]EISA 2007 authorized the Smart Grid Advisory Committee and Task Force through 2020. An earlier (2003) DOE plan was called Grid 2030; the intention was to have 100 percent of electricity running through a smart grid by 2030.

[35]Seamless, end-to-end connectivity of the hardware and software throughout the transmission and distribution system to the electrical energy source.

seamless, uninterruptible balancing of electricity supply and demand, which could allow distributed renewable power generation to be broadly dispatchable. Dispatchability would improve intermittent renewables' compatibility with the reliability and operational requirements of the bulk power system. The result could help transform buildings into power plants and provide a more reliable, efficient, and clean electricity supply system.

Advanced Metering

Advanced metering—the use of electricity meters that provide detailed consumption profiles—is one technology for improving the intelligence of the grid that would be particularly important to increasing the use of distributed renewables. Unlike conventional metering, advanced metering would couple the cost of electricity generation with the price to the consumer. In the context of renewables integration, the ability to do time-of-day pricing and net metering would better enable the deployment of renewables, especially solar PV. Such meters also could communicate real-time information to the consumer for billing and pricing purposes. Because solar PV generation peaks close to the late-afternoon price peak, meters allowing time-of-day pricing could improve the cost-competitiveness of solar PV at the consumer end. Advanced metering also helps to create incentives to use energy at off-peak times when possible, thereby reducing demands on the transmission and distribution systems. Chapter 4 discusses the use of real-time pricing to encourage the development of renewables.

Furthermore, advanced metering technologies would enable net metering for those with on-site renewable generation. Net metering improves the integration into the grid of distributed renewable resources such as solar PV installed at residential and commercial facilities. It measures both the consumption of electricity and the excess energy produced on-site, and at least partly credits the consumer for excess generation produced by consumer-owned solar PV or other renewable electricity technologies.

Software/Modeling Support

New grid operating tools are also needed to incorporate renewable energy resources, including operating models and system impact algorithms that address the transient behavior of renewable energy; improved operators' visualization techniques and new training methodologies; and advanced simulation tools that can provide an accurate understanding of grid behavior. These grid operating tools would also assist system planners in designing reliable power systems for this new environment. Better forecasting algorithms would allow better use of temporally varying resources such as wind energy. The objective of this work is to improve the forecasting of wind and its use in electricity markets (Ahlstrom et al., 2005; Hawlins and Rothleder, 2006; Smith, 2007).

Reactive Dynamic Power

The demand that some renewables place on ancillary services, such as reactive power and dynamic voltage control, also must be considered. Reactive power is the portion of electricity that establishes and maintains the electric and magnetic fields of alternating current (AC) equipment. Because wind and solar power produce direct current (DC), reactive power must be provided in the DC-to-AC conversion process, a requirement that is complicated by the variable/intermittent nature of these renewable energy sources: the reactive power must be equally dynamic to keep pace. Many early wind machines were induction generator wind turbines with a constant frequency and so required reactive power to be supplied from the grid. Although newer machines have solved this problem, voltage stability remains an issue. The European Transmission System Operators (ETSO) recently completed a study on the ancillary services required by wind power as the amount of installed wind capacity in Europe increased from 41 GW in 2005 to an expected 67 GW in 2008 (ETSO, 2007). In particular, the ETSO study looked at the effects of variable power output on the electricity grid and the ability of various wind turbine types to provide system service needed for the stable operation of an electricity grid. Another study describes technologies used to provide reactive power for a large wind farm and the interactions of the wind farm, reactive power compensation, and the power system network (Muljadi et al., 2004).

FINDINGS

The most critical elements of the panel's findings on renewable electricity generation technologies are highlighted below.

Over the first timeframe through 2020, wind, solar photovoltaics and concentrating solar power, conventional geothermal, and biopower technologies are technically ready for accelerated deployment. During this period, these technologies could potentially contribute a much greater share (up to about an additional

10 percent of electricity generation) of the U.S. electricity supply than they do today. Other technologies, including enhanced geothermal systems that mine the heat stored in deep low-permeability rock and hydrokinetic technologies that tap ocean tidal currents and wave energy, require further development before they can be considered viable entrants into the marketplace. The costs of already-developed renewable electricity technologies will likely be driven down through incremental improvements in technology, "learning curve" technology maturation, and manufacturing economies of scale. Despite short-term increases in cost over the past couple of years, in particular for wind turbines and solar photovoltaics, there have been substantial long-term decreases in the costs of these technologies, and recent cost increases due to manufacturing and materials shortages will be reduced if sustained growth in renewable sources spurs increased investment in them. In addition, support for basic and applied research is needed to drive continued technological advances and cost reductions for all renewable electricity technologies.

In contrast to fossil-based or nuclear energy, renewable energy resources are more widely distributed, and the technologies that convert these resources to useful energy must be located at the source of the energy. Further, extensive use of intermittent renewable resources such as wind and solar power to generate electricity must accommodate temporal variation in the availability of these resources. This variability requires special attention to system integration and transmission issues as the use of renewable electricity expands. Such considerations will become especially important at greater penetrations of renewable electricity in the domestic electricity generation mix. **A contemporaneous, unified intelligent electronic control and communications system overlaid on the entire electricity delivery infrastructure would enhance the viability and continued expansion of renewable electricity in the period from 2020 to 2035.** Such improvements in the intelligence of the transmission and distribution grid could enhance the whole electricity system's reliability and help facilitate integration of renewable electricity into that system, while reducing the need for backup power to support the enhanced utilization of renewable electricity.

In the third time period, 2035 and beyond, further expansion of renewable electricity is possible as advanced technologies are developed, and as existing technologies achieve lower costs and higher performance with the maturing of the technology and an increasing scale of deployment. Achieving a predominant (i.e., >50 percent) penetration of intermittent renewable resources such as wind and solar into the electricity marketplace, however, will require technologies that are largely unavailable or not yet developed today, such as large-scale and distributed cost-effective energy storage and new methods for cost-effective, long-distance electricity transmission. Finally, there might be further consideration of an integrated hydrogen and electricity transmission system such as the "SuperGrid" first championed by Chauncey Starr (Starr, 2002), though this concept is still considered high-risk.

REFERENCES

Ahlstrom, M., L. Jones, R. Zavadil, and W. Grant. 2005. The future of wind forecasting and utility operations. Power and Energy Magazine, IEEE 3(6):57-64.

Bain, R.L. 1993. Electricity from biomass in the United States: Status and future direction. Bioresource Technology 46:86-93.

Bett, A.W., F. Dimroth, G. Stollwerck, and O.V. Sulima. 1999. III-V compounds for solar cell applications. Applied Physics 69:119-129.

Beaudry-Losique, J. 2007. Biomass R&D Program and Biomass-to-Electricity. Presentation at the first meeting of the Panel on Electricity from Renewable Resources, September, 18, 2008. Washington, D.C.

Bossel, U., and B. Eliasson. 2003. Energy and the Hydrogen Economy. Arlington, Va.: Methanol Institute. Available at http://www.methanol.org/pdf/HydrogenEconomy Report2003.pdf.

Bowen, J.L., and C.M. Christensen. 1995. Disruptive technologies: Catching the wave. Harvard Business Review 73(1):43-53.

Bridgewater, A.V. 1995. The technical and economic feasibility of biomass gasification for power generation. Fuel 74:631-653.

Bullough, C., C. Gatzen, C. Jakiel, M. Koller, A. Nowi, and S. Zunft. 2004. Advanced adiabatic compressed air energy storage for the integration of wind energy. Proceedings of the European Wind Energy Conference, EWEC 2004, November 22-25, 2004, London. Brussels: U.K. European Wind Energy Association. Available at http://www.2004ewec.info.

Christensen, C.M. 1997. The Innovators Dilemma: When New Technologies Cause Great Companies to Fail. Cambridge, Mass.: HBS Press.

DOE (U.S. Department of Energy). 2005. Basic Research Needs for Solar Energy Utilization: Report on the Basic Energy Sciences Workshop on Solar Energy Utilization. Washington, D.C.

DOE. 2007a. Workshop for Enhanced Geothermal Systems Technology Evaluation: Summary Report. Workshop for Enhanced Geothermal Systems, Technology Evaluation, June 7-8, 2007. Washington, D.C.

DOE. 2007b. Enhanced Geothermal Systems Reservoir Creation Workshop: Summary Report. Enhanced Geothermal Systems Reservoir Creation Workshop. Houston, Tex.

DOE. 2007c. Enhanced Geothermal Systems Reservoir Management and Operations Workshop: Summary Report. Enhanced Geothermal Systems Reservoir Management and Operations Workshop. San Francisco.

DOE. 2007d. Enhanced Geothermal Systems Wellfield Construction Workshop: Summary Report. Enhanced Geothermal Systems Wellfield Construction Workshop. San Francisco.

DOE. 2007e. National Solar Technology Roadmap: Wafer-Silicon PV. Office of Energy Efficiency and Renewable Energy. Washington, D.C.

DOE. 2007f. National Solar Technology Roadmap: Film-Silicon PV. Office of Energy Efficiency and Renewable Energy. Washington, D.C.

DOE. 2007g. National Solar Technology Roadmap: Concentrator PV. Office of Energy Efficiency and Renewable Energy. Washington, D.C.

DOE. 2007h. National Solar Technology Roadmap: CdTe PV. Office of Energy Efficiency and Renewable Energy. Washington, D.C.

DOE. 2007i. National Solar Technology Roadmap: CIGS PV. Office of Energy Efficiency and Renewable Energy. Washington, D.C.

DOE. 2007j. National Solar Technology Roadmap: Organic PV. Office of Energy Efficiency and Renewable Energy. Washington, D.C.

DOE. 2007k. National Solar Technology Roadmap: Sensitized Solar Cells. Office of Energy Efficiency and Renewable Energy. Washington, D.C.

DOE. 2007m. National Solar Technology Roadmap: Nano-Architecture PV. Office of Energy Efficiency and Renewable Energy. Washington, D.C.

DOE. 2008. 20% Wind Energy by 2030: Increasing Wind Energy's Contribution to U.S. Electricity Supply. Office of Energy Efficiency and Renewable Energy. Washington, D.C.

EIA (Energy Information Agency). 2001. Biomass for Electricity Generation. Washington, D.C.: U.S. Department of Energy, EIA.

EIA. 2007. Renewable Energy Annual, 2005. Washington, D.C.: U.S. Department of Energy, EIA.

EIA. 2008. Annual Energy Review 2007. DOE/EIA 0834(2007). Washington, D.C.: U.S. Department of Energy, EIA.

EPA (U.S. Environmental Protection Agency). 2008. An Overview of Landfill Gas Energy in the United States. U.S. Environmental Protection Agency Landfill Methane Outreach Program. Washington, D.C.

EPRI (Electric Power Research Institute). 2007a. Assessment of Waterpower Potential and Development Needs. Palo Alto, Calif.

EPRI. 2007b. Compressed Air Energy Storage (CAES) Scoping Study for California. Sponsored by California Energy Commission. Sacramento, Calif.

Ernst, B., B. Oakleaf, M.L. Ahlstrom, M. Lange, C. Moehrlen, B. Lange, U. Focken, and K. Rohrig. 2007. Predicting the wind. IEEE Power & Energy Magazine 5(6):78-89.

ETSO (European Transmission System Operators). 2007. European Wind Integration Study (EWIS) Towards a Successful Integration of Wind Power into European Electricity Grids. Brussels. Available at http://www.etsonet.org/upload/documents/Final-report-EWIS-phase-I-approved.pdf.

Fletcher, E.A. 2001. Solar thermal processing: A review. Journal of Solar Energy Engineering 123:63-74.

Gyuk, I. 2008. Energy storage for a greener grid. Presentation at the third meeting of the Panel on Electricity from Renewable Resources, January 16, 2008. Washington, D.C.

Hawlins, D., and M. Rothleder. 2006. Evolving role of wind forecasting in market operation at the CAISO. Pp. 234-238 in Power Systems Conference and Exposition, 2006 (PSCE '06). Washington, D.C.: Institute of Electrical and Electronics Engineers.

IEEE (Institute of Electrical and Electronics Engineers). 2005. November/December Issue: Working with Wind—Integrating Wind into the Power System. IEEE Power & Energy Magazine 3(6).

IEEE. 2007a. November/December Issue: Wind Integration, Driving Policies, and Economics. IEEE Power & Energy Magazine 5(6).

Jones, A.T., and W. Finley. 2003. Recent developments in salinity gradient power. Pp. 2284-2287 in OCEANS 2003: Celebrating the Past, Teaming Toward the Future. Columbia, Md.: Marine Technology Society.

King, D.L., W.E. Boyson, and J.A. Mratochvil. 2004. Photovoltaic Array Performance Model. Photovoltaic System R&D Department. Albuquerque, N.Mex.: Sandia National Laboratories.

Kroposki, B. 2007. Renewable energy interconnection and storage. Presentation at the first meeting of the Panel on Electricity from Renewable Resources, September, 18, 2008. Washington, D.C.

Mancini, T., P. Heller, B. Bulter, B. Osborn, S. Wolfgang, G. Vernon, R. Buck, R. Diver, C. Andraka, and J. Moreno. 2003. Dish Stirling systems: An overview of development and status. Journal of Solar Energy Engineering 125:135-151.

McKenna, J., D. Blackwell, C. Moyes, and P.D. Patterson. 2005. Geothermal electric power supply possible from Gulf Coast, Midcontinent oil field waters. Oil & Gas Journal (Sept. 5):3440.

Miles, A.C. 2008. Hydropower at the Federal Energy Regulatory Commission. Presentation at the third meeting of the Panel on Electricity from Renewable Resources, January 16, 2008. Washington, D.C.

Mills, D., P. Le Lievre, and G.L. Morrison. 2004. Lower temperature approach for very large solar power plants. Proceedings of the 12th International Symposium on Solar Power and Chemical Energy Systems (SolarPACES '04), Oaxaca, Mexico. Available at http://www.ausra.com/pdfs/LowerTempApproach_Mills_2006.pdf.

Minerals Management Service. 2006. Wave energy potential on the U.S. Outer Continental Shelf. Technology white paper. Renewable Energy and Alternate Use Program, Minerals Management Service. Washington, D.C.: U.S. Department of the Interior.

MIT (Massachusetts Institute of Technology). 2006. The Future of Geothermal Energy: Impact of Enhanced Geothermal Systems (EGS) on the United States in the 21st Century. Cambridge, Mass.

Muckerman, J.T., D.E. Polyansky, T. Wada, K. Tanaka, and E. Fujita. 2008. Water oxidation by a ruthenium complex with noninnocent quinone ligands: Possible formation of an O–O bond at a low oxidation state of the metal. Inorganic Chemistry 47(6):1787-1802.

Muljadi, E., C.P. Butterfield, R.Yinger, and H. Romanowitz. 2004. Energy storage and reactive power compensator in a large wind farm. Paper presented at 42nd AIAA Aerospace Sciences Meeting and Exhibit, January 5-8, 2004, Reno, Nevada. AIAA 2004-352. Reston, Va.: American Institute of Aeronautics and Astronautics. Available at http://pdf.aiaa.org/preview/CDReadyMASM04_665/PV2004_352.pdf.

NAS-NAE-NRC (National Academy of Sciences-National Academy of Engineering-National Research Council). 2009a. America's Energy Future: Technology and Transformation. Washington, D.C.: The National Academies Press.

NAS-NAE-NRC. 2009b. Liquid Transportation Fuels from Coal and Biomass: Technological Status, Costs, and Environmental Impacts. Washington, D.C.: The National Academies Press.

NREL (National Renewable Energy Laboratory). 2008. About Geothermal Electricity. Golden, Colo. Available at http://www.nrel.gov/geothermal/geoelectricity.html.

NYISO (New York Independent System Operator). 2008. Forecasts Sufficient Electricity Supply for Summer 2008. NYISO press release. Rensselaer, N.Y.

O'Regan, B., and M. Grätzel. 1991. Low-cost, high-efficiency solar cell based on dyesensitized colloidal TiO_2 films. Nature 353:737-740.

Parsons, B., M. Milligan, E. DeMeo, B. Oakleaf, K. Wolf, M. Schuerger, R. Zavadil, M. Ahlstrom, and D.Y. Nakafuji. 2006. Grid impacts of wind power variability: Recent assessments from a variety of utilities in the United States. Conference Paper NREL/CP-500-39955. National Renewable Energy Laboratory. Washington, D.C.: U.S. Department of Energy. July.

Perkins, C., and A.W. Weimer. 2004. Likely near-term solar-thermal water splitting technologies. International Journal of Hydrogen Energy 29:1587-1599.

Perkins, C., and A.W. Weimer. 2009. Solar-thermal production of renewable hydrogen. AIChE Journal 55(2):286-293.

RAB (Renewables Advisory Board). 2008. Marine Renewables: Current Status and Implications for R&D Funding and the Marine Renewable Deployment Fund. United Kingdom.

Rastler, D. 2008. Electric energy storage briefing. Presentation at the fourth meeting of the Panel on Electricity from Renewable Resources, March 11, 2008. Washington, D.C.

SECO (State Energy Conservation Office). 2008. Texas Wind Energy. Available at http://www.seco.cpa.state.tx.us/re_wind.htm.

SEIA (Solar Energy Industries Association). 2004. Our Solar Power Future—The U.S. Photovoltaic Industry Roadmap Through 2030 and Beyond. Washington, D.C.

SERI (Solar Energy Research Institute). 1989. Ocean Thermal Energy Conversion: An Overview. Golden, Colo.

Smith, C. 2007. Integration of wind into the grid. Presentation at the second meeting of the Panel on Electricity from Renewable Resources, December 6, 2007. Washington, D.C.

Starr, C. 2002. National energy planning for the century: The continental SuperGrid. Nuclear News 45(2):31-35.

Steinfeld, A. 2005. Solar thermochemical production of hydrogen: A review. Solar Energy 78(5):603-615.

Surek, T. 2001. Photovoltaics: Energy for the new millennium. Physics and Society 30(1). Available at http://www.aps.org/units/fps/newsletters/2001/january/aaajan01.html.

Thresher, R., M. Robinson, and P. Veers. 2007. To capture the wind. IEEE Power & Energy Magazine 5(6):34-46.

WGA (Western Governors' Association). 2006a. Clean and Diversified Energy Initiative: Geothermal Task Force Report. Washington, D.C.

WGA. 2006b. Clean and Diversified Energy Initiative: Solar Task Force Report. Washington, D.C.

Wilburn, D., and D. Bleiwas. 2004. Platinum-Group Metals—World Supply and Demand. U.S. Geological Survey (USGS) Open-File Report 2004-1. Washington, D.C.: USGS.

Williams, M., and M. Zhang. 2008. Challenges to Offshore Wind Development in the United States. MIT-USGS Science Impact Collaborative. Cambridge, Mass.

Williams, R.H., and E.D. Larson. 1996. Biomass gasifier gas turbine power generating technology. Biomass and Bioenergy 10:149-166.

Wiltsee, G. 2000. Lessons Learned from Existing Biomass Power Plants. Golden, Colo.: National Renewables Energy Laboratory. February. Available at http://www.nrel.gov/docs/fy00osti/26946.pdf.

4 Economics of Renewable Electricity

The previous chapters established the availability of renewable resources and outlined the technology options for converting those resources into electricity. This chapter explores the challenges and opportunities for bringing substantial renewable electricity generation to market to serve future U.S. electricity needs. Given the experience with renewables over the past 20–30 years, there is an inherent understanding that the economics of renewables have not been favorable. The economics of renewables is about profitability, and profitability depends on three drivers: (1) the market price or value of renewable electricity; (2) the costs of renewables relative to those of other energy resources; and (3), importantly, policies to promote renewables and environmental goals (particularly climate and energy security policies) that raise costs of using fossil fuels and/or subsidize costs of renewables.

The economic future for renewables depends on how market price, costs, and policy evolve. This chapter examines these drivers, the factors that underlie them, and issues associated with making predictions about them and their effects on the success of renewables in the marketplace. It sets out the fundamentals of the electricity market, explores technical and regional issues that affect renewables economics, and outlines the many entities engaged in renewable generation and what they bring to the table. The chapter concludes by summarizing and analyzing cost estimates for the renewable technologies with the greatest likelihood of contributing significantly to electricity generation in the next decade. The goal is not only to compare the costs of various technology options and how they will evolve over time, but also to clarify how markets and government actions can affect the near-term deployment of renewables.

This chapter focuses on the renewable technologies that are closest to market and for which assessments of current and future costs are thus more readily available. These include biomass, wind, concentrating solar power, solar photovoltaics, and geothermal (hydrothermal), but exclude traditional hydropower, because the potential for future extraction of this resource is limited, as noted in Chapter 2. The chapter also excludes hydrokinetics and enhanced geothermal technologies, which are still in the early stages of technological development. The costs presented here come from the wealth of data obtained from projects built in the recent past.

THE VALUE OF RENEWABLES

Predicting the economics of future renewable generation involves predicting the cost of generation from alternative sources and the value of electricity delivered to the marketplace. The competitive value would be the wholesale price of electricity for grid scale resources and something close to the retail price of electricity for distributed renewable resources.[1] These prices define the value of adding renewables to the mix. The ability to predict electricity price is key to making predictions about future market penetration of renewable sources of electricity.

The value of generation from renewables will vary geographically and by time of day, because the marginal generator,[2] which sets the electricity price, varies with location and over the course of the day with fluctuations in total electricity demand and available supply. Construction of more transmission facilities will increase the value of renewables by reducing transmission constraints between regions with abundant renewable resources and those with abundant load (Vajjhala et al., 2008).

[1]In his analysis of the value of electricity produced by solar PV installations on household and business rooftops, Severin Borenstein (2008b) points out that, although the value to a consumer of not having to purchase electricity may be the retail price of the purchases avoided, the avoided cost to society from installing PV on one's rooftop is less than the full retail price, which includes payments for recovery of past costs, including the California Energy crisis, and sunk costs of past high-priced electricity contracts.

[2]To meet electricity demand at lowest cost, system operators tend to dispatch electricity generators in the order of their variable cost of generation, which includes fuel and operating and maintenance costs. The marginal generator is the last generator, and therefore typically the highest-cost generator, that is dispatched to meet electricity demand at any point in time.

The importance of relative costs means that efforts to understand how future expected declines in renewables cost are likely to affect renewables penetration will depend on future predictions of the market price of electricity. An analysis of the accuracy with which past studies from the 1970s and 1980s of several different renewable technologies—including wind, solar photovoltaics (PV), concentrating solar power (CSP), geothermal, and biomass—predicted future costs and future penetrations finds that these past studies performed reasonably well at predicting future cost declines but did not accurately predict market penetrations (McVeigh et al., 2000). McVeigh's analysis shows that predictions consistently overestimated the expected retail price of electricity in future years. The renewable technologies included in the study had, for the most part, large reductions in cost over time, but these reductions were matched or exceeded by declines in the real cost of supplying electricity with fossil fuels, and thus renewables did not achieve predicted increases in penetration. This suggests that the challenge of predicting future costs of renewables may be exceeded by the challenge of predicting future market conditions that will confront those technologies, which will be equally if not more important in determining the ability of renewables to penetrate the market.

In addition to selling electric energy, most wholesale electricity markets also have an additional source of revenue from capacity payments. Capacity payments are made to encourage some generation to be readily available to meet changes in demand and ensure a high level of reliability in delivered electricity despite unforeseen outages. Requirements for the amount of capacity required vary regionally, but the value directly correlates to the expected performance of the unit when needed for generation. For dispatchable fossil generation and renewables, the capacity value is the highest, usually based on close to 100 percent of the unit's rated capacity. For other renewables, the capacity value is typically lower to reflect the intermittent availability of the resource. The capacity value of a given renewable technology is regionally specific owing to how the capacity value is determined and the relative alignment between resource and load. Although intermittent, the capacity value of grid-scale solar would typically be higher than that of wind, because there is often better correlation between electricity demand and when the sun is shining. Solar resource availability is more predictable than wind is, though clouds do have a serious impact on solar flux. In a region where the wind resource availability does not correlate well with periods of system load, the capacity value may be as low as 8–10 percent of the rated capacity of the unit (ERCOT, 2007; GE Energy Consulting, 2005). In areas where resource or transmission availability allows for better correlations with load, renewables will qualify for higher capacity payments. Capacity payments do not lower costs, but they affect the economics of renewables, because they provide an additional incentive to increase dispatchability.

Another source of value for most renewables[3] is that their operation typically does not contribute to air pollution through emissions of NO_x and SO_2 and greenhouse gas (GHG) emissions, particularly emissions of CO_2.[4] Substituting renewable generation for fossil-fuel generation could reduce air pollution and greenhouse gas emissions. These benefits would depend on the type of fossil generation displaced, the emission controls on the fossil generation, the resulting emissions rate of that fossil generation, and the form of environmental regulation governing pollutants.[5] For pollutants subject to an emissions cap, as is the case for SO_2 nationally or CO_2 in states participating in the Regional Greenhouse Gas Initiative, there will not be reduced emissions or environmental benefits. Emissions caps are both a ceiling and a floor on the level of emissions, as emissions reductions at one facility will be made up by increases at another facility, unless the cap is reduced or is no longer binding, which could occur with a dramatic increase in renewable generation.

If emissions are capped and emission trading is allowed, there could be an important effect on emission allowance markets and thus on the costs of electricity production from fossil fuels with greater penetration of renewables. Greater use of renewables could reduce demand for emission allowances for SO_2 and NO_x and other capped pollutants, which could reduce their allowance price. To the extent that renewables displace natural gas, at least initially, this effect is likely to be small for pollutants like SO_2 and NO_x. However, the effect could be larger for pollutants like CO_2 if they were capped, though it is a value that would accrue to everyone who has to purchase allowances and not just to the utility that is adopting more renewables.

Most emissions of CO_2 from electricity generators in the United States are not capped. Increasing renewables generation to replace fossil-fuel generation

---

[3]With the exception of hydrothermal, which emits SO_2 and CO_2, and biopower, which emits NO_x and CO_2.

[4]There are emissions associated with the manufacture of different renewable technologies. These life-cycle effects are discussed in Chapter 5.

[5]Greater reliance on intermittent renewables like wind or solar could increase the need for spinning reserves from fossil generators, and increased operation of these generators in spinning mode or at less than full capacity could reduce the CO_2 and NO_x emissions benefits.

would reduce CO_2 emissions, at least relative to business-as-usual emissions. Identifying the extent of those reductions requires some caution. The effects would vary by location, based on the composition of the existing generation fleet and the types of new non-renewable generators and fuels that might otherwise be put in place to meet future electricity needs. These reductions in CO_2 emissions would have value to society, and renewable generators might be able to capture some of that value if they could identify consumers willing to pay a premium for CO_2-free electricity or green power.

COSTS AND ECONOMICS OF RENEWABLE ELECTRICITY

Cost is the principal barrier to the widespread adoption of renewable technologies. Generating electricity using renewable energy technologies is more costly than generating it with fossil fuels, especially coal, which supplies about half of the electricity generated in the United States each year. More transmission infrastructure in key locations would also be required for a dramatic increase in power supplied by renewables. Recent increases in renewables market penetration, particularly new wind power, have largely been in response to policies like the federal renewable energy production tax credit and state renewables portfolio standards. These policies seek to close the cost gap in the short term by subsidizing renewable generation. By encouraging greater market penetration, these policies enable reductions in long-term costs through increased scale and learning in manufacturing and in the use of the technology.

To achieve greater market penetration, renewables would have to undergo cost reductions at a rate greater than the rate of cost improvement by technologies that set the market price of electricity, including natural-gas- and coal-fired generation. These reductions might result from major breakthroughs in technology, improvements in manufacturing, or improved operating performance of equipment, such as higher capacity factors for wind turbines. Likewise, increases in the costs of fossil generation could have an impact on the relative competitiveness of renewables, though the magnitude might not be as great if cost increases also improved the competitiveness of energy efficiency options and nuclear generation.

Estimates abound of present and future costs for particular types of renewables and other sources of generation. Comparability of these estimates depends on the underlying assumptions and the types of costs captured in summary measures. The next sections discuss the types of costs associated with constructing and operating renewable generating facilities, important assumptions underlying those costs, and how they can be used to construct summary measures of the cost of supplying energy.

Cost Concepts and the Levelized Cost of Energy

Developing a particular technology to generate electricity incurs costs for the capital equipment, such as the wind turbine and its tower, or solar panels; the land or property, if necessary for installation; and operating and maintaining the equipment. Some costs vary with the amount of electricity generated, and some costs are fixed. When a technology requires a fuel, such as biomass generation (biopower), the cost of the fuel would be a part of the variable operating and maintenance cost.

Capital costs do not vary with the amount of electricity generated by the facility and are typically stated in dollars per kilowatt ($/kW). Capital costs generally vary with the size of the facility or installation, with economies of scale or volume discounts on equipment orders favoring larger enterprises. Coal-fired and nuclear generating facilities exhibit economies of scale, and larger plants tend to have lower average cost of generation than smaller plants have. For renewables such as wind and solar PV, economies of scale can be greater at the equipment manufacturing stage than at the electricity-generating site, and increased capacity does not decrease the average cost of generation as much as it does for fossil and nuclear plants. Capital costs can also vary across sites, depending on land cost and the costs of installation or construction of the facility.

Fixed operating and maintenance (O&M) costs are also stated in dollars per kilowatt, but unlike capital costs, they are an ongoing expense associated with some unit of time ($/kW-year). Typically technologies are characterized by their annual fixed O&M costs. This category includes costs such as wages, materials, and land lease payments.

Variable O&M costs are typically expressed as dollars per megawatt-hour ($/MWh). Fuel costs can be expressed as dollars per unit of mass of the fuel ($/ton), dollars per unit of heat content of the fuel ($/Btu), or $/MWh. The last formula takes into account the efficiency of the technology in converting British thermal units (Btu) of heat input into megawatt-hours (MWh) of electricity.

In comparing the costs of generating electricity for different renewable technologies and for fossil fuels and nuclear technologies, cost estimates are typically converted into a levelized cost of energy (LCOE), which is expressed in $/MWh.

The initial cost of the capital equipment and installation constitutes a large portion of the cost of generating electricity, particularly for renewables, which have no fuel costs, with the exception of biopower generation. Converting this large up-front cost to cost per megawatt-hour requires making assumptions about the lifetime and capacity factor of the equipment,[6] as well as the discount rate and the timing of returns on that capital. For intermittent technologies such as CSP, solar PV, and wind power, the capacity factor can vary considerably, depending on the location and the quality of the resource (e.g., wind speed and constancy for wind turbines, and hours of sunlight with no cloud cover for CSP and PV); likewise, the LCOE will vary depending on the capacity factor at a particular installation and location.

The cost of fuel plays an important role in calculating levelized cost for biopower. Biopower is typically a baseload technology with a high capacity factor. On an annual basis its fixed equipment costs could be recovered over many hours of operation. However, the hours of operation and the amount of electricity generated by biopower would depend on the cost of fuel, which accounts for about one-third of the total LCOE from biopower (Venkataraman et al., 2007). The cost of biomass fuel is uncertain and would depend on competing demands for crops and other agricultural inputs, including demands for biofuels from the transportation sector.

Costs Beyond Generator Costs

The costs of purchasing, installing, and operating a specific power plant might not be the total costs to the system and to electricity consumers of deploying a new renewable generation facility. Costs that might be missing from the traditional levelized cost measure include the costs of new infrastructure necessary to connect the renewable generator to the grid and to ensure continued quality of power supply. Other costs include up-front costs for approval of siting the new facility and costs for appraising the resource at the site, as well as costs of obtaining financing and environmental permits.[7]

---

[6]The capacity factor is defined as the ratio (expressed as a percent) of the electricity output of a plant to the electricity that could be produced if the plant operated at its nameplate capacity.

[7]Levelized cost estimates also typically exclude the costs of the ultimate disposal of the generation equipment at the end of its useful life. Disposal may be complicated and costly for some types of equipment that contain hazardous chemicals that require special disposal procedures.

Transmission

While fossil fuels may be transported from the mine or the wellhead to an electric generation facility, renewable generating plants must be located at or near the resource. There might be some degree of greater flexibility in location for biopower, but not much. It can be costly to ship biomass fuels, given the relatively low energy density compared to fossil fuels. Thus, biopower facilities are typically located close to sources of fuel.

Wind and some solar resources often are located at some distance from the existing transmission grid, and would require new transmission lines to transport the power to the centers of electricity demand or load. As with any new generation, the cost of constructing additional transmission lines should be included in the cost of supplying electricity from renewable resources. A recent report looked at 40 transmission studies covering a broad geographic area on the costs associated with the transmission requirements for wind power (Mills and Wiser, 2009). The transmission costs associated with wind ranged from $0 to $1500/kW, and the majority were less than or equal to $500/kW, with a median of $300/kW. These numbers correspond to $0–79/MWh, with the majority below $25/MWh, and a median of $15/MWh. Intermittent renewables generation requires an additional consideration. Because of low capacity factors, dedicated transmission lines sized to transmit the full amount of power produced during peak generation hours would be unused or underused some of the time. Siting additional peaking capacity along a new transmission corridor could potentially leverage the available transmission capacity during periods of underuse by the renewables.

A caveat to the preceding discussion is that distributed renewables, such as distributed PV, might end up closer to the load than conventional generation and could lead to less need for investment in transmission. To really achieve substantial benefits in terms of avoiding investment in transmission infrastructure may require substantial amounts of distributed renewables investment in particular locations.

Intermittency

At sufficiently high capacities of solar and wind generation, the costs of intermittency could extend beyond costs associated with dedicated transmission facilities to affect the operation of the interconnected transmission grid. More generation from intermittent resources will require additional or alternative resources to help track load, provide voltage support, and meet needs for capacity reserves.

These include demand for second-by-second electricity load balancing service, or regulation; load following within the hour; and unit commitment of generators to be available at particular times of the day or week. Renewable electricity must be used when generated because the electricity can be generated only when the resource is available. Typically, fossil-fuel generators that are easily dispatchable, such as natural gas combustion turbines, supply these ancillary services. As renewables generation increases and fossil generators are curtailed, renewable generation technologies themselves or additional system assets, such as storage, will be needed to meet the increased need for ancillary services, at some additional cost. When system managers have improved tools and technology for predicting resource availability, it will be easier to determine the need for additional generation resources to back up intermittent renewables. Smart Grid technologies, which allow system managers to manage supply and demand in real time, could also mitigate some of the costs of renewable intermittency. An upgrade and expansion of the electricity grid will be necessary no matter what happens with renewables, given the age of the grid and the anticipated growth in electricity demand.

Studies in the past five years looked at the costs of integrating wind into the grid, as summarized in Figure 4.1 (Smith, 2007; Wiser and Bolinger, 2008). These

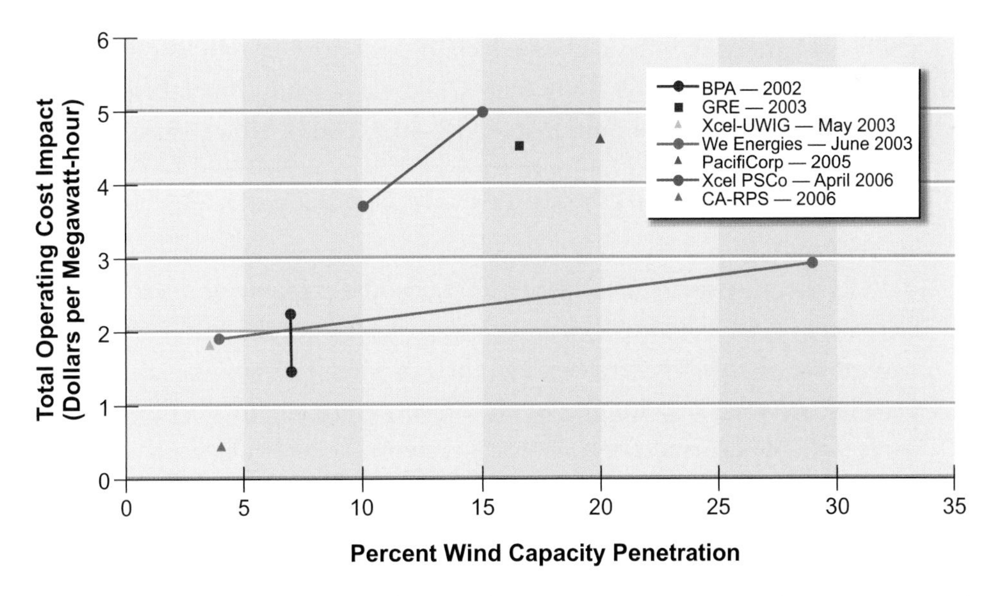

FIGURE 4.1 *Summary of wind plant ancillary services costs from various studies looking at the cost of regulation service, load following, unit commitment, and natural gas. Source: Developed from data in Smith (2007) and in Wiser and Bolinger (2008).*

studies examined the costs of regulation service, load following, unit commitment, and natural gas and found that the incremental costs per megawatt-hour range from about $1.50 to almost $5.00. A study on using wind to serve 50 percent of demand showed that the incremental costs are $10–20/MWh, including transmission, storage, and backup generation (DeCarolis and Keith, 2004). The European Wind Energy Association conducted a study of more than 180 sources and determined that additional costs range from $1.50 to $10.20/MWh for market penetration levels of 10 percent and from $2.80 to $11.50 for higher penetration levels (EWEA, 2005). Typically the predicted costs are higher in studies that focus on higher market penetration of wind. In the studies on different levels of penetration, the costs were higher with the higher levels of penetration, but the incremental effect of increased penetration varied across studies. Generally, where the average cost of wind generation would be about $80/MWh, the impact of grid integration costs appeared to be less than 15 percent where wind produced 20 percent or less of total electricity generation.

Energy Storage

Energy storage could mitigate the impact of intermittent renewables. Today there is very little storage in the United States, as high costs, low efficiencies, and technological uncertainty precluded storage from becoming economically viable.[8] Costs for battery and other storage technologies are generally about two to five times higher than the cost target that would make them competitive (less than about $200/kWh for a 4-hour system) (Rastler, 2008). However, technologies might be called on in the future to store electricity generated from intermittent renewable resources if their combined market penetration would rise to 20 percent and beyond.

Efficient, cost-effective energy storage could promote grid-scale renewable electricity. Wind and solar system operators have limited control over the amount and timing of power generation, and their production does not line up well with demand requirements. Storage would allow a grid operator to align the dispatch curve with the demand curve, a process referred to as load shifting. In addition to generating revenue when the wholesale market is at its peak, the ability to draw

[8]The exception is pumped hydroelectric storage, of which there was 21,461 MW of capacity nationwide in 2006 (EIA, 2007a). However, it is widely acknowledged that there is little chance for additional pumped storage because most of the viable pumped hydro opportunities have been exhausted. For this reason, this section omits pumped storage.

on storage would obviate the need for some peaker generators at the margin. Storage would also alleviate the reliability concerns associated with wind and solar. When renewables provide less than forecasted output, the operator has to turn to the spot market or bring on idle combined cycle natural gas generators to make up the difference. Conversely, when renewables provide too much power, those holding day-ahead gas contracts might not realize the value of their contracts. The market penalizes renewables for this uncertainty and, while recent studies have shown that this might not matter until intermittent renewables reach penetration levels in excess of 20 percent, this uncertainty may have to be addressed if they are to extend any further (DOE, 2008).

Storage would also mitigate some site limitations of renewable electricity and help reduce the size and increase the utilization of transmission lines installed for renewable sources. Small-scale domestic storage could also change the economics of distributed wind and solar generation, providing homes with energy security while perhaps making it possible to sell stored energy or capacity back to the grid. As plug-in hybrid electric vehicles (PHEVs) become a reality, households could store the energy they generate right on their vehicles. The National Renewable Energy Laboratory (NREL) found that PHEVs could enable increased penetration of wind energy (Short and Denholm, 2006).

Figure 4.2 displays how some of these storage technologies compare in terms of cost of energy and cost of power. At the grid-scale level (greater than 10 MW), compressed air energy storage (CAES) appears to be the most economical now, though the practicality of CAES also depends on the availability of suitable sites. Iowa Energy Storage Park (IESP), a 268 MW system, is scheduled to come on line in Iowa in 2011. Projected costs for IESP are $200–250 million, or $746–933/kW, and the system is designed to go from idle to full output in under 15 minutes. The Texas State Energy Conservation Office estimated total overnight capital costs of a new CAES system at $605/kW. Development and fixed O&M costs were listed at $28.00/kW and $14.07/kW, respectively, and variable O&M costs were estimated to be $1.50/MWh (Ridge Energy Storage, 2005). Batteries are modular and non-site specific, which makes them ideal for distributed generation. The quick, cheap response time also makes batteries ideal for providing backup power, or uninterruptible power supply (UPS). Yet despite broad application in other sectors, batteries are still very expensive, as shown in Figure 4.2.

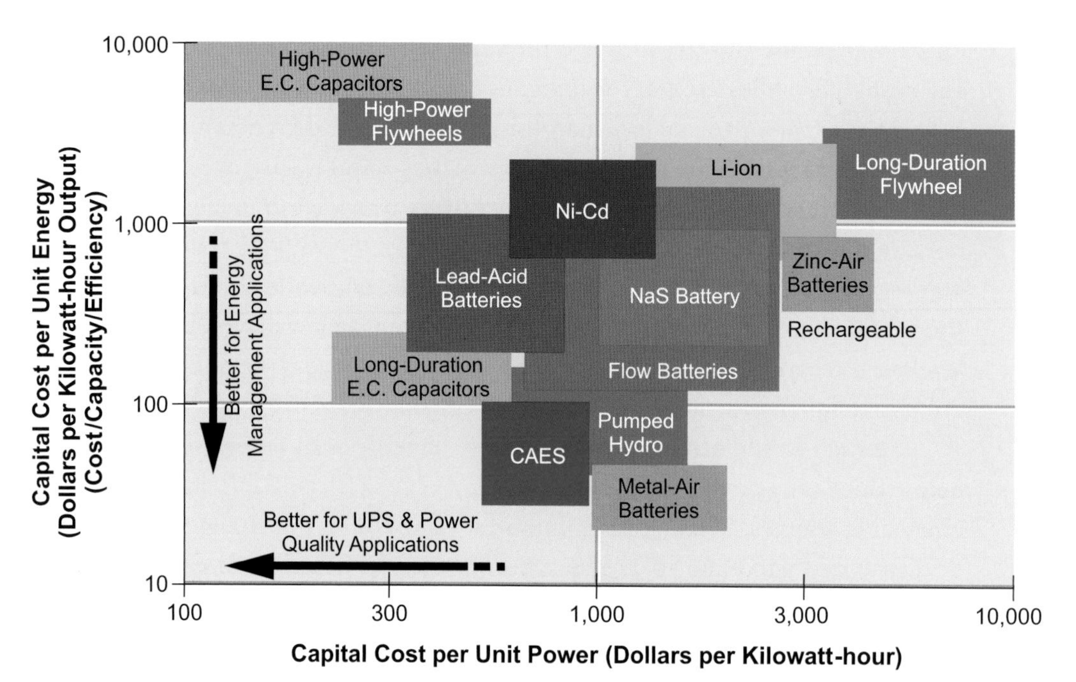

FIGURE 4.2 *Storage technolgies and costs of energy and power.*
Source: Electric Power Research Institute; presented in Rastler, 2008.

Financing Costs

Another element of cost that should be included when evaluating the overall competitiveness of any generation project is the cost of money. Because electricity generation projects are capital-intensive and have long lifetimes, access to capital and the rates at which it is paid back are key components of project cost. The magnitude of these costs differs, depending on the type of generation financed. For example, a renewable project that does not require fuel has a much larger portion of its costs associated with the initial capital expense of the plant than a gas-fired power project that will have greater operating expenses throughout its lifetime, even if both have similar LCOEs. These costs are project-specific, based on circumstances related to the project's financing strategy, the maturity of the technology, and risk factors discussed in Chapter 6.

Financing of renewables projects differs from that of fossil-fueled plants. Although the total magnitude of capital may be smaller for a renewable project than for a fossil project, the capital intensity relative to operating costs is much higher for renewables without fuel, such as wind and solar. There is more upfront risk in the renewables project's financial model. Further, the tax incentives that could subsidize some renewables might not be directly available to the project developer and could require more complex financing structures to access the benefit. The project financing structure, such as the debt-to-equity ratio; the types and costs of loans, depending on the risk profile; and the magnitude and timing of returns to financing entities can have an impact of as much as $15/MWh on a wind project's levelized cost of energy (Harper et al., 2007). Regional variability in the costs of land and logistical support and time variability in the selling price of electricity due to market forces complicate financing for renewables projects. For wind projects before the 2008–2009 economic crisis, the cost of tax equity appeared to decline by approximately 3 percent and interest rate margins on debt transactions by approximately 0.5 percent (Wiser and Bolinger, 2008). This trend toward cheaper capital resulted directly from reduced project risks as the wind power industry matured and the available capital for wind projects increased. Economic events dramatically reversed this trend for all forms of power generation but have affected renewables to a greater degree, due to the reliance on investor tax capacity in order to realize the economic benefit of the production tax credit (PTC). The American Recovery and Reinvestment Act of 2009 attempted to address this issue by allowing non-solar renewable electricity facilities to elect a 30 percent investment tax credit in lieu of the PTC.

Because of their small scale and modularity, advantages that wind and solar PV projects have over fossil projects is the shorter time between purchase of the equipment and placing it on line and the ability to start up the first few generators while others are under construction (Bierden, 2007; Royal Academy of Engineering, 2004; Sheehan and Hetznecker, 2008). These two features reduce the magnitude of draws on cash flow and accelerate the repayment of debt.

Methodologies for Projecting Costs

An overview of the different approaches to projecting future costs places the cost projections in this report in context. The panel identified three methods for predicting future costs of renewables.

The first methodology predicts the levelized cost of energy that must be achieved from a particular renewable generating source to be competitive with other sources of electricity by some date in the future. This method requires estimating the future wholesale market price of electricity with which renewable resources must compete. These predictions omit consideration of uncertainties, the relationship between government policy and expenditures, and changes in the costs of using renewables to supply electricity. The Western Governors' Association (WGA) Solar Task Force took this approach and developed a series of referent market prices that depend on the assumed price of natural gas, the energy source typically setting the market price of electricity in the western states (WGA, 2006b). The higher the predicted price of natural gas, the lower the cost reduction hurdle for renewable technologies.

A second approach that is similar to the first involves the enunciation of cost and technical performance goals, such as availability factors, that those researching future developments of the technology, such as the Office of Energy Efficiency and Renewable Energy (EERE) at the DOE, expect the technology to achieve as a result of their research program. The idea behind this approach is to establish goals for a research and development (R&D) program and also to provide some benchmark expectations about technological improvement that could be used later to judge the performance of the research program after the fact. This approach is used in NREL's projections that it develops for DOE (NREL, 2007).

The Energy Information Administration (EIA) took a third approach to cost prediction in constructing a more formal model of how technical costs might evolve over time. Models can be based on a projection of past trends or a formal learning or technological improvement function. Cost reductions through learning are greater for new technologies than for mature technologies. EIA took the learning-curve approach to predict how costs would evolve as greater amounts of a particular technology penetrate the market in response to a combination of policies and electricity demand growth (EIA, 2007a). This is the approach underlying the cost estimates for EIA and the Electric Power Research Institute (EPRI) shown in Tables 4.A.1 and 4.A.2 in the annex at the end of this chapter.

Some might argue that present-day costs would be the best predictor of future costs, particularly in the short term. This approach might suffice for forecasting future costs for more mature renewable technologies such as wind, but might be less appropriate for nascent technologies. Another problem with this approach is that factors contributing to short-term cost increases, such as the recent increases in the cost of wind turbines and solar cells due to material shortages, might not be sustained into the future, as entry into the industry, greater availability of materials, and innovations might bring costs down.

POLICIES AND PRACTICES THAT AFFECT THE ECONOMICS OF RENEWABLE ELECTRICITY GENERATION

The United States and other nations have implemented policies to increase the market penetration of renewables. Typically these policies work by either subsidizing the cost of renewable generation (and thereby decreasing their relative costs) or by increasing the demand for renewable generation. Some policies provide for additional sources of revenue for renewable generators, such as through the sale of renewable energy credits. The form of the policy and the amount of the payment or subsidy for renewables are both determinants of the policy's effect on renewables penetration.

The panel considers three classes of policies in this section: (1) policies and practices targeted at renewable technologies; (2) environmental policies that raise the cost of using conventional technologies, thereby improving the relative cost competitiveness of renewables; and (3) other electricity market policies that could affect the economics of using renewables and their ability to penetrate the future market.

Policies to Promote Renewables

Both the federal government and a majority of the states have policies to promote the use of renewable technologies to supply electricity. Most of these policies are described in Chapter 1. Here, the panel focuses on a few policies and describes how they appear to affect the economics of renewable generation. Some policies target large central station facilities, while others focus on distributed renewables intended for personal consumption. The following review focuses on the major policies in terms of their potential capability and their relevance for renewables market penetration.[9]

Production Tax Credits

A renewable energy production tax credit (PTC) policy allows firms that generate electricity with eligible renewable technologies to offset their income tax liability

[9]In addition to the policies identified below, some states offer low-interest loans for renewables. However, generators that avail themselves of this type of state support may make a particular project ineligible for the renewable energy production tax credit discussed in the next section.

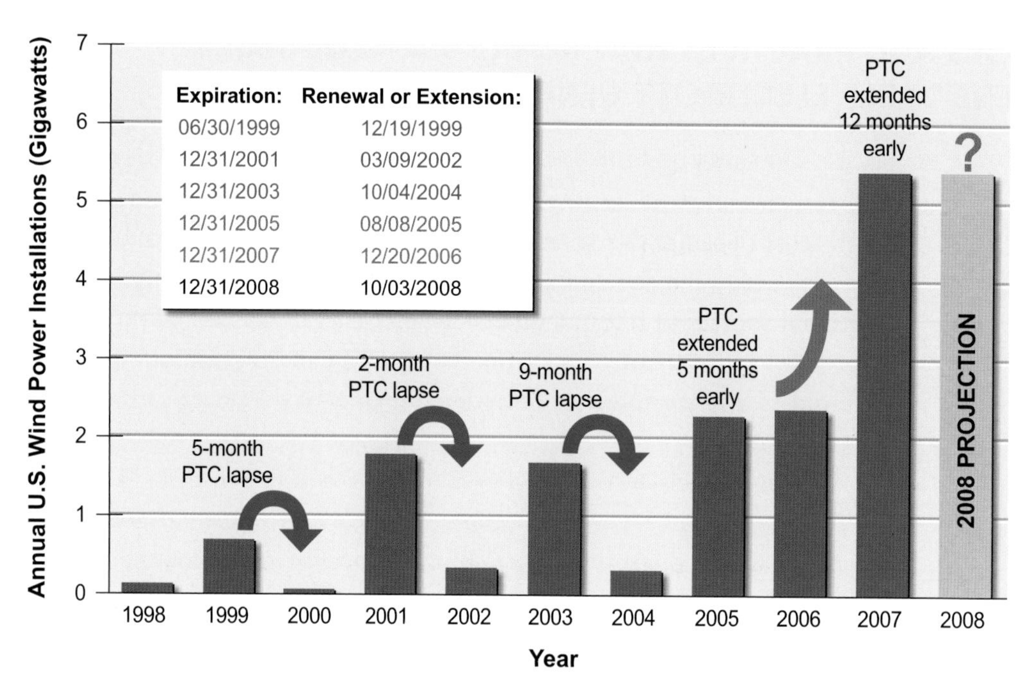

FIGURE 4.3 *Effects of production tax credit expiration and extension on wind power investment. Not shown are the almost 8,400 GW of installed wind power in 2008 and the extension of the PTC until 2012.*
Source: Lawrence Berkeley National Laboratory; presented in Wiser, 2008.

by the amount of the tax credit times the number of kilowatt-hours generated. The federal PTC applies to a range of renewable technologies, with some technologies, including wind, solar, closed-loop biomass, and geothermal, eligible for a larger tax credit than others, such as open-loop biomass, small hydroelectric, landfill gas, and municipal solid waste.[10] Generators are eligible for the tax credit for every kWh of electricity generated during their first 10 years of operation. The federal renewables PTC policy was recently extended until 2012 and beyond, as described in Chapter 1. Initially passed in 1992, this policy typically had only been approved for 1–2 years into the future and lapsed three times since its inception. As shown in Figure 4.3, the intermittency of this policy led to large fluctua-

[10]Several renewable technologies (wind, solar, geothermal, and small biomass generators) are also eligible for accelerated depreciation, which allows them to depreciate their capital over 5 years instead of the 20-year lifetime for most fossil generators (15 years for new nuclear). In addition, renewables may be eligible for a method of depreciation within the 5-year time period that allows more than half of the investment value to be depreciated in the first 2 years of use.

tions in demand for wind turbines as project developers raced to beat the deadline and then lost interest in new projects when the policy lapsed (Wiser, 2008).

In addition to the federal PTC, five states (Florida, Iowa, Maryland, Nebraska, and New Mexico) also offer PTCs that provide a tax credit for every kilowatt-hour of electricity generated. Another seven states offer direct payments for each kilowatt-hour of electricity generated by certain renewable technologies.

Both an investment tax credit (ITC) and a renewable PTC reduce the cost of generating electricity using renewables. The PTC is arguably more effective at getting performance out of a generator, because the level of the PTC subsidy depends directly on how much electricity the generator produces, whereas the ITC does not differentiate between a renewable that is productive and one that does not generate much electricity. However, using a tax credit based on electricity production may not be viable for distributed technologies owned and operated by end users. At present levels, the federal PTC reduces the cost of supplying renewables by 1–2¢/kWh, depending on the technology, divided by 1 minus the marginal income tax rate. The reason for the transformation is that the tax credit is equivalent to after-tax income. To earn 2¢/kWh in after-tax income, before-tax income would have to be greater than 2¢. Likewise, since a decrease in cost is the same as an increase in income, the decrease in before-tax costs that is equivalent to the 2¢ tax credit must be greater than 2¢. Thus, for companies in the 33 percent marginal tax bracket, this would be an increase of after-tax income of about 1.5 times the value of the tax credit, as that would be the increase in total revenue equivalent to the decrease in tax burden (assuming the affected firm still has a tax burden) (EIA, 2005).

Modeling analysis demonstrates the potential capability for an extended PTC to increase investment in and generation from eligible renewables. In response to a request from the House Committee on Ways and Means, EIA (2007b) analyzed the effects of both a 5-year and an indefinite extension of the PTC for wind generators only. This analysis considered the effects of a continued PTC for wind at 1¢, 1.5¢, and 1.9¢/kWh. The results suggest that extending the PTC at the current level for 5 years would lead to 30 percent more wind capacity in 2020 compared to the AEO 2007 forecast, and total wind generation equal to 1.5 percent of total generation in 2020, compared to a baseline share for wind of about 1 percent in 2020. Extending the PTC indefinitely would more than double the amount of wind capacity in 2020 compared to the AEO 2007 baseline and would almost triple it by 2030. These increases in investment and associated wind generation would have very little impact on the price of electricity to consumers, although there would be a cost to the U.S. Treasury ranging from $2 billion, with a 5-year extension of the present credit, to more than $20 billion with a permanent extension.[11]

According to the EIA analysis, extending the PTC for only 5 years would have a negligible effect on CO_2 emissions, largely because the displaced generation comes mostly from natural gas. However, with a permanent extension of the PTC, the investment in wind would displace more investment in new coal-fired generation, resulting in about a 2 percent reduction in CO_2 emissions from the electricity sector compared to the AEO 2007 forecast in 2020. Palmer and Burtraw in an earlier study (2005) extended the PTC indefinitely to both wind and biopower and demonstrated a more substantial increase in renewables generation by 2020, and nearly 5 percent lower CO_2 emissions in 2020 compared to a no-extension base case. Despite this substantial reduction in CO_2 emissions, their work suggested that the PTC is not as cost-effective in limiting CO_2 emissions as other policies, including renewable portfolio standards (RPS) or a CO_2 cap and trade approach, which could achieve a similar reduction in CO_2 emissions at much lower costs.

Renewable Portfolio Standards

As of 2008, the District of Columbia and 27 states have renewable portfolio standards (RPSs) that require a minimum percentage of electricity sold to customers within a state to be generated using renewable resources. An additional six states have voluntary programs. Details of all state RPSs can be found in Appendix D of the report. Figure 4.4 from the Lawrence Berkeley National Laboratories shows the timing of the adoption of RPS policies in different states and indicates that many states revised their policies after adoption, typically to make them more ambitious.

An RPS policy creates an increase in demand for electricity supplied by renewables and, in most cases, a demand for a complementary product that renewable generators can sell, a renewable energy credit (REC). RECs can be traded and bundled with the electricity that the generator produces or as an unbundled separate product. Revenue from the sale of RECs provides an additional incentive for renewable generators to supply electricity. Researchers at Lawrence Berkeley National Laboratory found that 60 percent of wind capacity

---

[11]The costs to the U.S. Treasury reported in this report are in real 2006 dollars and are not discounted to reflect the time value of money.

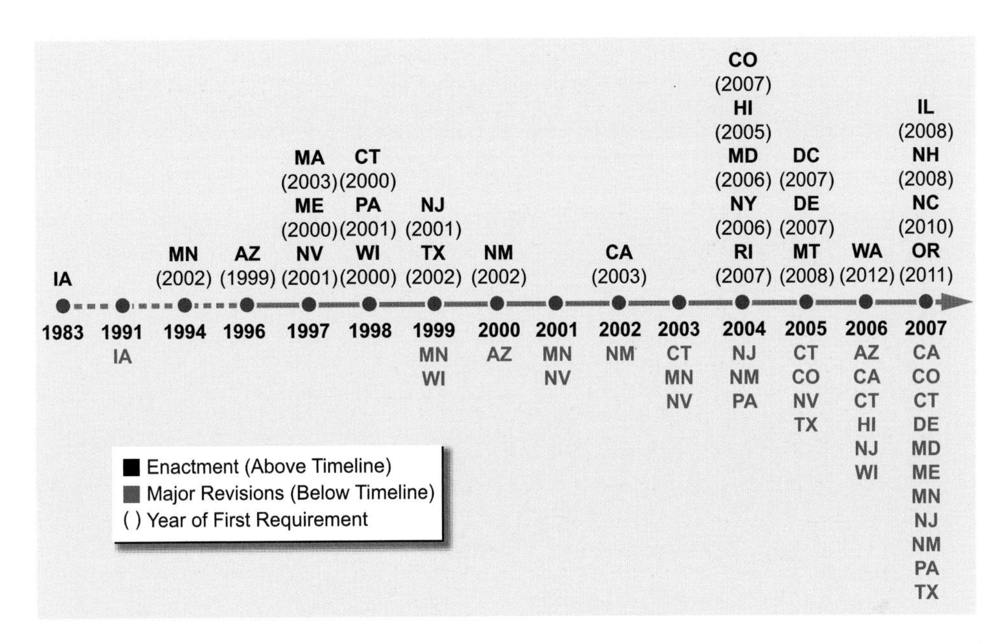

FIGURE 4.4 *Renewable portfolio standards policy adoption and modification across the states.*
Source: Wiser and Barbose, 2008.

constructed in 2006 was motivated at least partially by state RPS policies (Wiser and Bolinger, 2008).

The effects of a state RPS policy on the economics of renewable generation in particular and electricity supply more generally depend on features of the policy, including its stringency, what renewable technologies are included, site restrictions on renewables eligibility (e.g., limitations on out-of-state renewables), cost containment measures, and enforcement penalties. For example, a policy that caps the price of an REC at a low level or allows regulated electricity suppliers to make a small alternative compliance payment in lieu of meeting the RPS obligation would provide a weaker incentive for renewables development than would a policy that has no REC cap and stringent requirements for RPS compliance. State RPS policies differ dramatically in terms of features.[12] Figure 4.5 shows how REC prices in various state programs for tier-one resources—the most valuable and flexible resources—have evolved since early 2003, and how prices differ across states. In general, prices in New England tend to be higher than prices in Texas

[12]For an overview of the various ways in which state RPS policies differ across states see Cory and Swezey (2007).

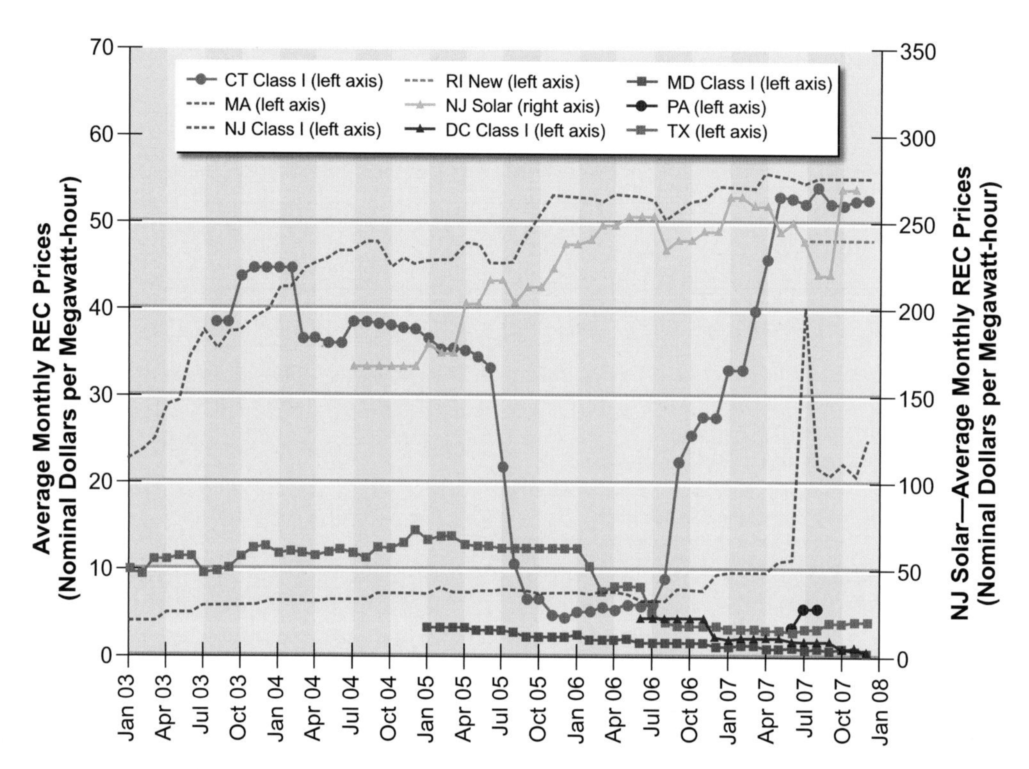

FIGURE 4.5 *Renewable energy credit (REC) prices in the renewables portfolio standard compliance market.*
Source: Wiser and Barbose, 2008.

and the mid-Atlantic states (with the exception of New Jersey solar). These differences reflect in part the difficulty in siting new renewable generators in New England. Fluctuations in REC prices over time within a state program reflect changes in state conditions. For example, large fluctuations in REC prices in Connecticut followed from changes in the states from which Connecticut electricity suppliers could purchase RECs and from imposing RPS requirements on municipal generators in 2007, which increased demand for Connecticut RECs. The states with the highest REC prices or with the highest RPS requirements in 2007 saw the largest impact on their electricity prices in that year.

A federally mandated RPS policy could reduce the differences in RPS policies across states. There have been unsuccessful attempts in Congress to pass a national RPS. EIA's analysis of a federal policy mandating a 25 percent RPS and a 25 percent renewable fuel standard by 2025 suggested that a federal RPS of 25 percent would result in REC prices between \$35 and \$50/MWh in 2025, depending on assumptions about fuel costs and technology improvement (EIA,

2008b). The 25 percent RPS policy would result in about 20 percent lower annual CO_2 emissions from the electricity sector than the AEO 2007 baseline in 2025. EIA's earlier analysis of a 15 percent RPS showed that annual CO_2 emissions from the electricity sector in 2030 could be 6.7 percent lower with such a policy than under a base case scenario (EIA, 2007c).[13]

Previous modeling of less stringent RPS policies suggested that the REC price is not a linear function of the level of the portfolio standard requirement. Instead, the REC price for a national policy seems to be an increasing function of the threshold, particularly for levels higher than 15 percent (Palmer and Burtraw, 2005). These models assumed that the federal standard would replace existing state-level RPSs. However, with a federally mandated policy in addition to existing state RPS policies, the price of a federal REC would tend to be lower and would depend on regional transmission capability to bring power from resource-rich areas to regions with high levels of electricity demand (Vajjhala et al., 2008).

Green Power Marketing

Green power marketing is the direct marketing of power generated by renewable resources and supplied directly to end users of electricity. Green power refers to all types of renewables except hydropower. Green power marketing typically focuses on selling power from new facilities. Green power marketing can occur in competitive electricity markets or as an elective tariffed service that regulated utilities offer rate-paying customers. According to researchers at NREL, voluntary purchases of green power represented about 32 percent of total green power generation in 2005 and about 36 percent of total green power generated in 2006 (Bird and Swezey, 2006; Swezey et al., 2007; Bird et al., 2007). In 2006, voluntary green power markets supplied about 12 billion kWh of generation out of a total demand of more than 3600 billion kWh.

As more and more states adopt RPS policies, the question arises as to whether these mandatory policies will replace voluntary markets for green power. A study of the relationships between mandatory RPS programs and voluntary green power markets found potential for overlap in the form of double-counting, which is selling renewable kilowatt-hours to voluntary markets and using the

---

[13]This analysis looks at a 15 percent RPS that is phased in by 2020, but because the proposal includes multiple credits for a subset of technologies such as wind and a price cap on RECs of 1.9¢/kWh, the actual percentage of renewables that is achieved is closer to 9 percent of total generation by 2020.

same renewable kilowatt-hours to comply with an RPS (Bird and Lockey, 2007). Most states prohibit double-counting; in some states some amount is allowed. In others, there are no rules, and it is difficult to know if voluntary markets are producing more renewable generation. One benefit of voluntary green power markets is that excess renewable generation beyond that required for RPS compliance may be sold into the green power market, providing a way to manage timing inconsistencies and lumpiness in renewable resource development.[14]

Voluntary and compliance REC markets differ in that most compliance REC markets are limited in scope, while voluntary green power markets can be national in scope. This means that the relationships between prices of compliance RECs and voluntary RECs are often unrelated. The effects of an RPS on voluntary purchases of green power are difficult to identify, but analysis of data from four states that have both RPS policies and active green markets showed that voluntary green power sales have continued to increase after the adoption of the RPS policy (Bird and Lockey, 2007).

Renewable Feed-In Tariff

To encourage renewables, most European nations, including France, Germany, Spain, and Denmark, prefer not to set a relative quantity target as is done with an RPS. Instead, policy makers in these countries specify a minimum price, called a feed-in tariff, that utilities must pay generators for renewable electricity. The level of the feed-in tariff varies by technology and is calculated to ensure profitability of the generation regardless of its levelized cost of energy (LCOE). For example, solar power receives a much higher price than does wind. The tariff is guaranteed to be in place at a predetermined level for a statutorily defined time, enabling its benefit to be incorporated into the evaluation of renewable project financiers. To illustrate, the feed-in tariff in France for onshore wind provides €0.082/kWh for 10 years, followed by between €0.028 and €0.082/kWh for 5 subsequent years, depending on the site (IEA, 2006). Feed-in tariffs for solar PV tend to be substantially higher, ranging from a low level of €0.052/kWh in Estonia to as high as €0.60/kWh in Austria (Klein et al., 2006).

The feed-in tariff is typically funded by revenue collected from all electricity customers. The German government estimated that each German household

[14]Another way to do this would be to allow banking of RECs so that excess renewable generation this year could be used to comply with the stricter RPS in a future year.

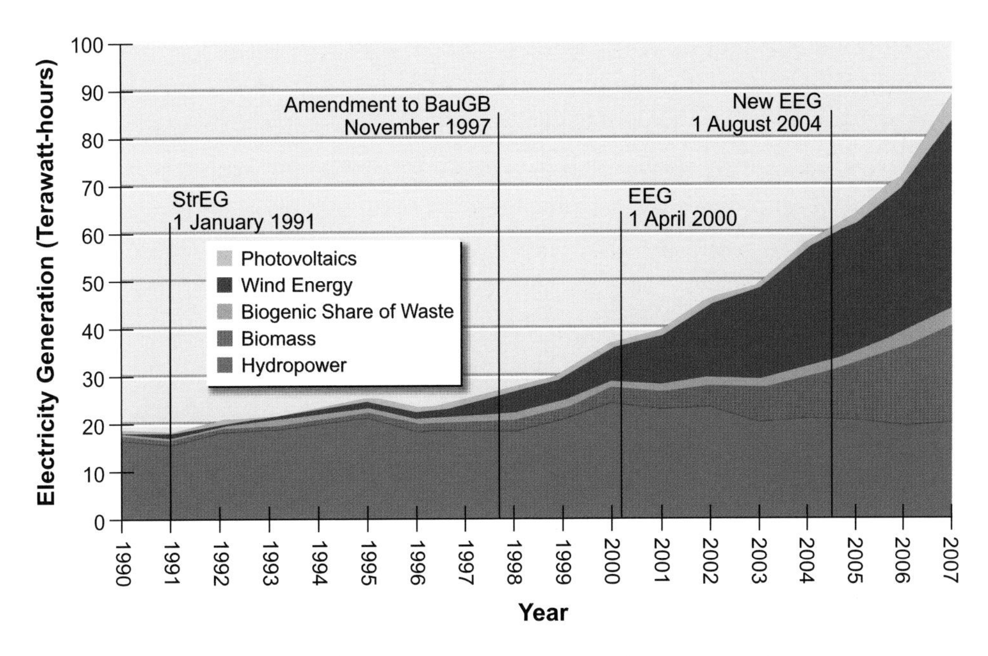

FIGURE 4.6 *Growth in electricity generation from renewable energy sources in Germany. Source: German Federal Ministry for the Environment, Nature Conservation, and Nuclear Safety, 2007.*

paid an additional €2.10/month to cover the costs of the 53.4 TWh eligible for the tariff (German Federal Ministry for the Environment, Nature Conservation, and Nuclear Safety, 2007). The correlation between growth in renewables penetration, shown in Figure 4.6, and tariff eligibility is clear. In Germany the feed-in tariff for solar declined by 5 percent each year to reflect expected reductions in cost of solar panels due to learning. However, a government proposal called for a more dramatic reduction in tariff levels to help lower the costs of this policy. The anticipated cost reductions from increased production of solar PV had not materialized, and instead, shortages of high-grade silicon used in solar cell production resulted in more than a 10-fold increase in silicon prices since 2003 (*Economist*, 2008). The German experience with feed-in tariffs demonstrates the complexity of trying to influence the economics of a particular technology through policy.

Environmental Policies

Policies such as the Title IV cap and trade program for SO_2 emissions raise the cost of fossil-fuel electricity generation and could potentially promote generation from renewables. This effect has been small for policies focused on criteria air pollutants and mercury. Because there is a lack of tested cost-effective ways to reduce CO_2 emissions directly from fossil generators, policies to cap emission of greenhouse gases may provide a stronger economic signal to adopt renewables. The success of that signal will depend on the stringency of the policy, the expected evolution of the policy over time, and the relative economics of demand-side alternatives.

Policies to Control Conventional Pollutants and Mercury

Policies to restrict emissions of SO_2, NO_x, and mercury from electricity generated from fossil fuels have had only a small effect on renewable generation. Fuel switching from high-sulfur coal to low-sulfur coal and installing post-combustion controls for reducing NO_x emissions, such as selective non-catalytic reduction (SNCR), are more cost-effective than is switching from fossil-fuel generation to renewable generation. Modeling suggests that policies restricting emissions may produce a small increase in renewable generation, particularly if mercury emissions are being restricted (Palmer et al., 2007; EIA, 2004). For example, EIA's study of the Clear Skies Act that placed caps on national emissions of SO_2, NO_x, and mercury found that capping emissions of these pollutants would result in 20 percent more generation by non-hydropower renewables in 2020 compared to the base case, raising the non-hydropower renewables generation's share of the total electricity used in the United States from 3.0 percent in the base case to 3.6 percent in 2020. The economics of renewables appears to depend more on the price of natural gas or coal than on the stringency of policies to limit SO_2, NO_x, and mercury.

Policies to Limit Emissions of CO_2

A host of approaches has been proposed for limiting emissions of CO_2. The electric utility sector and other large stationary sources seem to prefer a CO_2 cap and trade program, which has been adopted in the Regional Greenhouse Gas Initiative in the northeastern United States and has been implemented for large stationary and utility sources in the EU CO_2 Emissions Trading Scheme. Several pieces of federal legislation in the 110th Congress proposed cap and trade programs. A cap on CO_2 emissions might have a greater impact on renewables penetration than would caps on other pollutants, since direct abatement of CO_2 emissions is not currently feasible. Until CO_2 capture and sequestration becomes realistic and economic, reductions in CO_2 emissions will come from generating with different fuels, including renewables; more efficient operation of existing emitting facilities; and more efficient electricity consumption. A CO_2 cap would raise the cost of using fossil fuels, in turn raising the market price of electricity, which could have positive effects on the profitability of renewables. The effect of different prices of CO_2 on the LCOE of coal- and natural-gas-generated electricity is shown in Table 4.1.

The EIA provides insight into how climate policy might affect investment in renewables technologies in its study of the Bingaman-Specter proposal to cap economy-wide CO_2 emissions near the 2006 levels in 2020 and 1990 levels in 2030. This bill also includes a cap on the price of CO_2 emission allowances of $15 in 2020 that rises to $25 by 2030. Under this policy, investment in new non-hydropower renewable generating capacity would either double or triple in 2020 in response to the policy relative to baseline levels, depending on underlying cost and performance assumptions. However, the share of electricity provided by renewables including hydropower would only increase from 10 percent in 2007

TABLE 4.1 2020 Cost Projections of Electricity from Fossil Fuel with CO_2 Tax, from AEO 2009 Reference Case (in 2007 dollars)

	Levelized Cost of Energy (2007 $/kWh)				
Technology	No Tax	$10 Tax	$20 Tax	$50 Tax	$100 Tax
Coal					
Pulverized coal	0.083	0.095	0.107	0.145	0.206
IGCC	0.088	0.099	0.109	0.141	0.194
IGCC with sequestration	0.103	0.104	0.105	0.109	0.115
Natural Gas					
Combined cycle	0.083	0.088	0.092	0.105	0.127
Advanced combined cycle	0.079	0.083	0.087	0.099	0.120
Advanced combined cycle with sequestration	0.110	0.110	0.111	0.112	0.115
Combustion turbine	0.138	0.145	0.152	0.172	0.205
Advanced combustion turbine	0.121	0.127	0.132	0.149	0.176

Note: Taxes are denominated in dollars per short ton of CO_2; assumes that sequestration technology captures 90 percent of CO_2 emissions.
Source: EIA, 2008d.

to 15 percent by 2020 (EIA, 2008a); non-hydropower renewables would increase from 3 percent to approximately 9 percent of electricity generation. In analyzing the more stringent Lieberman-Warner proposal, EIA concluded that with a CO_2 allowance price of about $48 in 2025, the amount of generation from non-hydropower renewables would more than quadruple in 2025 relative to the no-climate-policy baseline scenario and would climb to more than 13 percent of total generation in that year (EIA, 2008a).

Some policies to reduce CO_2 could have the perverse effect of limiting the ability of those seeking to market renewable energy directly to consumers to make environmental claims about the emissions consequences of switching to renewable power. Because a cap on emissions of CO_2 is both a ceiling and a floor, increased generation from renewables would free up a CO_2 allowance for use elsewhere. To support voluntary renewables markets, several states participating in the Regional Greenhouse Gas Initiative plan to retire CO_2 allowances in connection with voluntary sales of green power to maintain the opportunity for credible green power claims.

Other Types of Policies

Time-of-Use or Real-Time Pricing

Time-of-use or real-time pricing of electricity can affect incentives for the installation of distributed renewables such as solar PV. If peak demand times or peak-load pricing coincide with the times of availability for solar PV or other distributed renewables, more widespread use of time-of-day or real-time pricing could encourage the development of renewables, particularly when peak period price is several times higher than base period price for electricity. For example, with real-time pricing, the value of electricity from PV in California is 30–50 percent higher than in the absence of real-time pricing (Borenstein, 2008a). The effect of a shift to real-time pricing on the value to electricity consumers of installing PV would depend on how electricity prices were set in the absence of real-time metering. For example, the tiered structure of flat electricity rates in the Southern California Edison territory, which led to higher electricity prices for heavy-use consumers, meant that the value of PV installation was higher for many customers under a flat-rate structure than with real-time pricing (Borenstein, 2007).

Advanced meters that keep track of electricity use in real time are a prerequisite for real-time pricing. A study by the Federal Energy Regulatory Commission (FERC, 2006) found that only about 6 percent of customers nationwide had the advanced meters necessary for real-time pricing, but in some parts of the country as many as 15 percent of the customers had these meters.

Conversely, time-of-use pricing could actually make grid-scale solar less attractive to investors. Time-of-use pricing would tend to flatten the load curve, reducing the market clearing wholesale price of electricity during peak hours. To the extent that these hours coincide with solar generation, this would reduce returns to solar investment.

Policies to Promote Biofuels in Transportation

The recent surge in policies to promote biofuels for transportation has already produced a large increase in demand for corn to make corn-based ethanol and has led to an increase in corn prices. Biomass electricity generation has competition from liquid biofuel development for feedstock inputs such as energy crops, agro-waste, and even wood pulp, which will ultimately raise the cost of biopower.

On the other hand, by-products from liquid biofuels production could provide a source of renewable fuel for electricity generation to supply at least some of the electricity needs of ethanol production facilities. Electricity generated using this biofuel production by-product may qualify for credit as a renewable source of generation under existing state RPS policies. To a degree, policies to promote biofuels, including ethanol and other liquid fuels, might also provide opportunities for some additional generation from biomass, albeit largely to serve the energy needs of ethanol production.

Energy Efficiency Policies

Conversations about the changes necessary to reduce the greenhouse gas intensity of electricity generation often refer to promoting renewables and promoting energy efficiency as complementary strategies. However, investment in energy efficiency and investment in renewables are two different ways of balancing demand and supply in energy markets. Policies to promote energy efficiency may conversely make it harder for renewables to compete in electricity markets. If efficiency programs are cost-effective, electricity prices would be lower than they would be without the program, though not necessarily lower than before the program. There would be less demand for investment in renewables, and investment would be less profitable, all else being equal. By reducing overall electricity demand, energy efficiency programs also reduce the minimum quantity of renewable generation required under an RPS. If demand reduction is significant enough to reduce the growth rate to zero, then excess capacity from existing fossil generation would become available and further reduce the marginal cost target that must be met by new renewable generation.

ESTIMATES OF CURRENT COSTS

Estimates of the cost of energy from new generating facilities indicate that the levelized costs of wind and other renewables are typically greater than the levelized cost of energy from generators fueled by coal or natural gas. Table 4.2 shows estimates of the national average levelized cost per megawatt-hour of new generation

TABLE 4.2 Levelized Cost of Energy (in 2007$/MWh) for New Plants Coming on Line in 2012, from AEO 2009, by Technology

Technology	Capacity Factor (%)	Capital Costs	Fixed O&M	Variable O&M/Fuel Costs	Transmission Costs	Total[a]
Pulverized coal	85	56.9	3.7	23.0	3.5	87.1 (58.1)
Conventional gas combined cycle	87	20.0	1.6	55.2	3.8	80.7 (72.7)
Conventional combustion turbine	30	36.0	4.6	80.1	11.0	131.7 (121.5)
Concentrating solar power	31	218.9	21.3	0.0	10.6	250.8 (166.1)
Wind	36	73.0	9.8	0.0	8.3	91.1 (84.9)
Offshore wind	33	171.3	29.2	0.0	9.0	209.5 (164.9)
Solar photovoltaic	22	342.7	6.2	0.0	13.2	362.2 (308.1)
Geothermal	90	76.7	21.6	0.0	4.9	103.3 (66.8)
Biopower	83	61.1	8.9	24.7	3.9	106.6 (84.0)

Note: Fuel cost imputed from AEO 2009 Early Release model solution (EIA, 2008d). AEO 2009 energy prices (2007$/million Btu) in 2012 are $1.91 for coal, $6.63 for natural gas, and $1.96 for biomass. O&M, operating and maintenance.

[a] Numbers for total LCOE from AEO 2008 (EIA, 2008c) shown in parentheses.

Source: EIA, 2008c,d.

facilities constructed in 2012 in the AEO 2009 from the EIA (2008d) by technology type.[15]

The levelized costs reported in the last column of Table 4.2 include capital and finance costs (including the cost of site development), variable O&M (including fuel), fixed O&M, and the cost of transmission necessary to connect the new facility to the grid.[16] The costs for renewables do not reflect the renewable PTC. However, they do reflect the effects of state RPS policies on the mix of wind resources and other renewables that are expected to come on line in response to those policies.

Table 4.2 shows that the three renewable technologies with the lowest cost of energy are geothermal, biopower, and wind. Pulverized coal and conventional gas combined cycle are less costly than all of the renewable technologies. According to the AEO 2009 results, the present $20/MWh level of the PTC would basically close the gap between the levelized costs of new wind and the LCOE of new coal plants, ignoring issues of relative dispatchability. However, the costs of other technologies, particularly solar PV, concentrating solar power, and offshore wind, would remain higher than the costs of other renewables, and additional subsidies or set-asides in RPS policies would be necessary for these technologies to penetrate the markets given existing costs.

Annex Table 4.A.1 shows the levelized costs of renewable sources of generation from EIA compared to those from the EERE Office at DOE (EPRI, 2007b); a recent report from Standard and Poor's (S&P) (Venkataraman et al., 2007); the inputs to the American Wind Energy Association (AWEA) and NREL 20 percent wind study (Black & Veatch, 2007); and the Solar Energy Industry Association (SEIA, 2004). While the estimates in Table 4.A.2 include the costs of installation and construction of transmission necessary to facilitate power delivery, Table 4.A.1 contains more generic estimates of costs relevant for today or for 2010, the first year reported by many sources.[17]

---

[15]The year 2012 was selected because that is the first year that a new baseload coal plant would be able to come on line in the forecast, owing to the lead times for constructing a new baseload coal plant.

[16]The cost estimates reported in Table 4.2 reflect EIA's assumption of no variable O&M costs for wind, solar, and geothermal. As shown in Table 4.A.1, other data sources, including Black & Veatch and Standard and Poor's (Venkataraman et al., 2007) include variable O&M costs for at least some of these technologies.

[17]For all cases in Table 4.A.1 where capital costs were available from the data sources, but comparable to LCOE estimates were not, those estimates were construed assuming a 20-year equipment life and a 7.5 percent discount rate.

The snapshot of costs presented in Table 4.A.1 does not reveal a number of important factors that affect the estimates of levelized costs. The next several paragraphs discuss important factors for several of the technologies considered in this study.

Wind Power

Table 4.A.1 estimates of levelized cost of energy for onshore wind in 2010 range from $0.029 to $0.10/kWh, with EIA estimates falling in the middle at $0.069/kWh. Most estimates of the capital cost of new wind facilities are in the $1750/kW range, close to 10 percent lower than the EIA estimates of nearly $1900/kW.[18] In addition, EIA estimates that average capacity factors are somewhat lower than recent forecasts from EPRI.

A single national average estimate of the levelized cost of wind does not communicate how wind costs depend on the capacity factor of new wind turbines, which in turn depends on wind class. Figure 4.7 shows estimates from DOE of the amount of wind capacity available at different levelized costs of energy, after netting out the PTC, and how the cost of electricity increases when moving from higher wind classes to lower wind classes and from onshore sites to offshore sites.

Capacity factors differ across the country, as shown in the regional differences for existing facilities in Figure 4.8. As is also shown in Figure 4.8, capacity factors for wind have been improving over time due to improvements in equipment performance, although this improvement may be offset as the lower cost sites are taken.

The costs of offshore wind are likely to be much more uncertain because currently there are no operating offshore facilities in the United States. As a result, we are several years from a point where we can be more certain about what offshore wind generation costs would look like in the future and how they would compare to the costs of other renewables.

Solar Photovoltaics

The cost of energy produced using solar PV technology is a function of the efficiency of the cell in producing electricity, which is typically 15 percent or less depending on the material system and the total cost of installation. The capital

---

[18]The AEO 2009 estimate is consistent with the expected installed costs of new wind facilities completed in 2008, which averaged about $1920/kW (Wiser and Bolinger, 2008).

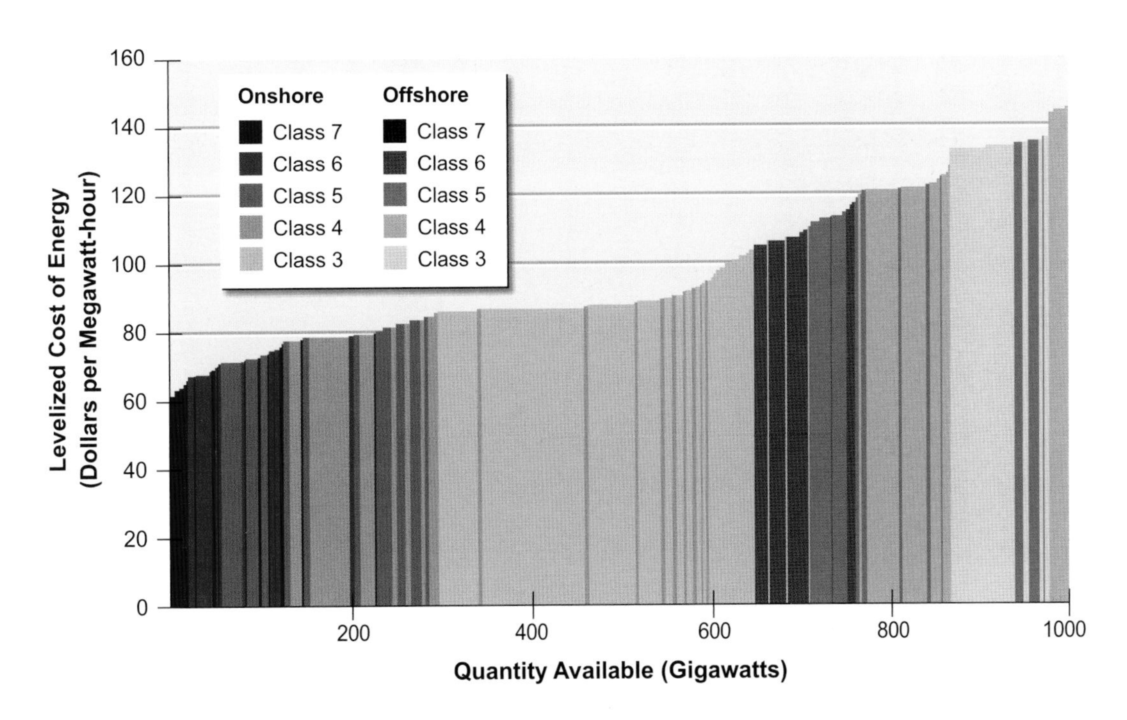

FIGURE 4.7 *Supply curve for wind accounting for the production tax credit and transmission costs (assuming $1600/MW-mile) but no other integration costs. Source: DOE, 2008.*

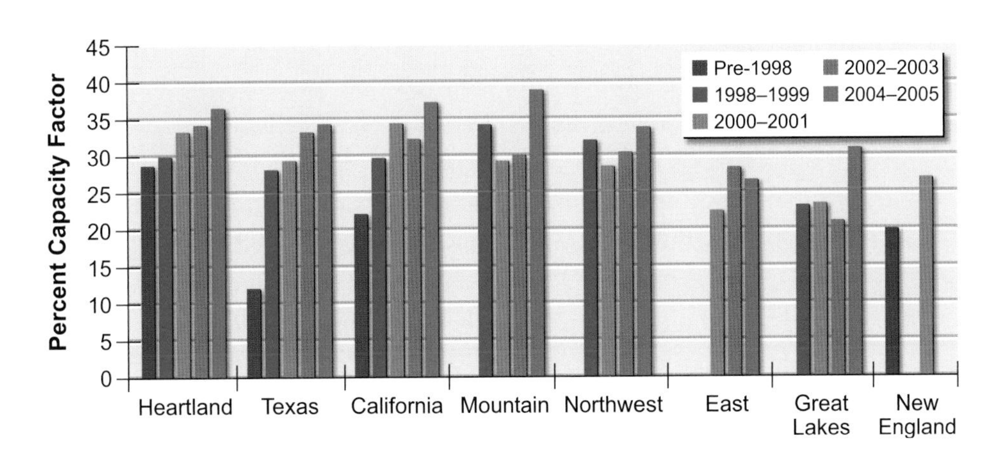

FIGURE 4.8 *Wind capacity factor in 2006 by region and vintage of wind facility. Source: Wiser and Bolinger, 2008.*

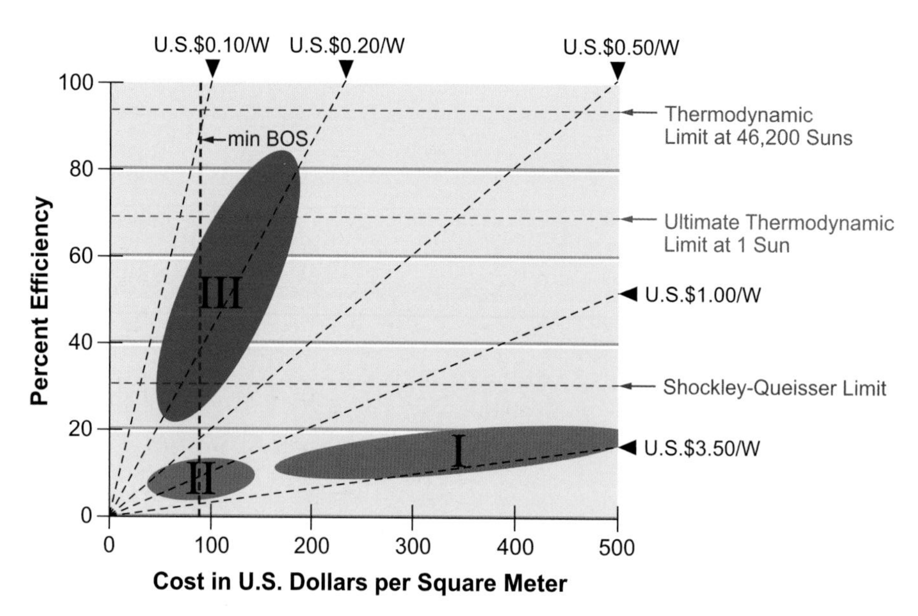

FIGURE 4.9 *PV power costs as function of module efficiency and cost. Source: Green, 2004. Copyright 2004 by Springer. Reprinted with permission of Springer Science+Business Media.*

cost of a PV cell module is typically expressed as dollars per peak watt of production ($\$/W_p$) and is determined by the ratio of the module cost per unit of area ($\$/m^2$) divided by the maximum amount of electric power delivered per unit of area (module efficiency multiplied by 1000 W/m^2, the standard insolation rate at 25°C). In Figure 4.9, this cost per peak watt is indicated by a series of dashed straight lines having different slopes. Any combination of area cost and efficiency on a given dashed line produces the same cost per peak watt indicated by the line labels. For example, present single-crystalline Si PV cells, with an efficiency of 10 percent and a cost of $\$350/m^2$, have a module cost of $\$3.50/W_p$. The area labeled I in Figure 4.9 represents the first generation (Generation I) of solar cells and covers the range of module costs for these cells. Areas labeled II and III in Figure 4.9 present the target module costs for Generation II (thin-film PV) and Generation III PV cells (advanced future structures) that are still in development.

In addition to module costs, a PV system also has costs associated with the non-photoactive parts of the system, called balance of system (BOS) costs, which are in the range of $\$250/m^2$ for Generation I cells. The total cost of current PV systems is about $\$6/W_p$. Taking into account the cost of capital, interest rates, depreciation, system lifetime, and the available annual solar irradiance integrated over the year (i.e., considering the diurnal cycle and cloud cover, which reduce the peak power by a factor of about 1/5), the $/W$_p$ figure of merit can be converted to $/kWh by the following simple relationship: $1/W$_p$ ~ $0.05/kWh. This calculation leads to a present cost for grid-connected PV electricity of about $0.30/kWh. The estimates of levelized energy costs for PV generally are distributed around the 30¢/kWh level, as shown in Annex Table 4.A.2. The one exception was a 2004 SEIA study of levelized costs that predicted the cost of energy from PV would fall to about $0.14/kWh by 2010, in the absence of aggressive policies to promote the technology, and to $0.08/kWh with such policies in place.

The costs of supplying electricity from rooftop PV installations will vary across different locations and depend on factors such as the cost of land, options for orienting the installation (particularly on rooftops), and amount of energy produced in a particular location. A study of the factors affecting supply curves for solar PV from rooftops used data on building stock, rooftop orientation, solar insolation, and other factors to construct relative supply functions for solar PV for three U.S. electric interconnections as shown in Figure 4.10 (Denholm and Margolis, 2008). These supply curves relate to the system with the greatest yield, which results from the best orientation in the most productive location. The sup-

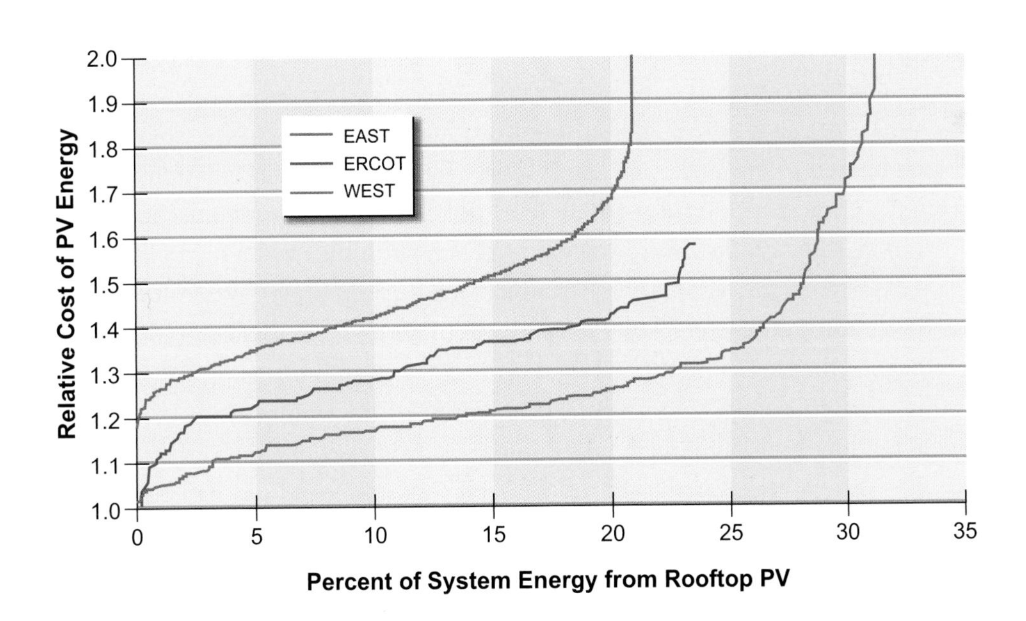

FIGURE 4.10 *Fractional energy PV rooftop supply curves for three U.S. interconnections. Source: Denholm and Margolis, 2008.*

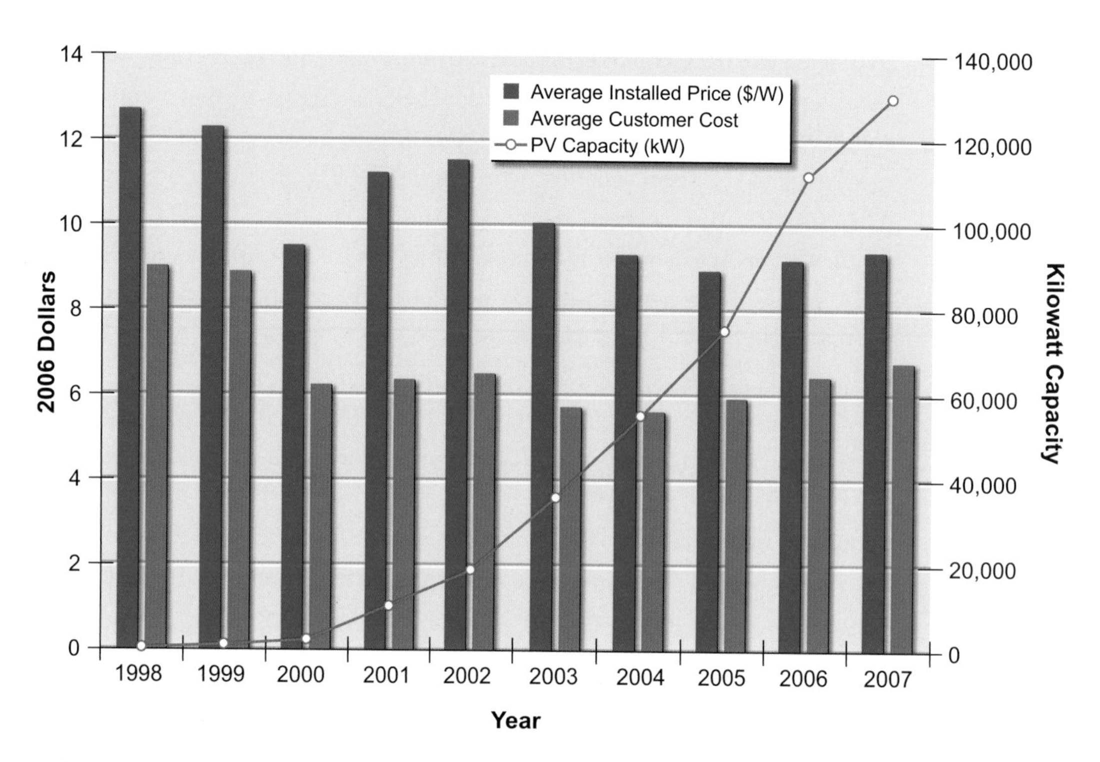

FIGURE 4.11 *Price, customer cost after subsidy, and number of PV installations per year in California under California Energy Commission incentive programs.*
Source: CEC, 2008.

ply curves show the higher costs of producing electricity using solar PV in the east compared to the west, and the resource limits in different locations.

Largely as a result of state-level policies to promote the use of solar PV, the number of installations is growing. As shown in Figure 4.11, in California about 130 MW of the cumulative PV capacity installed by 2007 was under incentive programs administered by the California Energy Commission (CEC), more than double the total amount installed under these programs as of 2004. This increase in capacity coincided with the 2006 launch of the California Solar Initiative with a funding level of about $3.3 billion for subsidy payments available to new solar PV installations. Data from the CEC on total costs and costs to customers of PV installations suggest that costs per kilowatt for consumers rose slightly over this period, a period of only slight increases in consumer costs per watt of PV installations, due in part to the subsidies afforded by the California policy (CEC, 2008). The CEC PV database contains information on about one-third of the total amount of PV capacity installed in California.

Most of the solar PV installations appear to be taking place in regions that have aggressive pro-solar policies. According to solarbuzz.com, in 2006 California accounted for 63 percent of the grid-connected PV market, and New Jersey, which also has an aggressive policy to promote PV, accounted for 19 percent. In general, achieving grid parity, the point at which electricity from PV is equal to or cheaper than power from the electricity grid, would require a two- to three-times improvement for costs per kilowatt-hour for the whole system (PV modules, batteries, inverters, and other system components) as well as for installation and O&M costs (Cornelius, 2007).

Concentrating Solar Power

According to the EIA (2008d) AEO 2009 model runs, the levelized cost of generating electricity using concentrating solar power (CSP) is higher than the cost with wind, but lower than the cost of solar PV as shown in Table 4.2. If technological learning for CSP is a function of aggregate investment, as assumed by the EIA, then the economics of CSP may be improved by policies that promote investment in this technology and provide incentives for using it to generate electricity. Twelve states have set-asides for solar technologies in their RPS policies, and in nine of those states, which include Arizona, New Mexico, and Nevada, solar thermal generating technologies for hot water generation qualify for the set-aside (Wiser and Barbose, 2008). The set-aside typically requires that a specific portion of the RPS target must be met with a solar technology. Some policies also include a credit multiplier for generation from solar such that solar-produced electricity creates RECs at a ratio of greater than 1:1.

Estimates of the levelized cost of central station CSP have typically been around 16¢/kWh at the busbar (Venkataraman et al., 2007; WGA, 2006b; EIA, 2008c). Table 4.2 shows that EIA reported a much higher levelized cost of 25¢/kWh in the AEO 2009 forecast, reflecting increases in 2007–2008 in raw materials costs (EIA, 2008d). With the 16¢ cost as a starting point, the supply curve for CSP in the southwestern United States shown in Figure 4.12 displays costs at the busbar. The total supply curve in this graph is the horizontal sum of the individual supply curves for different levels of solar resource intensity. This cost curve is very flat at levels of around 16¢/kWh.

Figure 4.13 shows a supply curve that goes beyond the busbar and takes into account the costs of incremental transmission necessary to deliver power to load. This curve is based on assumptions about the portion of local load that

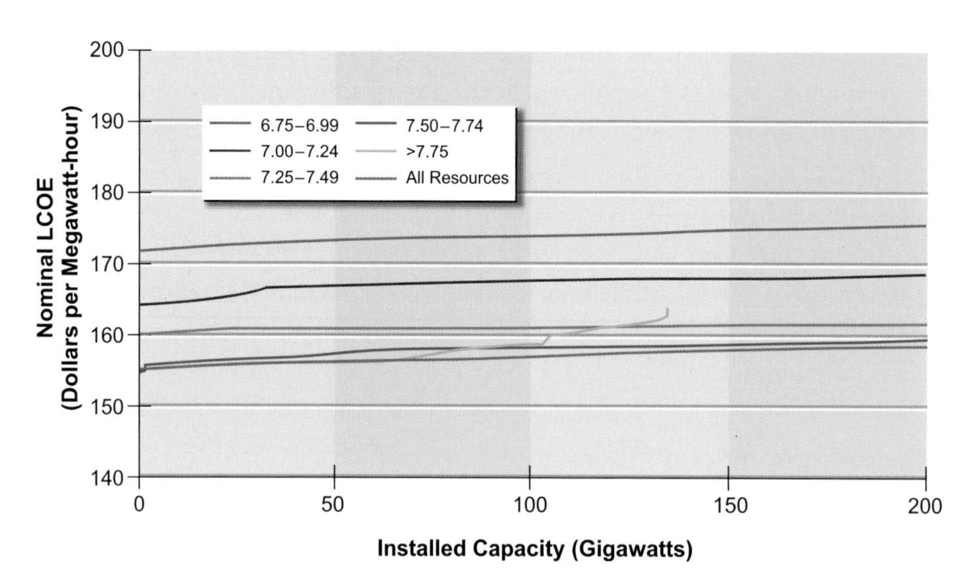

FIGURE 4.12 *Supply curves describe the potential capacity and current busbar costs in terms of nominal levelized cost of energy (LCOE) of concentrating solar power. Colored lines indicate different amounts of insolation measured in kilowatt-hours per square meter per day.*
Source: ASES, 2007. Used with permission of the American Solar Energy Society. Copyright 2007 ASES.

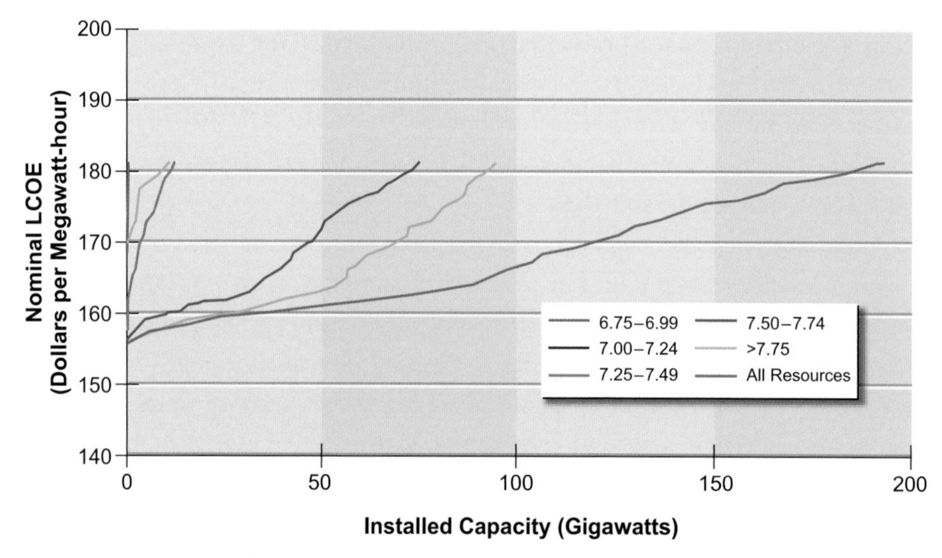

FIGURE 4.13 *Concentrating solar power supply curve based on 20 percent availability of city peak demand and 20 percent availability of transmission capacity. Colored lines indicate different amounts of insolation measured in kilowatt-hours per square meter per day.*
Source: ASES, 2007. Used with permission of the American Solar Energy Society. Copyright 2007 ASES.

could be served by solar power, the availability of transmission to move power from generation sites to load centers, and the cost of expanding transmission at $1000/MW-mile, lower than the $1600/MW-mile used in the DOE (2008) 20 percent wind study. As shown in Figure 4.13, the resulting aggregate supply curve for this region has a bit of slope to it, rising to approximately 18¢/kWh at or near 180 GW of generation. The basic message from the fairly flat slope of this supply curve is that at this time the constraining factor for concentrating solar power supply is not the amount of the resource, which is widely distributed and available abundantly in the southwest, but the costs of developing that resource.

How this cost picture might change over time depends on future adoption of renewable technologies. According to the WGA study, technology learning and economies of scale in manufacturing and installation indicate that the levelized cost of energy in 2015 for a parabolic trough technology would decrease by 50 percent with an increase of 4 GW of installed capacity (WGA, 2006b). The American Solar Energy Society (2007) anticipates further decreases in levelized cost of another 25 percent between 2015 and 2030. Research and development is also expected to have an important effect on costs. The DOE's Office of EERE (NREL, 2007) anticipates that both capital costs and capacity factors for CSP could improve dramatically through its R&D program for concentrating solar power, including storage capacity and location of new systems in the most productive sites. Levelized costs of energy at the busbar could decrease by 50 percent as soon as 2010, as shown in Table 4.A.1, though this sounds quite optimistic.

Geothermal Power

Most of the economic U.S. hydrothermal resources are located in the western states. Recent studies sponsored by the WGA (WGA, 2006a) identify approximately 13,600 MW of geothermal potential in the west that could be developed economically, at busbar costs of up to 20¢/kWh in $2005, and 5,600 MW that reasonably could be developed by 2015 at costs of less than 10¢/kWh in $2005. Both cost estimates omit the renewable PTC that would reduce the costs of developing these resources.

The WGA report and one conducted by the CEC (GeothermEX Inc., 2004) were used to update the geothermal supply curves in NEMS (Smith, 2006). The supply curves are limited to the 80 most likely sites to be developed and extend to include 8 GW of new capacity. The NEMS geothermal supply curve, shown in Figure 4.14, is similar to the supply curves found in the WGA report. According

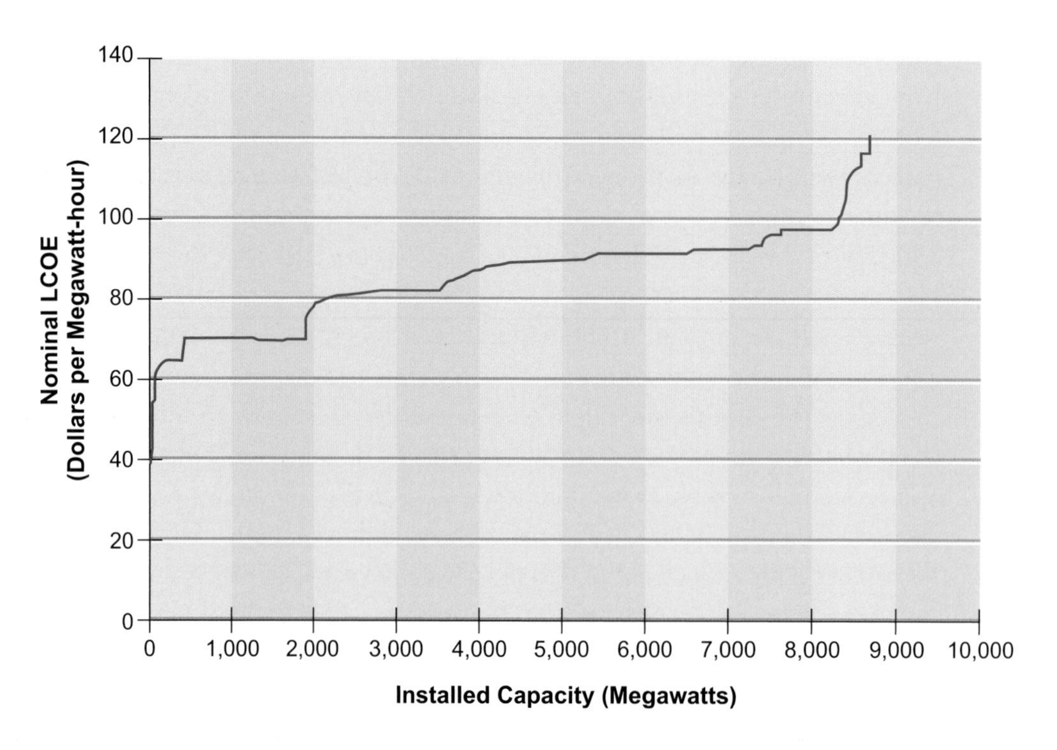

FIGURE 4.14 *Geothermal supply curve.*
Source: Developed from data supplied by Smith, 2006.

to EIA, this supply curve, added to the NEMS model with the development of AEO 2007, would not capture all potentially economic geothermal resources, but it is an important start and likely does capture the most economic resources available (Smith, 2006). Enhanced geothermal systems (EGS) may offer greater opportunity in the future, but, as indicated in Chapter 3, this technology is too early in its development to reliably estimate its cost.

Existing geothermal generating capacity is closer to 2.5 GW (EIA, 2008c). One hurdle to the development of geothermal resources is that, like wind, they may be located far from load and require new transmission lines to facilitate delivery. However, geothermal energy provides constant, baseload power, which is an advantage over solar and wind.

Biopower

The costs of new biopower generation will depend on two important factors: the generation technology and the cost of the fuel. In its NEMS model, EIA assumed that any new biopower generation would use gasification with a combined cycle technology. These generators have high capital costs and lower heat rates than a conventional boiler. However, none of these types of biopower generators are now in commercial operation in the United States, so it is difficult to know how the predicted costs would compare to actual experience.[19] In its *Technology Assessment Guide*, EPRI reported costs for both stoker and circulating fluidized bed boilers, technologies that are well suited to the small scale of most biomass plants and that can handle the fuel well EPRI (2007a). Capital costs, including interest during construction and project specific costs, would be on the order of $3400/kW for each technology with capacity factors of 85 percent. The levelized cost of energy would depend on fuel costs, but the EPRI summer study reports a cost of approximately 9.6¢/kWh for a fluidized bed generator in 2010, which, assuming similar fuel costs of $34/MWh (about $2.70/million Btu of high heat value), would yield a levelized cost estimate for a stoker of 9.4¢/kWh.[20]

The costs of biomass fuels are also subject to uncertainty and potential volatility. Much of the existing biopower generation occurs as self-generation at facilities that have a ready source of fuel (such as pulping operations, paper mills, or forest products plants). Expanding capabilities beyond these generators could involve shipping fuel, which can become quite costly, which suggests that future biopower generation capability would be located close to fuel sources and would use more economical biomass fuels that are concentrated locally and do not face substantial competition for their use.

This uncertainty about fuel costs is reflected in the different estimates of levelized costs of biopower reported in Annex Table 4.A.1. The fuel cost assumptions in the recent EIA forecasts are substantially lower than those assumed by other sources, including EPRI and S&P. These lower costs are a major factor in the substantially lower levelized cost of energy in the EIA numbers, which are about 85 percent lower than those provided by other sources.

One option for greater use of biomass fuel is co-firing the fuel with coal. Biomass co-firing of up to 10 to 15 percent of fuel on a heat-input basis is a potential way of reducing the CO_2 emissions associated with coal-fired generation (NREL, 2006). The costs of making a coal-fired generation facility available for

[19]NEMS includes this technology instead of biomass combustion in boilers under the expectation that gasification, once commercial, could trump biomass combustion and the need for the model to be somewhat parsimonious in including different technologies.

[20]See Table 4.A.1 for all the components of cost.

co-firing could be substantial and involve large investments in new fuel-handling equipment. Certain types of boiler configurations are more amenable than others. Even though co-firing counts as renewable generation under many state RPS policies, co-firing has not increased much in response to state RPS policies (McElroy, 2008). Placing a cap on CO_2 emissions may be necessary to drive coal plants to start making the investment necessary to co-fire, and then only when the facility can identify an economic source of biomass fuel.

COSTS IN 2020

Table 4.A.2 in the annex at the end of this chapter provides estimates from a few different sources of the levelized costs in 2020 of a range of different renewable technologies. These sources include the EIA AEO 2009 reference case forecast, the EERE Office at DOE, EPRI, the SEIA, U.S. DOE's 20 percent wind study, and the cost estimates published in a 2007 report edited by the American Solar Energy Society. Table 4.A.2 also includes levelized cost of energy projections for a number of fossil generation technologies based on the AEO 2009 forecasts. These forecasts all include the effect of learning on reducing capital costs, where the potential cost reductions from learning vary across technologies. The projections from EIA also include the effect of moving along the supply curve, such as when less accessible or lower quality wind resources are tapped for wind electricity generation.

Table 4.A.2 shows a wide range of forecasts on the future of renewables costs. Most of the forecasts envision renewables as continuing to be more costly than the EIA forecasts of generation using conventional coal and gas technologies. The exceptions are the EERE forecasts that envision substantial improvements in costs for concentrating solar power and wind, and the SEIA forecast for solar PV. The differences between the program scenarios and the baseline scenarios for the EERE forecasts show how full funding of renewable energy research at DOE is expected to affect the future costs of renewable generation.

The different forecasting groups and scenarios also envision different rates of change in levelized costs of energy over the next decade as shown in Figure 4.15, which compares forecasts of costs for 2020 and 2010 for several sources for four of the technologies. This graph shows that EERE and EPRI summer study forecasts envision large decreases in the costs of wind generation between 2010 and 2020, whereas the levelized costs in EIA forecasts increase as a result of the cost

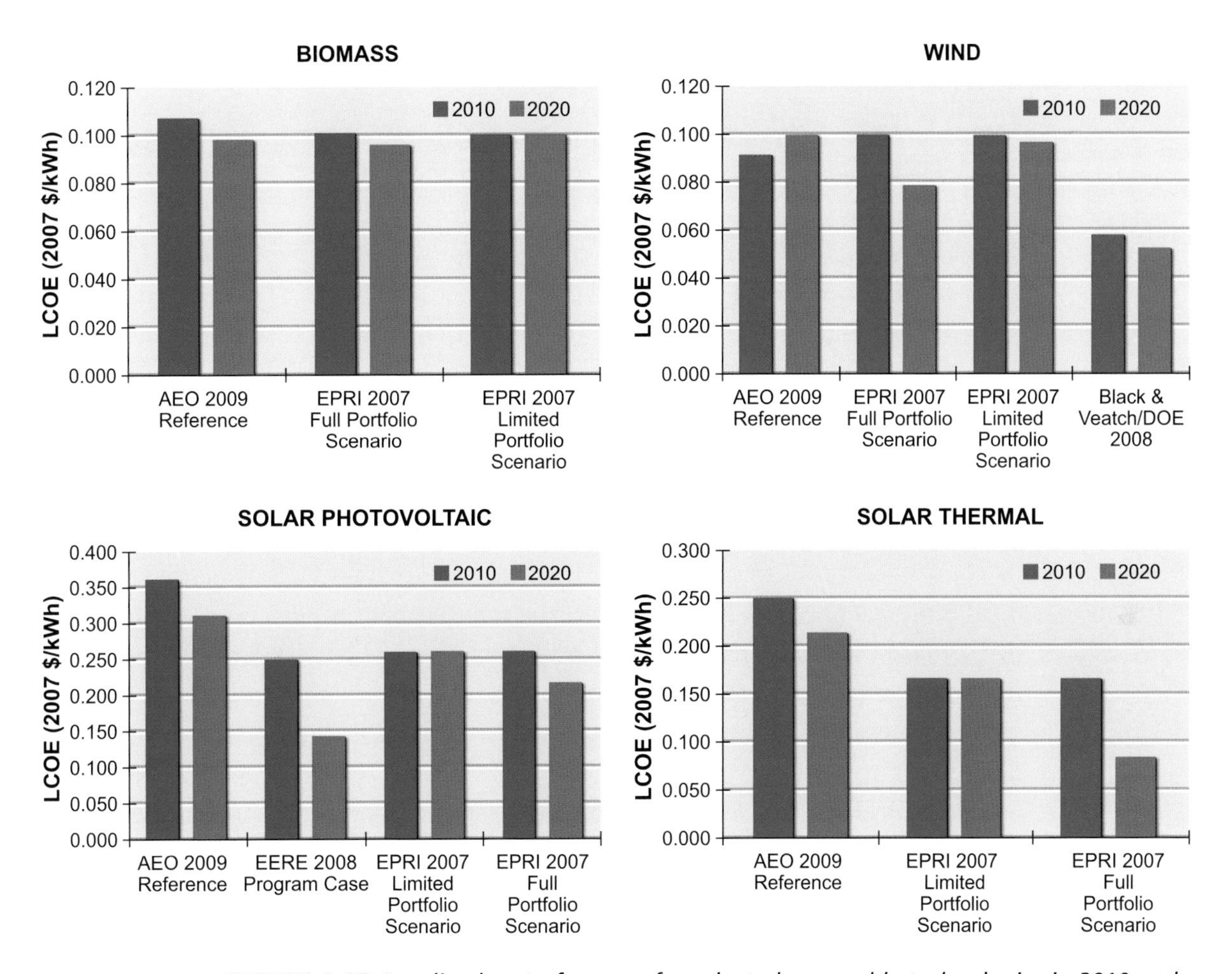

FIGURE 4.15 *Levelized cost of energy for selected renewable technologies in 2010 and 2020 from various sources. Note that AEO 2009 numbers are for 2012.*

increases inherent in tapping increasingly difficult sites, which are not reflected in the estimates reported by the other studies.

Differences in cost projections for wind turbines appear to be at least partly due to differences in assumptions about capacity factors. The predictions from EERE are largely the result of improvements in engineering resulting from research and development in this technology and greater deployment. In the AEO projections, capacity factor predictions for 2020 are based on where the wind resource would be developed in that year. The model presumably would have used up the better sites for the least-cost development of resources in earlier years. Incorporating resources found in higher wind class regions, as suggested in Figure 4.7, would likely lead to lower capacity factors at new facilities after the better wind sites are taken.

The DOE 20 percent wind study assumed that, as a result of technology improvements, capacity factors would improve between 2005 and 2030 by 10–18.7 percent, with faster rates of improvement anticipated in the lower wind resource regions (DOE, 2008). Most of this improvement is expected by 2020. This study also assumed that capital costs of new onshore wind generators would fall by 5 percent between 2005 and 2020, and that new offshore wind generators would see capital cost decreases of just over 10 percent during the same period. This study also anticipated a marked decline in variable and fixed O&M costs between 2005 and 2020, particularly for offshore installations.

CSP and PV also have a wide range of future cost predictions, representing the large degree of uncertainty and differing opinions about how solar costs are likely to evolve over this decade. PV is expected to remain more expensive than CSP, although the SEIA forecasts dramatic improvement in the cost of distributed PV, and EERE anticipates decreases in PV costs, too. EERE also projects potential cost improvements for solar thermal projects. But unlike other forecasters, EERE predicts substantially lower costs in the near term, suggesting differences in what goes into their cost measures. The DOE Solar Energy Technologies Program report envisions declines in CSP costs of about 50 percent from present levels, similar to the aggressive technology case from the EPRI summer study.

Analysis of the evolution of PV costs suggests that the prices of PV modules have followed a historical trend along the "80 percent learning curve." That is, for every doubling of the total cumulative production of PV modules worldwide, the price has dropped by approximately 20 percent. This trend is illustrated in Figure 4.16 (Surek, 2005). The final data point for 2003 corresponds to about $3.50/W$_p$ and a cumulative PV capacity of 3 GW. A major reduction in the projected future cost of PV modules depends on the introduction of thin films, concentrator systems, and new technologies. The graph projects the path of future costs under historical learning rates as well as with slower and faster rates of learning.

ANALYSIS OF 2020 COST PROJECTIONS

Cost projections for renewables can vary widely within a particular resource type and technology, depending on the source and the assumptions. For this report, the panel did not have access to all of the underlying assumptions, ren-

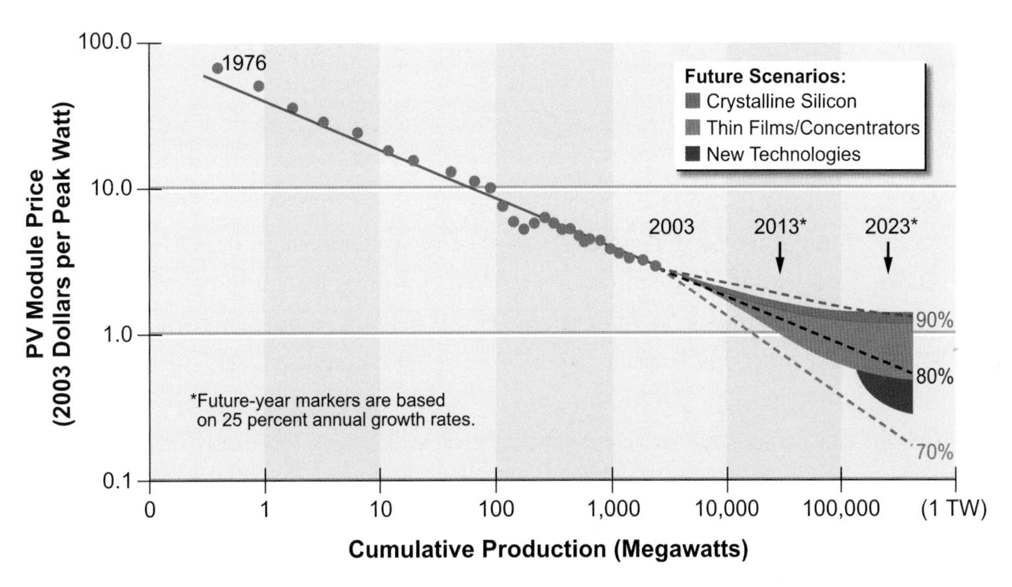

FIGURE 4.16 *Learning curve for PV production.*
Source: Surek, 2005.

dering impossible a total breakdown of the differences among different sources. The panel did have sufficient information to compare the assumptions and the general approaches to developing projections of costs and to make substantive comments.

The panel is aware of the difficulties of predicting future energy prices, especially after observing the cost of a barrel of light crude oil vary from a low of $35 per barrel to a high of almost $140 per barrel within the span of 12 months. Such volatilities should humble any group attempting to predict the future. The panel considers these alternate cost estimates to come from reputable sources and to be representative of the range of cost estimates for renewables in the future. Taking all of these factors into consideration, the panel supported the idea that the upper cost bounds in these estimates are in line with the panel's estimate of a reasonable upper bound, while some of the lower cost estimates tended to represent aspirational goals. Volatilities in energy and commodities markets and other large-scale macroeconomic factors are difficult to predict, and therefore the panel does not label any projection as more correct than another.

Nevertheless, the panel considers it essential to understand why these estimates vary. Indeed, while the most recent AEO 2009 shows a significant increase in PV costs over AEO 2008 (see Table 4.2), recent reports indicate a drastic decrease in solar PV due to increased supply and decreased demand (Patel, 2009). One key factor is the assumed cost of capital equipment today, and whether or not it reflects recent increases in costs for material and labor, wind turbines, solar panel components, and construction in general. Material and construction cost increases in the past few years have resulted in substantially higher costs of new facilities and have tended to more than offset reductions in capital costs for wind turbines and solar PV installations from technology learning over that short time horizon. Some studies, such as the DOE 20 percent wind penetration study and AEO 2009, more fully reflect these recent cost increases, while others, including EERE, only partially inflate the costs, possibly in trying to capture just the portion of the recent cost increases anticipated to be long-lived, rather than short-run spikes in costs that will likely be resolved over time.

Assumptions about technology learning and its likely evolution independent of scale of deployment are also key to projections about future costs. The EIA NEMS model assumes that substantial cost reductions will occur over time for CSP and solar PV technologies as a result of learning both from greater deployment and from research and development. The model assumes high initial learning rates, with initial capital costs projected to decrease by at least 20 percent by 2025, and potentially more if the technology is widely adopted. The NEMS model also assumes that gasification technology forecasted for new biopower installations will realize substantial cost reductions as installations of this technology, either for use with coal or biomass, grow. On the other hand, the model treats wind as a fairly mature technology and assumes fairly low rates of improvement in costs in the future, with a 1 percent reduction in capital costs for each doubling of installed capacity.

These different learning assumptions help explain the differences in cost improvements over time among the different technologies within the EIA projections. The fact that other studies, such as the DOE (2008) 20 percent wind study, are more bullish on the possibility of improvements in cost of wind technology explains some of the differences in projections of levelized costs across studies. The 20 percent wind study also assumes larger improvements in capacity factors of wind turbines over time than the maximum potential improvements in performance allowed in the NEMS model. These differences in assumptions contribute to differences in the projections of future levelized costs.

For studies that assume technology learning is a function of the level of installed capacity, the underlying assumptions about policies to promote renewables and how those policies and market dynamics will play out in the future renewables market will affect projections of the LCOE. The approach to modeling state-level RPS policies and assumptions about the future of the PTC will influence the predictions of future costs, and possibly performance, of renewables. For AEO 2009, assumptions include regional compliance with relevant state RPS requirements and expiration of the PTC at the end of 2009 for wind and 2010 for other technologies. Changes in these policies will affect gains from capacity-related learning. For example, the recent extension of the PTC beyond the 2009 or 2010 expiration dates could increase learning-related improvements in costs. On the other hand, substantial increases in investment in energy efficiency and associated reductions in electricity demand could lower projected renewables penetrations by 2020 and reduce the gains from capacity-related learning. The DOE (2008) 20 percent wind study imposes only external assumptions about technology learning that are a function of time, but not of overall penetration, and so the level of penetration before 2020 does not matter for learning, but it does matter for determining where each region will find itself in 2020 on the relevant wind supply curves.

In general, projections that are more in the form of goals for renewable technology performance tend to be more optimistic than projections based on learning, whether the learning is from projections of past trends or as a function of anticipated deployment, which is in turn determined partly based on policy assumptions. Thus, the SEIA and EERE projections for the costs of solar technologies tend to be more optimistic than those from other sources.

Cost projections for biopower depend on the assumed technology. For its NEMS model projections, EIA assumes that all new investment in biopower will use an integrated gasification combined cycle (IGCC) technology that has a very low heat rate and relatively low capital costs, whereas EPRI, in its 2007 summer study, assumes that new biopower is generated using a stoker boiler technology that has somewhat higher capital costs and a much higher heat rate. The comparability of these two cost assumptions is unclear, given the complexity of the IGCC technology relative to the proven stoker boiler technology. The fact that biopower from IGCC technology is a relatively untested technology in the United States at this time is also not reflected in these different estimates, and thus the cost estimates from EIA are based on engineering projections and not on actual experience with the technology.

FINDINGS

The most critical elements of the panel's findings on the economics of renewable electricity generation are highlighted below.

A principal barrier to the widespread adoption of renewable electricity technologies is that electricity from renewables (except for electricity from large-scale hydropower) is more costly to produce than electricity from fossil fuels without an internalization of the costs of carbon emissions and other potential societal impacts. Policy incentives, such as renewables portfolio standards, the production tax credit, feed-in tariffs, and greenhouse gas controls, thus have been required, and for the foreseeable future will continue to be required, to drive further increases in the use of renewable sources of electricity.

Unlike some conventional energy resources, renewable electricity is considered manufactured energy, meaning that the largest proportion of costs, external energy, and materials inputs, as well as environmental impacts, occur during manufacturing and deployment rather than during operation. In general, the use of renewable resources for electricity generation involves trading the risks of future cost increases for fossil fuels and uncertainties over future costs of carbon controls for present fixed capital costs that typically are higher for use of renewable resources than for use of fossil fuels. Except for biopower, no fuel costs are associated with renewable electricity sources. Further, in contrast to coal and nuclear electricity plants, in which larger facilities tend to exhibit lower average costs of generation than do smaller plants, for renewable electricity the opportunities for achieving economies of scale are generally greater at the equipment manufacturing stage than at the generating site itself.

The future evolution of costs for generation of electricity from renewable resources will depend on continued technological progress and breakthroughs. It will also depend on the potential for policies to create greater penetration and to accelerate the scale of production largely an issue of long-term policy stability and policy clarity. Markets will generally exploit the lowest-cost resource options first, and thus the costs of renewables may not decline in a smooth trajectory over time. For example, in the case of wind power, the lowest-cost resources are generally available at the most accessible sites in the highest wind class areas. Development of these prime resources will thus entail significant resource cost shifts as markets adjust to exploit next-tier resources. **At present, onshore wind is an economically favored option relative to other (non-hydroelectric) renewable resources, and**

hence wind power is expected to continue to grow rapidly if recent policy initiatives continue into the future.

Although some forecasts show that biopower will play an important role in meeting future RPS targets, the degree of competition with and recent mandates for use of liquid biofuels for providing transportation fuel and, of course, the use of biomass for food, agricultural feed, and other uses will impact the prospects for greater use of biomass in the electricity market. The future of distributed renewable electricity generation from sources such as residential photovoltaics will depend on how its costs compare to the retail price of power delivered to end users, on whether prices fully reflect variations in cost over the course of the day, and on whether the external costs of fossil-based electricity generation are increasingly incorporated into its price.

Formulation of robust predictions about whether the price of electricity will meet or exceed the price required for renewable sources to be profitable and what their resulting level of market penetration will be remain a difficult proposition. Comparisons between past forecasts of renewable electricity penetration and actual data show that, while renewable technologies generally have met forecasts of cost reductions, they have fallen short of deployment projections. Further, the profitability and penetration of electricity generated from renewable resources may be sensitive to investments in energy efficiency, especially if efficiency improvements are sufficient to meet growth in the demand for electricity or to lower the market-clearing price of electricity. If the financial operating environment for fossil-fuel and other in-place sources of electricity remains unchanged, then the competitiveness of renewable electricity may be affected more than that of other electricity sources. However, at this time, the deployment of renewable electricity is being driven by tax policies, in particular by the renewable production tax credit, and by renewable portfolio standards.

In particular, the panel finds that:

- Onshore wind is an economic option that could scale to a material penetration by 2020, and will likely see rapid growth if recent policy initiatives continue into the future. Biopower is also forecast to play an important role in meeting future RPS targets, but greater use in the future electricity market will depend on competition from demand for liquid biofuels for transportation.
- Some kind of incentive (RPS with high REC price, PTC, or feed-in tariff) will be needed to increase renewables' use as long as external costs

of fossil generation, particularly those associated with greenhouse gas emissions, are not incorporated into the costs of using those technologies and fuels.

- Renewables policies will be more effective if they are stable and predictable than if they cycle on and off or have a highly uncertain future, as has been the case with the federal PTC.
- Intermittency of wind is manageable using current technology, and storage is not required for levels of market penetration expected within the 2020 timeframe.
- The evolution of renewables' costs will depend on technological breakthroughs, the potential for policies to achieve greater market penetration and technological learning, and the rapidity with which low-cost resources, such as the most accessible sites in the highest wind class areas, are exhausted.
- Investment in energy efficiency, which can lower the market-clearing price of electricity, could diminish the future profitability of renewable electricity generation.
- A key determinant of the future success of renewables in penetrating the market is the value that renewables suppliers will obtain for their generation in the electricity marketplace, which is largely determined by the wholesale market price of electricity for grid-scale renewables and the retail price of electricity for distributed renewables. Predicting that price and the resulting level of renewables market penetration has been and continues to be a difficult proposition.
- Projections of levelized costs of energy in 2020 for wind across all data sources are generally no higher than EIA projections of levelized costs for coal integrated gasification combined cycle with carbon capture and storage or natural gas combined cycle with carbon capture and storage. However, the full costs of integrating wind power into the grid, which are not typically reflected in levelized cost projections, could lead to a change in the relative cost ranking of these technologies.

REFERENCES

ASES (American Solar Energy Society). 2007. Tackling Climate Change in the U.S.: Potential Carbon Emission Reductions from Energy Efficiency and Renewable Energy by 2030. Washington, D.C.

Bierden, P. 2007. The process of developing wind power generators. Presentation at the second meeting of the Panel on Electricity from Renewable Resources, December 6, 2007. Washington, D.C.

Bird, L., and B. Swezey. 2006. Green Power Marketing in the United States: A Status Report. 9th Edition. NREL/TP-640-40904. Golden, Colo.: National Renewable Energy Laboratory. November.

Bird, L., and E. Lockey. 2007. Interaction of Compliance and Voluntary Renewable Energy Markets. NREL/TP-670-42096. Golden, Colo.: National Renewable Energy Laboratory. October.

Bird, L., L. Dagler, and B. Swezey. 2007. Green Power Marketing in the United States: A Status Report. 10th Edition. NREL/TP-670-42502. Golden, Colo.: National Renewable Energy Laboratory. December.

Black & Veatch. 2007. 20 Percent Wind Energy Penetration in the United States—A Technical Analysis of the Energy Resource. Black & Veatch Project #144864, prepared for the American Wind Energy Association, Walnut Creek, Calif.

Borenstein, S. 2007. Electricity Rate Structures and the Economics of Solar PV: Could Mandatory Time-of-Use Rates Undermine California's Solar Photovoltaic Subsidies? Center for the Study of Energy Markets Working Paper 172. University of California, Berkeley. September. Available at http://www.ucei.berkeley.edu/PDF/csemwp172.pdf.

Borenstein, S. 2008a. The Market Value and Cost of Solar Photovoltaic Electricity Production. Center for the Study of Energy Markets Working Paper 176. University of California, Berkeley. January.

Borenstein, S. 2008b. Response to Critics of "The Market Value and Cost of Solar Photovoltaic Electricity Production." Available at http://faculty.haas.berkeley.edu/borenste/SolarResponse.pdf.

CEC (California Energy Commission). 2008. Energy Renewable Programs. Available at http://www.energy.ca.gov/renewables/emerging_renewables/index.html.

Cornelius, C. 2007. DOE Solar Energy Technologies Program. Presentation at the first meeting of the Panel on Electricity from Renewable Resources, September 18, 2007. Washington, D.C.

Cory, K.S., and B.G. Swezey. 2007. Renewable Portfolio Standards in the States: Balancing Goals and Implementation Strategies. Technical Report NREL/TP- 670-41409. Golden, Colo.: National Renewable Energy Laboratory. December.

DeCarolis, J.F., and D.W. Keith. 2004. The economics of large-scale wind power in a carbon constrained world. Energy Policy 34:395-410.

Denholm, P., and R. Margolis. 2008. Supply Curves for Rooftop Solar PV-Generated Electricity for the United States. Technical Report, NREL/TP-6A0-44073. Golden, Colo.: National Renewable Energy Laboratory. November.

DOE (U.S. Department of Energy). 2008. 20% Wind Energy by 2030: Increasing Wind Energy's Contribution to U.S. Electricity Supply. Office of Energy Efficiency and Renewable Energy. Washington, D.C.: U.S. DOE.

Economist. 2008. German lessons: An ambitious cross-subsidy scheme has given rise to a new industry. Vol. 387(8574), pp. 67-68, April 5.

EIA (Energy Information Administration). 2004. Analysis of S. 1844, the Clear Skies Act of 2003; S. 843, the Clean Air Planning Act of 2003; and S. 366, the Clean Power Act of 2003. SR/OIAF/2004-05. Washington, D.C.: U.S. Department of Energy, EIA. May.

EIA. 2005. Production Tax Credit for Renewable Energy Generation, Issues in Focus, AEO2005. Washington, D.C.: U.S. Department of Energy, EIA.

EIA. 2007a. Assumptions to the Annual Energy Outlook for 2007. DOE/EIA-0554(2007). Washington, D.C.: U.S. Department of Energy, EIA.

EIA. 2007b. Analysis of Alternative Extensions of the Existing Production Tax Credit for Wind Generators. Memorandum to the Committee on Ways and Means, U.S. House of Representatives. Washington, D.C.: U.S. Department of Energy, EIA.

EIA. 2007c. Impacts of a 15-Percent Renewable Portfolio Standard. SR/OIAF/2007-03. Washington, D.C.: U.S. Department of Energy, EIA.

EIA. 2008a. Energy Market and Economic Impacts of S. 1766, The Low Carbon Economy Act of 2007. SR/OIAF/2007-06. Washington, D.C.: U.S. Department of Energy, EIA.

EIA. 2008b. Energy and Economic Impacts of Implementing Both a 25-Percent Renewable Portfolio Standard and a 25-Percent Renewable Fuel Standard by 2025. SR/OIAF/2007-05. Washington, D.C.: U.S. Department of Energy, EIA.

EIA. 2008c. Annual Energy Outlook 2008 with Projections to 2030. DOE/EIA-0383(2008). Washington, D.C.: U.S. Department of Energy, EIA.

EIA. 2008d. Annual Energy Outlook 2009, Early Release. DOE/EIA-0383(2009). Washington, D.C.: U.S. Department of Energy, EIA.

EPA (U.S. Environmental Protection Agency). 2008. EPA Analysis of the Lieberman-Warner Climate Security Act of 2008, S. 2191 in the 110th Congress. Washington, D.C.

EPRI (Electric Power Research Institute). 2007a. Renewable Energy Technical Assessment Guide. Palo Alto, Calif.

EPRI. 2007b. The Power to Reduce CO_2 Emissions: The Full Portfolio. Palo Alto, Calif.

ERCOT (Electric Reliability Counsel of Texas). 2007. ERCOT Target Reserve Margin Analysis. Sacramento, Calif.: Global Energy Decisions, Inc.

EWEA (European Wind Energy Association). 2005. Large Scale Integration of Wind Energy in the European Power Supply: Analysis, Issues and Recommendations. Brussels.

FERC (Federal Energy Regulatory Commission). 2006. Assessment of Demand Response and Advanced Metering Staff Report. Docket AD06-2-000. Washington, D.C.

GE Energy Consulting. 2005. The Effects of Integrating Wind Power on Transmission System Planning, Reliability, and Operations. Prepared for New York State Energy Research and Development Authority, Albany, N.Y.

GeothermEx, Inc. 2004. New Geothermal Site Identification and Qualification. Report to the California Energy Commission, Public Interest Energy Research Program, Sacramento, Calif.

German Federal Ministry for the Environment, Nature Conservation, and Nuclear Safety. 2007. Renewable Energy Sources in Figures—National and International Development. Bonn, Germany.

Green, M.A. 2004. Third Generation Photovoltaics: Advanced Solar Energy Conversion. Berlin: Springer.

Harper, J.P., M.D. Karcher, and M. Bolinger. 2007. Wind Project Financing Structures: A Review and Comparative Analysis. Technical Report LBNL-63434. Berkeley, Calif.: Lawrence Berkeley National Laboratory. September.

IEA (International Energy Agency). 2006. Renewables Database. Renewable Energy Feed-in Tariffs (III). Paris. Available at http://www.iea.org/textbase/pm/?mode=re&id=3846&action=detail.

Klein, A., A. Held, M. Ragwitz, G. Resch, and T. Faber. 2006. Evaluation of Different Feed-In Tariff Design Options—Best Practice for the International Feed-In Cooperation. Fraunhofer Institute Systems and Innovation Research, Economics Group, Karlsruhe, Germany.

McElroy, A.K. 2008. Lukewarm on co-firing. Biomass Magazine. February. Available at http://www.biomassmagazine.com/article.jsp?article_id=1429&q=&page=all.

McVeigh, J., D. Burtraw, J. Darmstadter, and K. Palmer. 2000. Winner, loser, or innocent victim: Has renewable energy performed as expected? Solar Energy 68(3):237-255.

Mills, A., and R. Wiser. 2009. The Cost of Transmission for Wind Energy: A Review of Transmission Planning Studies. Berkeley, Calif.: Lawrence Berkeley National Laboratory.

NREL (National Renewable Energy Laboratory). 2006. Power Technologies Energy Data Book. NREL/TP-620-39728. Golden, Colo. August. Available at http://www.nrel.gov/analysis/power_databook/.

NREL. 2007. Projected Benefits of Federal Energy Efficiency and Renewable Energy Programs. NREL/TP-640-41347. Golden, Colo. March. Available at http://www1.eere.energy.gov/ba/pba/pdfs/41347.pdf.

Palmer, K., and D. Burtraw. 2005. Cost effectiveness of renewable energy policies. Energy Economics 27:873-894.

Palmer, K., D. Burtraw, and J.S. Shih. 2007. The benefits and costs of reducing emissions from the electricity sector. Journal of Environmental Management 83:1.

Patel, S. 2009. PV sales in the U.S. soar as solar panel prices plummet. Power Magazine, March 1.

Rastler, D. 2008. Electric energy storage briefing. Presentation at the fourth meeting of the Panel on Electricity from Renewable Resources, March 11, 2008. Washington, D.C.

Ridge Energy Storage and Grid Services, L.P. 2005. The Economic Impact of CAES of Wind in TX, OK, and NM. Final Report for Texas State Energy Conservation Office, Austin, Tex. June 27.

Royal Academy of Engineering. 2004. The Costs of Generating Electricity. London.

SEIA (Solar Energy Industries Association). 2004. Our Solar Power Future—The U.S. Photovoltaic Industry Roadmap Through 2030 and Beyond. Washington, D.C.

Sheehan, G., and S. Hetznecker. 2008. Utility Scale Solar Power. IEEE Power & Energy Magazine 5 (October).

Short, W., and P. Denholm. 2006. A Preliminary Assessment of Plug-in Hybrid Electric Vehicles on Wind Energy Markets. NREL Technical Report. Golden, Colo.: National Renewable Energy Laboratory. April.

Smith, J.C. 2007. Integrating wind into the grid. Presentation to the America's Energy Future Panel on Electricity from Renewable Resources, Washington, D.C. December 7.

Smith, R.K. 2006. EIA geothermal supply curve. Memorandum. Washington, D.C.: U.S. Department of Energy, EIA. September 15.

Surek, T. 2005. Crystal growth and materials research in photovoltaics: Progress and challenges. Journal of Crystal Growth 275:292-304.

Swezey, B., J. Aabakken, and L. Bird. 2007. A Preliminary Examination of the Supply and Demand Balance for Renewable Electricity. NREL/TP-670-42266. Golden, Colo.: National Renewable Energy Laboratory. October.

Vajjhala, S., A. Paul, R. Sweeney, and K. Palmer. 2008. Green corridors: Linking interregional transmission expansion and renewable energy policies. Discussion Paper 08-06. Washington, D.C.: Resources for the Future, Inc. March.

Venkataraman, S., D. Nikas, and T. Pratt. 2007. Which power generation technologies will take the lead in response to carbon controls? S&P Credit Research. New York: Standard and Poor's. May 11.

WGA (Western Governors' Association). 2006a. Clean and Diversified Energy Initiative: Geothermal Task Force Report. Denver, Colo.

WGA. 2006b. Clean and Diversified Energy Initiative: Solar Task Force Report. Denver, Colo.

Wiser, R. 2008. The development, deployment, and policy context of renewable electricity: A focus on wind. Presentation at the fourth meeting of the Panel on Electricity from Renewable Resources, March 11, 2008. Washington, D.C.

Wiser, R., and G. Barbose. 2008. Renewables Portfolio Standards in the United States: A Status Report with Data Through 2007. Berkeley, Calif.: Lawrence Berkeley National Laboratory.

Wiser, R., and M. Bolinger. 2008. Annual Report on U.S. Wind Power Installation, Cost and Performance Trends: 2007. DOE/GO-102008-2590. Washington, D.C.: U.S. Department of Energy.

ANNEX

TABLE 4.A.1 Current Cost Assumptions for Renewable Technologies (in 2007 Dollars)

Technology	Source	Case/Scenario	Overnight Capital Cost ($/kW)[a]
Biopower			
Biopower–IGCC	EIA (2008d)	Input table	3766
Biopower–Stoker	EPRI (2007a)		3520
Biopower–50 MW fluidized bed	EPRI (2007a)		3629
Biopower	Venkataraman et al. (2007)		2596
Concentrating Solar Power			
Concentrating solar	NREL (2007)	Program	3645
Concentrating solar	EIA (2008d)	Reference	5021
Concentrating solar	EPRI (2007b)	Limited and full portfolio	
Concentrating solar–trough	EPRI (2007a)		3271
Concentrating solar	Venkataraman et al. (2007)		4153
Concentrating solar–trough	ASES (2007)		
Photovoltaic			
Photovoltaic	NREL (2007)	Program	4050
Photovoltaic–distributed	EPRI (2007b)	Limited and full portfolio	
Photovoltaic flat plate	EPRI (2007a)		5487
Photovoltaic 2-axis	EPRI (2007a)		8876
Photovoltaic–distributed	SEIA (2004)	Baseline	
Photovoltaic–distributed	SEIA (2004)	Roadmap	
Photovoltaic–central	EIA (2008d)	Input table	6038

Capacity Factor (%)	Variable O&M (+ Fuel Costs) ($/MWh)	Fixed O&M ($/kW)	Levelized Cost of Energy ($/kWh)[b,c]
83	6.71 (+ $15)[d]	64.45	0.080
85	3.74 (+ $35)[e]	91.79	0.0977[b,f]
85	4.26 (+ $35)[e]	94.49	0.101[b,f]
85	7.27 (+ $28)[e]	166.13	0.090
65	8.10	0.00	0.071[g]
31	0.00	56.7	0.200
34			0.170
34	0.00	60.2[f]	0.130
43	31.20	34.3	0.170
			0.160–0.190
21	0.00	10.4	0.220[g]
			0.260
25	0.00	19.5	0.251[b,f]
32	0.00	46.6	0.330[b,f]
			0.150
			0.080
22	0.00	11.7	0.320

continued

TABLE 4.A.1 Continued

Technology	Source	Case/Scenario	Overnight Capital Cost ($/kW)[a]
Wind			
Onshore wind	EIA (2008d)	Input table	1923
Onshore wind	NREL (2007)	Baseline	1052
Onshore wind	NREL (2007)	Program	927
Onshore wind	EPRI (2007b)	Limited and full portfolio	
Onshore wind	EPRI (2007a)	Class 6, 100 MW	1820
Onshore wind	Venkataraman et al. (2007)		1765
Onshore wind	Black & Veatch (2007)	20% wind energy study	1713
Onshore wind	Midwest ISO	MTEP 2008 reference	1983
Offshore wind	EIA (2008d)	Input table	3851
Offshore wind	Black & Veatch (2007)	20% wind energy study	2388
Conventional			
Pulverized coal	EIA (2008d)	Input table	2058
Conventional gas combined cycle	EIA (2008d)	Input table	962
Conventional combustion turbine	EIA (2008d)	Input table	670

[a] Fuel cost per megawatt-hour reported by source.
[b] Calculated from inputs based on a 20-year economic life and real cost of capital of 7.5 percent.
[c] Levelized costs here are generic and do not include site-specific development costs or cost of facilitating delivery.
[d] Fuel cost imputed from AEO 2009 Early Release model solution. AEO 2009 Energy Prices (2007$/million Btu) in 2012 are $1.91 for coal, $6.63 for natural gas, and $1.96 for biomass.
[e] Fuel cost per megawatt-hour imputed from EPRI summer study levelized cost and TAG specifications for CFB biomass plant.
[f] This estimate comes from a personal communication with Steve Gehl of EPRI.
[g] EERE numbers are for 2010.
[h] Depending on wind class.

Capacity Factor (%)	Variable O&M (+ Fuel Costs) ($/MWh)	Fixed O&M ($/kW)	Levelized Cost of Energy ($/kWh)[b,c]
36	0.00	30.3	0.069[c]
45	0.00	26.2	0.033[g]
46	0.00	25.3	0.029[g]
32.5			0.100
42	0.00	72.7	0.068[b,f]
33	0.00	26.0	0.073[c]
35–50[h]	5.70	11.9	0.064–0.047[c,h]
34	0.00	16.5	0.071[c]
34	0.00	89.5	0.157[c]
37–52[h]	15.60	18.7	0.094–0.071[c,h]
85	4.64 (+ $16.7)[d]	27.53	0.050
87	2.09 (+ $45.1)[d]	12.48	0.060
30	3.60 (+ $69.3)[d]	12.11	0.100

TABLE 4.A.2 2020 Cost Projections and Comparisons (in 2007 Dollars)

Technology	Source	Case/Scenario	Overnight Cost (/kW)[a]	Capacity Factor (%)
Conventional Sources				
Pulverized coal	EIA (2008d)	Reference	1985	85
IGCC	EIA (2008d)	Reference	2233	85
IGCC with sequestration	EIA (2008d)	Reference	3171	85
Combined cycle	EIA (2008d)	Reference	928	87
Advanced combined cycle	EIA (2008d)	Reference	892	87
Advanced combined cycle with sequestration	EIA (2008d)	Reference	1729	87
Combustion turbine	EIA (2008d)	Reference	647	30
Advanced combustion turbine	EIA (2008d)	Reference	587	30
Renewables				
Biopower				
Biopower	EIA (2008d)	Reference	3390	83
Biopower–Stoker	EPRI (2007b)	Full portfolio		85
Biopower–Stoker	EPRI (2007b)	Limited portfolio		85
Biopower	ASES (2007)	WGA Biomass Task Force		90
Geothermal				
Geothermal	EIA (2008d)	Reference	1585	90
Concentrating Solar				
Concentrating solar	NREL (2007)	Program case	2860	72
Concentrating solar	EIA (2008d)	Reference	4130	31
Concentrating solar	EPRI (2007b)	Limited portfolio		34
Concentrating solar	EPRI (2007b)	Full portfolio		34
Photovoltaic				
Photovoltaic	NREL (2007)	Program	2547	21
Photovoltaic	EPRI (2007b)	Full portfolio		
Photovoltaic	EPRI (2007b)	Limited portfolio		

Levelized Cost ($/kW)	Total Capital Cost ($/MWh)	Variable O&M/ Fuel Costs ($/MWh)	Fixed O&M ($/MWh)	Transmission Cost ($/MWh)	Levelized Cost of Energy ($/kWh)[b]
195	52.30	23.06	3.70	3.61	0.083
					[0.079]
219	60.64	18.59	5.19	3.61	0.088
					[0.084]
311	69.54	23.26	6.19	4.01	0.103
					[0.099]
91	18.63	59.21	1.64	3.88	0.083
					[0.079]
88	17.98	55.46	1.54	3.88	0.079
					[0.075]
170	34.64	68.84	2.61	3.93	0.110
					[0.106]
63	33.55	88.49	4.61	11.41	0.138
					[0.127]
58	30.71	75.21	4.01	11.41	0.121
					[0.110]
333	61.62	22.81	8.86	4.14	0.097
					[0.093]
					0.096
					0.101
					~0.080[c]
156	75.44	0.00	22.22	5.00	0.103
					[0.098]
	4.47	0.00			0.050
405	180.02	0.00	21.30	11.00	0.212
					[0.201]
					0.170
					<0.083[c]
250	135.81	0.00	5.59		0.141
					0.220
					0.260

continued

TABLE 4.A.2 Continued

Technology	Source	Case/Scenario	Overnight Cost (/kW)[a]	Capacity Factor (%)
Photovoltaic	EIA (2008d)	Reference	5185	22
Photovoltaic–distributed	SEIA (2004)	Baseline		
Photovoltaic–distributed	SEIA (2004)	Roadmap		
Photovoltaic–distributed	ACES (2007)	DOE's Solar America Initiative	2.50/W_p installed cost	
Wind				
Onshore wind	EIA (2008d)	Reference	1896	35
Onshore wind	NREL (2007)	Baseline	1076	46
Onshore wind	NREL (2007)	Program	916	49
Onshore wind	EPRI (2007b)	Full portfolio		42
Onshore wind	EPRI (2007b)	Limited portfolio		33
Onshore wind	Black & Veatch (2007); DOE (2008)	DOE 20% wind study	1630	38–52 (depending on wind class)
Offshore wind	EIA (2008d)	Reference	3552	33
Offshore wind	Black & Veatch (2007); DOE (2008)	DOE 20% wind study	2232	38–52 (depending on wind class)

Note: Reflects the base capital cost from AEO 2009 (EIA, 2008d), Table 39, adjusted for learning. This figure does not reflect taxes and depreciation, which are included in the total capital cost.

[a]The overnight cost includes the effects of technological learning but does not include other project costs, which are reflected in the levelized cost estimated.

[b] [ ] contain AEO estimates of busbar levelized cost of energy, i.e., without transmission-related costs.

[c] Cost estimate is for 2015.

[d] Interpolated between reported targets for 2015 and 2030.

Source: Based on data in ASES (2007); Black & Veatch (2007); EIA (2008d); EPRI (2007b); NREL (2007); and SEIA (2004).

Levelized Cost ($/kW)	Total Capital Cost ($/MWh)	Variable O&M/ Fuel Costs ($/MWh)	Fixed O&M ($/MWh)	Transmission Cost ($/MWh)	Levelized Cost of Energy ($/kWh)[b]
509	292.84	0.00	6.21	13.69	0.313
					[0.299]
					0.110
					0.050
					0.075–0.010[d]
186	81.38	0.00	9.95	8.66	0.100
					[0.091]
		0.00	27.10		0.033
		0.00	23.40		0.027
					0.078
					0.097
160	48.04–35.1	4.85	3.64–2.66		0.057–0.043
348	154.36	0.00	26.72	9.31	0.191
					[0.181]
219	64.1–46.29	4.87–3.52	4.62–3.33		0.074–0.053

5 Environmental Impacts of Renewable Electricity Generation

Environmental impacts are an inherent part of electricity production and energy use. Electricity generated from renewable energy sources has a smaller environmental footprint than power from fossil-fuel sources, which is arguably the major impetus for moving away from fossil fuels to renewables. However, although the types and magnitude of environmental effects differ substantially from fossil-fuel sources and from one renewable source to another, using renewables does not avoid impacts entirely. An understanding of the relative environmental impacts of the various electric power sources is essential to the development of sound energy policy.

This chapter reviews and compares the environmental impacts of various fossil-fuel and renewable sources of electricity. It applies life-cycle analyses in discussing impacts that occur typically on regional or larger scales, such as air, water, and global warming pollution. This chapter then addresses local impacts that are often considered and assessed as part of the siting and permitting processes.

LARGE-SCALE IMPACTS FROM LIFE-CYCLE ASSESSMENT

Life-cycle assessment (LCA) attempts to estimate the overall energy usage and environmental impact from the energy produced by a given technology by assessing all the life stages of the technology: raw materials extraction, refinement, construction, use, and disposal. Here, LCA is used to compare the relative impacts of various fossil-fuel-based and renewable sources of electricity. To place all analyses on a common footing, impacts are expressed in terms of emission or usage rate per kilowatt-hour (kWh). Finally, it should be noted that developing complete LCAs of electricity sources is beyond the scope of this panel. There are, however, a wide range of earlier assessments, and these form the basis of this section.

A major complication in comparing LCAs is that there is no set standard for carrying out such analyses. While it is the goal in using LCAs to cover technologies from cradle to grave in a systematic way, there is variability in the assumptions, boundaries, and methodologies used in these assessments. Therefore, caution should be used in comparing LCAs; each is an approximation of a technology's actual impact. Discussion of the attributes and assumptions used in life-cycle analysis is found in Appendix E.

The renewable energy technologies are wind, solar, geothermal, hydroelectric, tidal, biopower, and storage. Appendix F contains LCA studies for coal, natural gas, and nuclear technologies as a benchmark against which to assess the performance of renewables. LCA information for solar energy is limited to photovoltaic (PV) technologies, and no LCA studies were reviewed for concentrating solar power (CSP) technologies such as solar trough, power tower, or dish–engine technologies. No LCA information is included for enhanced geothermal systems. The life-cycle impacts considered here include net energy usage; atmospheric emissions of greenhouse gases expressed in units of carbon dioxide (CO_2) equivalents (CO_2e);[1] atmospheric emissions of sulfur dioxide (SO_2), nitrogen oxides (NO_x), and particulate matter;[2] land use; and water withdrawals and consumption. To provide a sense of the variability of the LCAs found in the literature, the maximum, minimum, and average energy usage and environmental impact for each technology are shown in figures discussed below in this chapter.

Energy

Energy input and output calculations, the basic building blocks for any life-cycle evaluation of greenhouse gas emissions, can be used to evaluate the energy inten-

---

[1]Equivalent carbon dioxide emissions (CO_2e) are the amount of greenhouse gas emissions expressed as carbon dioxide, taking into account the global warming potential of non-carbon dioxide greenhouse gases (e.g., methane and nitrous oxides).

[2]All energy technologies are included in the CO_2 section even if CO_2 emissions were low. Other pollutants with emissions less than 100 mg/kWh are not included in the data and discussion. Studies used to compile CO_2 data often make up a different data set from the studies used to compile other emissions. Often LCA studies focus only on CO_2. This required building another data set for other emissions.

sity and resource consumption of the energy technology itself. The literature is replete with assessments of life-cycle energy usage from renewable and non-renewable sources of electricity. However, these assessments adopt a wide range of energy metrics, making internal comparisons problematic. Spitzley and Keoleian (2005) describe eight distinct energy metrics defined in the literature.

Energy metrics should therefore be used with cautions and caveats. No single metric defines the ideal energy generation technology without an accompanying statement of the core value for assessment. For example, a metric such as capacity factor will effectively measure for intermittence or dispatchability. A metric such as price per unit of energy produced measures economic value according to conventional accounting, financing, and cost-accounting assumptions.

This review focuses on two of the more commonly used energy metrics: (1) net energy ratio (NER), which quantifies how much net energy a technology produces over its life cycle, and (2) energy payback time, which defines how long it takes for a given energy technology to recoup the lifetime energy invested in its development once the technology starts generating electricity. These metrics offer insight into the overall energy and environmental performance of generation technologies, especially in making macro-level resource acquisition and development decisions.

Net Energy Ratio

The NER is defined as the ratio of useful energy output to the grid to the fossil-fuel energy consumed during the lifetime of the technology. As such, it is critical to assessing whether or not a renewable energy source reduces our use of fossil fuel.

Renewable energy sources generally have an NER value greater than one. For fossil-fuel energy technologies, the NER is commonly referred to as the life-cycle efficiency. However, there is some inconsistency in the literature on how the NER is defined when the energy technology itself is based on a fossil fuel. In these cases, some researchers include only indirect (external) energy inputs and not the (primary) energy inherent in the fuel (Meier, 2002; White, 2006; Denholm and Kulcinski, 2003). However, this interpretation of the ratio is not an accurate reflection of the total resource consumption of the energy technology in question. For example, the energy consumed by combusting coal in a coal-fired plant is not included in this alternate use of the term. In cases where the primary energy of the fuel is not included in the energy inputs, the NER is more accurately defined as an external energy ratio (EER). The EER is also widely referred to in the literature as the energy payback ratio.

For renewable energy sources such as wind and solar, the NER and EER are very similar, since the energy technology's use of fuel (e.g., wind or solar radiation) does not deplete the energy resource. For the purposes of this text, the ratio is referred to as the EER when primary fossil energy inputs are not included.

Figure 5.1 illustrates the range of NERs and EERs found in the literature. NER values are influenced by a number of factors, including plant capacity factor, plant life expectancy, choice of plant materials (e.g., steel versus concrete for wind towers), and fuel mix during material construction. For wind and solar technologies, the location and the strength of the resource at that location also constitute an important variable. For example, a wind farm sited in a location with higher average wind speeds will generate more energy than will a wind farm sited at a

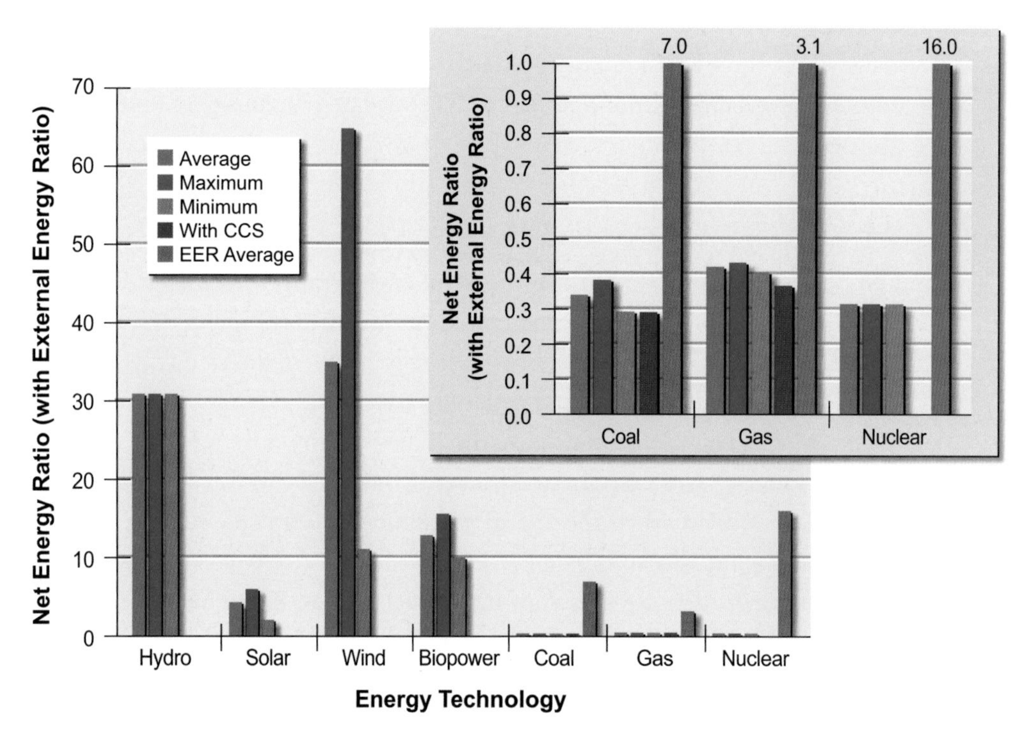

FIGURE 5.1 *Net energy ratio (NER) and external energy ratio (EER) for various renewable and non-renewable energy sources.*
Source: Developed from data provided in Denholm (2004), Denholm and Kulcinski (2003), Meier (2002), Pacca et al. (2007), Spath et al. (1999), Spath and Mann (2000), Spitzley and Keoleian (2005), and White (1998, 2006).

location with lower average wind speeds. In the same way, solar installations in areas with greater solar radiation will typically have higher NERs. Additional factors for PV technologies include position of module, solar conversion efficiency of module, and manufacturing energy intensity.

Figure 5.1 shows that NERs for renewable technologies tend to be higher than for conventional energy technologies, because they consume fewer resources. Of the technologies reviewed, wind has the highest NERs, with values that range from 11 to 65. The lower values tend to be for relatively small wind farms with low-capacity turbines and slower winds. Net energy ratios of 47 and 65 were reported for two large wind farms with higher-capacity turbines and higher average wind speeds. The NER for wind is very dependent on assumptions related to the frequency of blade and turbine replacement, because so much life-cycle energy is consumed in material manufacturing for this technology.

Figure 5.1 also indicates a relatively high NER for hydroelectric power, but this should be interpreted with caution, as it is based on only one LCA study (with a NER of 31) for a large reservoir facility in the United States with a 50-year lifetime. NERs for biopower reported here range from 10 to 16, based on analysis of four power plants that use cropping to supply biomass. Biopower from waste would be expected to have higher NERs, but no LCAs for this fuel stock appear to be available at this time. No NER data were reviewed for geothermal, tidal, or energy storage technologies.

While the NERs for solar PV plotted in Figure 5.1 tend to be relatively low, rapid innovation should improve this ratio in the coming years. For example, Pacca et al. (2007) developed an optimal case for multicrystalline and thin-film (a-Si) PV technologies (using the highest possible solar insolation and conversion efficiency, the least possible manufacturing energy, and maximum plant life) to evaluate the future potential of this technology and found that PV NERs improved to 43 and 132, respectively.

Unlike renewable sources, conventional energy technologies have NERs of less than 1. Their EERs, however, tend to be comparable to or even greater than the NERs for solar PV and biopower. Of the three non-renewable sources of energy considered here, nuclear has the highest average EER.

Energy Payback Time

The energy payback time (EPBT) is a measure of how much time it takes for an energy technology to generate enough useful energy to offset energy consumed during its lifetime. As such it provides an indication of the temporal fossil-fuel needs and emissions as an energy infrastructure is transformed from a carbon-intensive to a low-carbon system.

In the LCA literature, the EPBT is most commonly applied to wind and PV technologies as an additional measure of the economic viability of these newer technologies. Wind EPBT of 0.26 and 0.39 years were reported for two large wind farms with higher-capacity turbines and higher average wind speeds (Schleisner, 2000). The lower value is for a land-based wind farm, while the higher value reflects the additional materials needs for offshore installations. EPBT values for PV range from 7.5 years to less than 1 year. As illustrated in Figure 5.2, this range in EPBT for PV largely reflects a downward trend in time as each successive generation becomes less energy intensive. The EPBT of less than 1 year is from analysis of a hypothetical future generation of PV.

The length of the EPBT has important implications for how long it will take to displace fossil-fuel sources of energy with renewable sources. Consider a simple example. Suppose it takes four units of fossil-fuel energy to produce one unit of energy with a renewable energy technology (such as a wind turbine), and suppose that the unit of renewable technology displaces one unit of fossil-fuel energy.

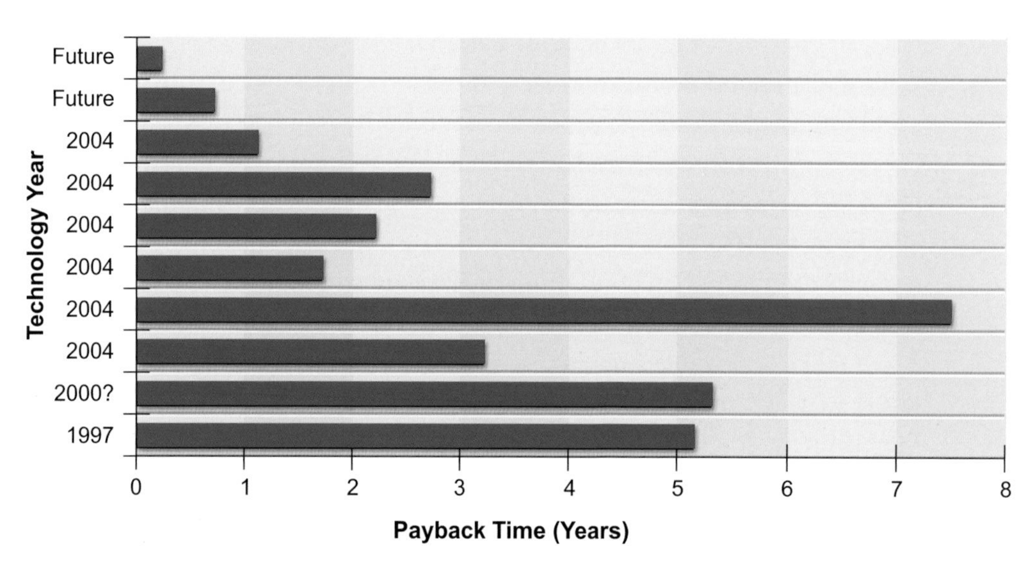

FIGURE 5.2 *Estimated PV energy payback time decreases as a function of the vintage year of the technology.*
Source: Developed from data provided in Fthenakis et al. (2006), Keoleian and Lewis (2003), Meier (2002), and Pacca et al. (2007).

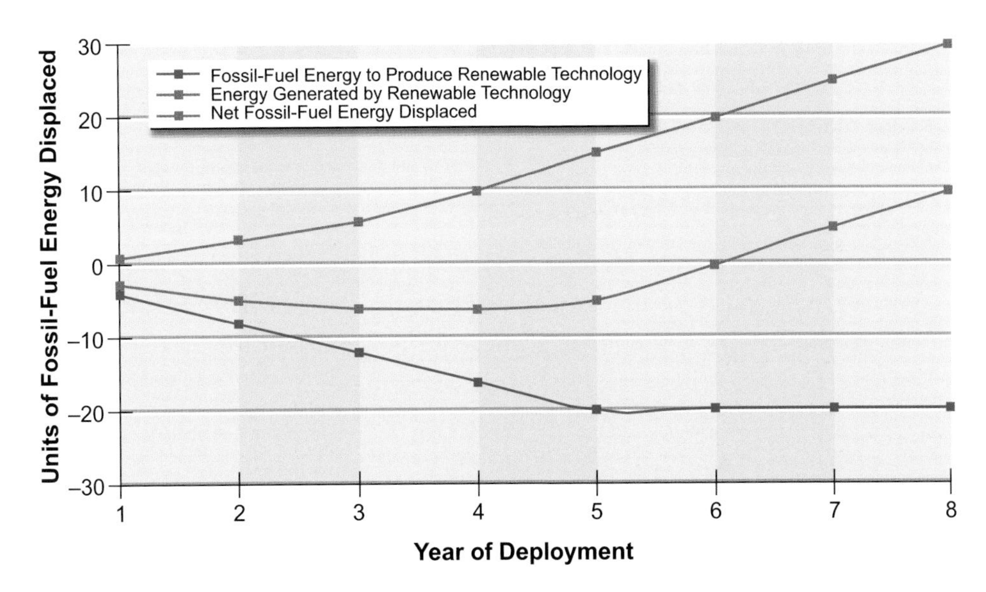

FIGURE 5.3 *Simple illustrative example of total fossil-fuel energy expended (red), renewable energy generated (green), and net fossil-fuel energy displaced (blue) for a scenario when one unit of a renewable technology with an energy payback time of 4 years is deployed each year over a 5-year period.*

Thus, the EPBT for the technology is 4 years. The renewable technology does not begin to displace fossil-fuel energy used per year until 4 years after its initial deployment.

However, the preceding example omits the reality that low-carbon technologies will be deployed over time, so that the energy costs of each successive installation accumulate and effectively extend the time it takes before the energy benefits of the renewable technology are realized. For example, suppose that one unit of the renewable technology discussed above is deployed each year for a period of 5 years. In this scenario, the break-even point between the expenditure of fossil-fuel energy and displacement of the same does not occur until 1 year after the completion of the deployment or 6 years after the first unit is deployed (see Figure 5.3). By the same token, large-scale deployment of renewable technologies with long EPBTs, such as PV, will likely not begin to provide a net displacement of fossil-fuel energy until some years after the deployment has begun. Since CO_2 emission reductions depend on displacing fossil-fuel energy, this means that the greenhouse gas emissions reductions from using renewable energy may not be realized for quite some time after the deployment begins. On the other hand, in terms of greenhouse gas emissions, adding new capacity using renewables is preferable to adding new capacity using CO_2-emitting fossil-fuel sources regardless of the EPBT because of the lifetime commitment to fossil-fuel use made by such plants.

Greenhouse Gas Emissions

Concern about climate change and greenhouse gas (GHG) emissions is a major driver in the push for use of renewable energy sources. This section reviews the LCAs of GHG or CO_2e for relevant renewable and non-renewable sources of electricity. Figure 5.4 illustrates the range of estimates of CO_2e emissions that appear in the literature. Table 5.A.1 (in the annex at the end of the chapter) provides a compilation of studies that estimate life-cycle emission of GHG in CO_2e.

Not surprisingly, renewables are estimated to have significantly less CO_2e emissions than coal and gas; most estimates of emissions from nuclear power use are similar in magnitude to those from the use of renewables. Adding carbon capture and storage (CCS) to coal and gas systems, however, significantly reduces the relative advantage renewables have in terms of carbon and energy savings. This relative advantage is also modestly reduced by adding energy storage to a renewable technology.

Solar Photovoltaic

Of the renewable technologies included in this review, solar PV technologies have the highest CO_2e emissions, ranging from 21 to 71 g CO_2e/kWh. CO_2e emissions from PV are sensitive to innovations in conversion efficiencies and to the energy mix used to generate electricity during manufacturing. Older systems have conversion efficiencies as low as 5 percent. In 2007, efficiencies had increased to 8–13 percent depending on the type of PV used. A study of newer PV systems dating from 2004–2006 by Fthenakis et al. (2008) puts CO_2e emissions at the lower end of the range (21–54 g CO_2e/kWh). By 2010 conversion efficiencies for CdTe PV are expected to increase from 9 percent to 12 percent, and efficiencies for crystalline silicon modules are expected to increase to 16 percent in the next few years, lowering emissions even further (Fthenakis and Kim, 2007).

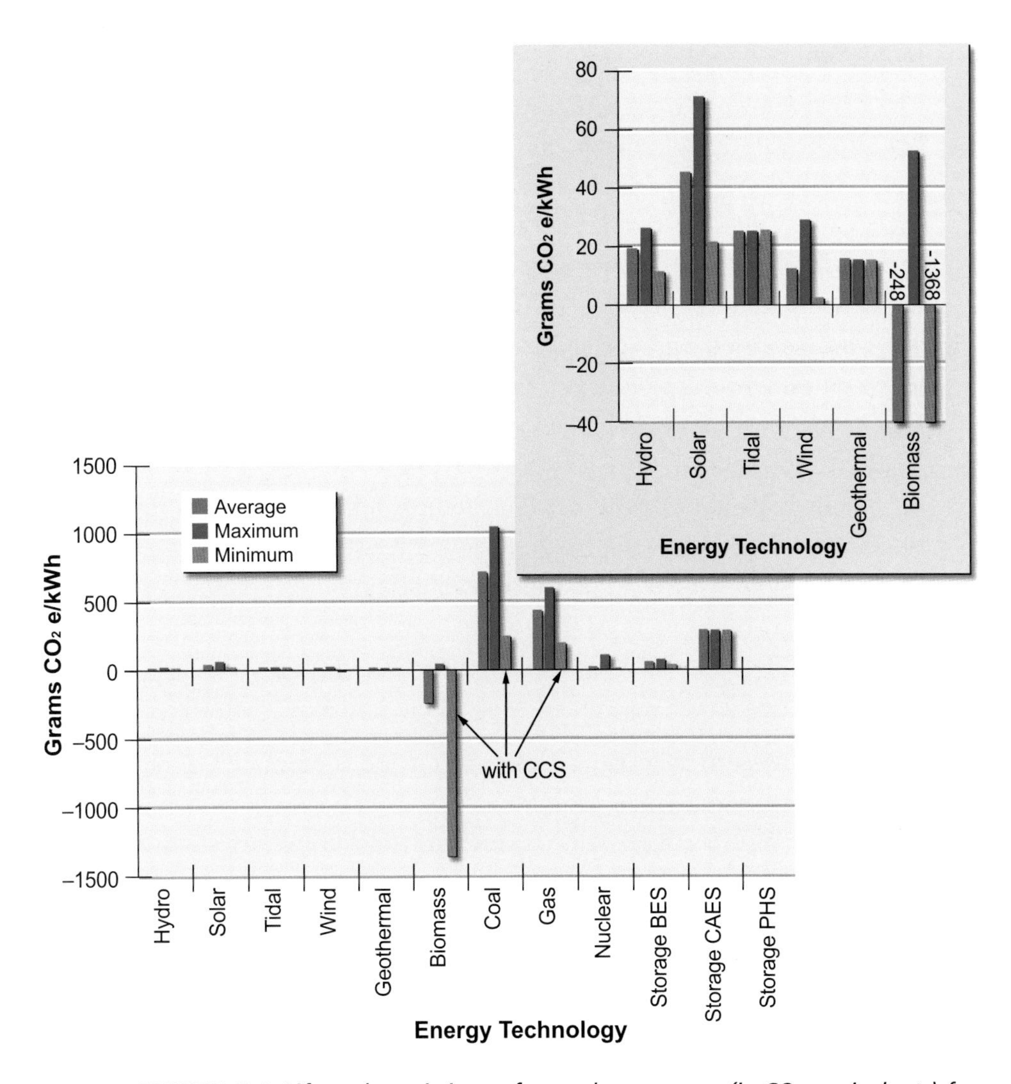

FIGURE 5.4 *Life-cycle emissions of greenhouse gases (in CO_2 equivalents) for various sources of electricity. Average, maximum, and minimum emissions are shown for each technology based on a review of the literature. Note that the inset provides a smaller scale and more details for sources that are not distinguishable in the main figure. Note: Values for biomass, coal, and natural gas include data for carbon capture and storage (CCS).*

Source: Developed from data provided in Berry et al. (1998), Chataignere et al. (2003), Denholm (2004), Denholm and Kulcinski (2003), European Commission (1997a,b,c,d), Frankl et al. (2004), Fthenakis and Kim (2007), Hondo (2005), Mann and Spath (1997), Meier (2002), Odeh and Cockerill (2008), Spath et al. (1999), Spath and Mann (2000, 2004), Spitzley and Keoleian (2005), Storm van Leeuwen and Smith (2008), Vattenfall AB (2004), and White (1998, 2006).

Biopower

The CO_2e emissions from biopower are affected not only by the feedstocks used but also by the yield, fertilizer, and fuel used to cultivate and harvest the feedstock, as well as the specifics of the power plant itself (Mann and Spath, 1997). Most CO_2e values for biopower range from 15 to 52 g CO_2e/kWh for biomass derived from cultivated feedstocks. Spath and Mann (2004) claim that biopower can actually lead to "negative" CO_2e emissions (i.e., act as a greenhouse gas sink). Their estimate of a negative emission of –410 g CO_2e/kWh for biopower was based on using waste residues as the feedstock and giving credit for the avoided GHG emissions that would have occurred as a result of normal waste disposal. Negative emissions of –667 g CO_2e/kWh and –1368 g CO_2e/kWh were estimated for biopower combined with carbon capture and storage using crops and residues, respectively. However, none of these studies considered CO_2 emissions from initial land conversion, which can be considerable (Searchinger et al., 2008; Fargione et al., 2008).

Wind

Among the renewable energy technologies, wind is estimated to be among the lowest life-cycle emitters of greenhouse gases, with emissions ranging from 2 to 29 g CO_2e/kWh. The high value corresponds to a wind farm with a 20 percent generating capacity (Hondo, 2005). This capacity factor is lower than the range of capacity factors (24–40 percent) used in other studies. The two lowest values of 1.7 and 2.5 g CO_2e/kWh are for two larger wind farms (with 50 or more 500-kW turbines) set in an area with good wind production (Class 6 and 4 wind areas, respectively) (Spitzley and Keoleian, 2005). While wind speed is a key factor in determining life-cycle CO_2e emissions, other variables such as generation capacity per unit of materials are also important. For example, Berry et al. (1998) found that a U.K. wind farm with 103 lower-capacity turbines (250 kW) located in an area with higher average wind speeds (Class 7) emitted 9 g CO_2e/kWh. This result, while still very low, is more than three times higher than that seen for the U.S. farm with 50 500-kW turbines located in an area with Class 4 winds.

In spite of producing very low life-cycle carbon emissions, wind is often discounted as a viable source of electricity because of its intermittent availability. Addressing this limitation, Denholm (2004) evaluated CO_2 emissions from wind generation with different storage options. The study found that a combination of wind and pumped hydropower storage (PHS) emitted only 24 g CO_2e/kWh, which is within the range of CO_2 emissions for wind technology alone. A combination of wind and compressed air energy storage (CAES) technology showed a higher value of 105 g CO_2e/kWh, but still far less than emissions seen with fossil-fuel electricity generation. The life-cycle data from Denholm (2004) demonstrate that current technologies for storage are capable of overcoming the limitations of wind generation intermittency without significant carbon emissions.

Geothermal

The total for CO_2e emissions from geothermal electricity generation incorporates the emissions associated with production of the facility and emissions during operation. The latter emissions depend on both the reservoir gas composition and whether the gas is vented to the atmosphere during electricity generation. In 2003, only 14 percent of geothermal facilities were closed-loop binary systems that did not vent gases to the atmosphere (Bloomfield et al., 2003). The analysis presented here considers hydrothermal plants and does not discuss enhanced geothermal systems.

The panel's review found only one LCA study of geothermal technologies that considered emissions from both facility construction and operation. Hondo (2005) reported a value of 15 g CO_2e/kWh for a double-flash geothermal facility. Other data from non-LCA literature show a range of CO_2e emissions from 0 to 740 g CO_2e/kWh for reservoir emissions only.

Hydropower

Most studies conclude that the life-cycle emissions of CO_2e from conventional hydropower technologies are quite small. For example, Hondo (2005) reported a value of 11 g CO_2e/kWh for a river system with a small reservoir. Spitzley and Keoleian (2005) evaluated a large-capacity, efficient U.S. reservoir system located in a semiarid region and estimated an emission rate of 26 g CO_2e/kWh that did not include emissions from flooded biomass. A limitation of most LCAs of hydroelectric generation is that they do not consider the CO_2 and CH_4 emissions that arise from the flooding of large quantities of biomass when the facility is first developed. Some studies suggest that these emissions may be significant for large and/or inefficient tropical hydroelectric projects that flood large quantities of biomass (Fearnside, 1995, 2002) or hydroelectric reservoirs sited on more temperately located peat lands (Gagnon and van de Vate, 1997). Ranges in the literature for carbon emissions from tropical reservoirs can be several hundred to several thousand grams CO_2e/kWh, but they do not reflect normalized life-cycle emissions. Gagnon et al. (1997) addressed this issue by deriving a theoretical life-cycle emission value of 237 g CO_2e/kWh for a hydroelectric reservoir located in Brazil. In this calculation, Gagnon et al. (1997) assumed that 100 percent of the flooded biomass would decay completely over 100 years and that 20 percent of the biomass carbon would be emitted as methane. This calculation does not include emissions from turbines and spillways. More study is needed of the impact of flooded biomass on life-cycle emissions associated with hydroelectric plants.

Hydrokinetic (Tidal/Wave)

No LCA data were reviewed for tidal or wave electricity-generating technologies, which are still very much in the pilot or demonstration stage. One source reported a value of 25 g CO_2e/kWh for the steel used to manufacture turbines for tidal generation installations (CarbonTrust, 2008). One would expect LCA emissions to be low and to occur primarily during material manufacturing and plant construction.

Storage

Storage is not a generating system, but it can be combined with generating technologies to provide backup power for intermittent and peak power needs. Storage options reviewed in the LCA literature included pumped hydropower storage, compressed air energy storage, and battery energy storage (BES) (Denholm and Kulcinski, 2003; Denholm, 2004). The estimate for PHS was a low 3 g CO_2e/kWh. When transmission and distribution (T&D) were included, the estimate increased to 6 g CO_2e/kWh. A variety of BES technologies were reviewed with values ranging from 33 to 81 g CO_2e/kWh. A subset of the BES data with values from 33 to 50 g CO_2e/kWh includes T&D. CAES had the highest emission values, 291 and 292 g CO_2e/kWh, primarily because it relies on natural gas.[3] The second example includes T&D.[4]

[3]Natural gas is used to reheat the air coming from the cavern in diabatic CAES.
[4]Most LCA studies cited here do not include T&D.

SO_2 Emissions

Figure 5.5 shows the range from the literature for life-cycle SO_2 emissions from power sources. Wind, hydropower, and nuclear technologies have extremely low life-cycle SO_2 emissions, less than 100 mg/kWh.

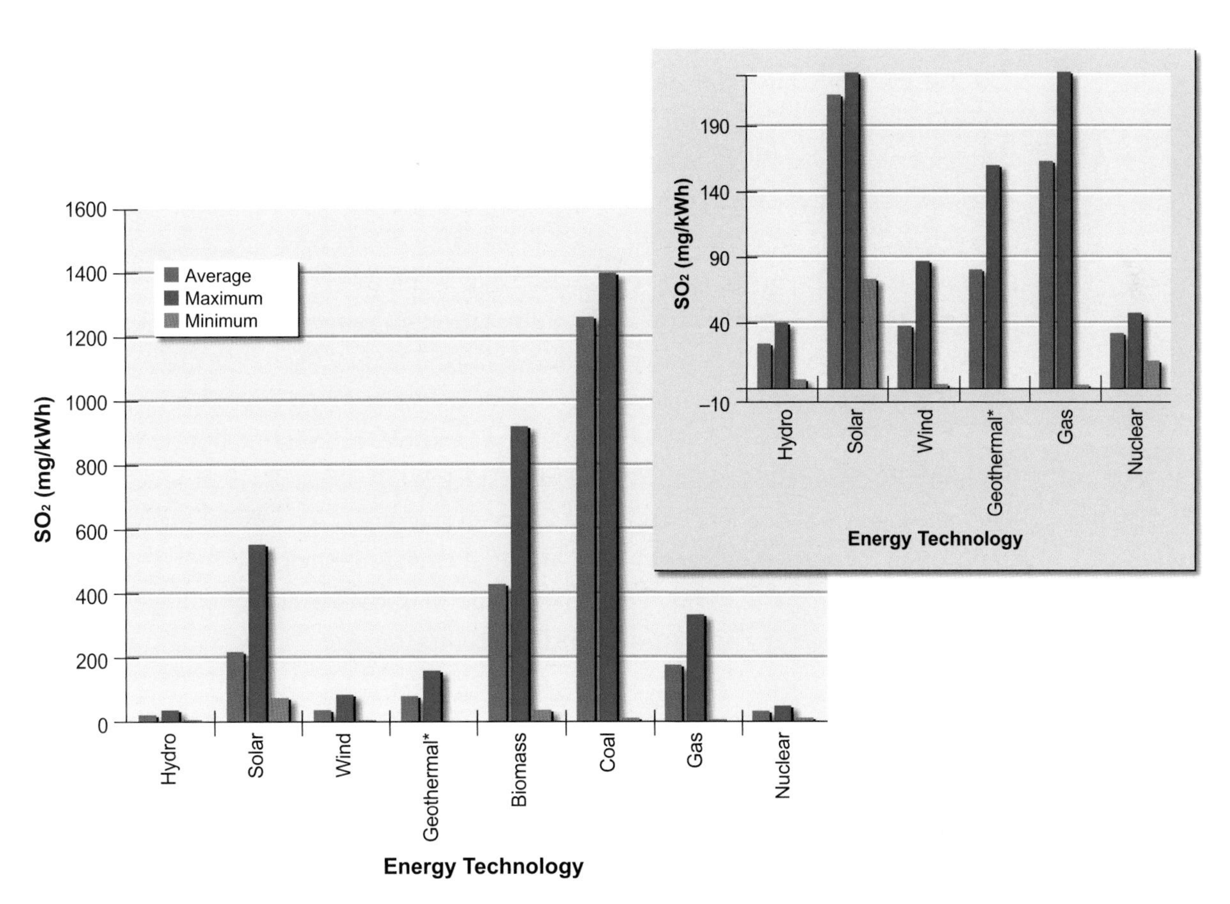

FIGURE 5.5 *Estimated life-cycle emissions of SO_2 in milligrams per kilowatt-hour for various renewable and non-renewable energy sources. No data on SO_2 emissions were found for tidal or energy storage technologies.*
Note: Asterisk indicates facility emissions only.
Source: Developed from data provided in Berry et al. (1998), Chataignere et al. (2003), Dones et al. (2005), European Commission (1997a,b,c,d), Frankl et al. (2004), Fthenakis et al. (2008), Green and Nix (2006), Mann and Spath (1997), Odeh and Cockerill (2008), Spath et al. (1999), Spath and Mann (2000), Spitzley and Keoleian (2005), and Vattenfall AB (2004, 2005).

Solar Photovoltaic

Rates of SO_2 emissions associated with electricity generation from PV are most affected by the energy intensity of the manufacturing process and the efficiency of the PV material, as well as the energy mix used to manufacture the PV material and the solar insolation at the site where the PV is installed. SO_2 emissions for PV installations in Europe range from 73 to 215 mg/kWh and include a range of PV technologies (single crystalline, multicrystalline, amorphous silicon, copper-indium-gallium-diselenide [CIGS] and CdTe) with conversion efficiencies of 6–14 percent, and insolation rates of 1700–1740 kWh m^2/yr over assumed lifetimes of 20–30 years. SO_2 emissions shown from studies in the United States have a wider range of values, from 158 to 540 mg/kWh. The high value of 540 mg SO_2/kWh is from an older U.S. PV installation with lower insolation rates and a greater reliance on coal for electricity generation compared to that of Europe (Spitzley and Keoleian, 2005). Fthenakis et al. (2008) compared 2004–2006 PV technologies for similar systems using the average U.S. and European inventory data and electricity mix. For the European cases, SO_2 emission values ranged from 73 to 146 mg/kWh, whereas for the U.S. cases the values ranged from 158 to 378 mg/kWh.

Interestingly, studies suggest that the most efficient PV material is not necessarily the best for minimizing emissions. For example, cadmium telluride (CdTe) technologies have the lowest conversion efficiencies (9 percent) yet produce lower SO_2 emissions because less energy is consumed during CdTe manufacturing than with other PV technologies that have higher conversion efficiencies (11.5–14 percent) (Fthenakis et al., 2008). This relationship may change as technology innovations decrease energy consumption during manufacturing.

Biopower

For biopower, reported values for SO_2 emissions range from 40 to 940 mg/kWh. Mann and Spath (1997) suggest that much of this variation arises from differences in power plant efficiency. The low end of the range, from 40 to 45 mg/kWh, is from two European studies cited by Mann and Spath (1997). The four remaining studies, with values ranging from 302 to 940 mg/kWh, are from the United States. Cases with results in the mid-range include two hypothetical integrated gasification combined cycle (IGCC) plants with different fuels. Both plants are based on models developed by the National Renewable Energy Laboratory (NREL). A plant using hybrid poplar as fuel has an estimated SO_2 emission rate of 302 mg/kWh (Mann and Spath, 1997), and a willow feedstock plant has an estimated SO_2 emission rate of 370 mg/kWh (Spitzley and Keoleian, 2005).[5] Mann and Spath (1997) include no special emission controls on combustion plants and assume that all SO_2 in the biomass is converted to emissions. The other U.S. examples include a direct-fired boiler and a high-pressure IGCC system, based on Electric Power Research Institute (EPRI) models, that emit SO_2 at rates of 930 and 940 mg/kWh, respectively (Spitzley and Keoleian, 2005). The base models developed by NREL and EPRI have very different emission profiles for plant combustion (Heller et al., 2004); the EPRI plant is assumed to emit approximately three times more SO_2 than is assumed for the NREL plant.

Geothermal

No LCA data were found that included SO_2 emissions for geothermal technologies. Data from Green and Nix (2006) show reservoir-only emissions ranging from 0 to 160 mg/kWh.

Emissions of Nitrogen Oxides

Figure 5.6 illustrates the range of life-cycle NO_x emissions estimated from various electrical generation technologies. Among these technologies, hydroelectric, wind, geothermal, and nuclear technologies have low estimated NO_x emissions (<100 mg/kWh) and are not discussed in detail in this section.

As a rule, energy sources based on combustion have significantly higher levels of NO_x emissions than do those that do not involve combustion. The NO_x produced from combustion arises from two sources: the oxidation and volatilization of the nitrogen contained in the fuel, and the high-temperature reactions involving atmospheric nitrogen and oxygen. The production of NO_x from atmospheric sources can be reduced or even completely eliminated by carrying out the combustion under high-oxygen conditions, so-called oxy-fuel combustion. Because of a lack of LCAs, the levels of NO_x emissions described here do not reflect the performance of these systems.

[5]Spitzley and Keoleian (2005) attributed incorrect SO_2 and NO_x emission data to the base model IGCC plants from Heller et al. (2004). SO_2 and NO_x emission results cited here have been corrected to be consistent with Heller et al. (2004).

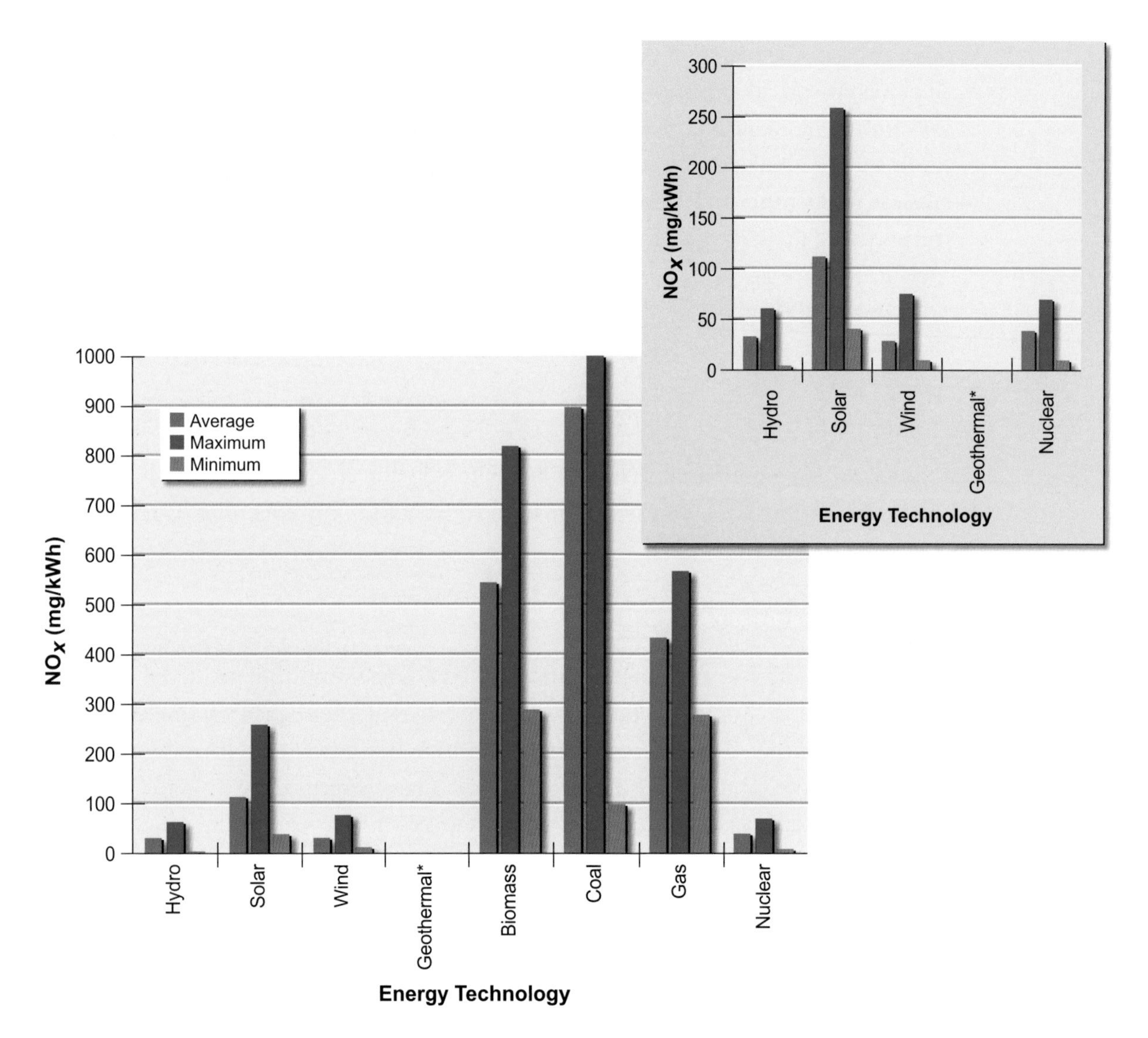

FIGURE 5.6 *Estimates of life-cycle emissions of NO_x from various technologies. No LCA data on emissions of NO_x were found for geothermal, tidal, or energy storage technologies.*
Note: Asterisk indicates facility emissions only.
Source: Developed from data provided in Berry et al. (1998), Chataignere et al. (2003), Dones et al. (2005), European Commission (1997a,b,c,d), Frankl et al. (2004), Fthenakis et al. (2008), Green and Nix (2006), Mann and Spath (1997), Odeh and Cockerill (2008), Spath et al. (1999), Spath and Mann (2000), Spitzley and Keoleian (2005), and Vattenfall AB (2004, 2005).

Solar

NO_x emission data for PV technologies range from 40 to 260 mg/kWh. This range largely reflects the differing mixes of grid energy used to produce the PV material as well as the conversion efficiencies and life expectancies of the PV facility. The high value of 260 mg/kWh is for an older U.S. PV installation with lower insolation; the greater reliance on coal for electricity generation in the United States as compared to Europe leads to greater life-cycle emissions in the United States (Spitzley and Keoleian, 2005). NO_x values from European studies ranged from 40 to 99 mg/kWh. A study by Fthenakis et al. (2008) demonstrates how the carbon intensity of the grid can affect emissions from PV technologies. They compared 2004–2006 PV technologies for similar systems using the average U.S. and European inventory data and electricity mix. For the European cases, NO_x values ranged from 40 to 82 mg/kWh, whereas for the U.S. cases reported values ranged from 79 to 188 mg/kWh.

Biopower

Of all the renewable electricity technologies, biopower can have the highest NO_x emissions, with estimates ranging from 290 to 820 mg/kWh. Mann and Spath (1997) found that NO_x emissions are most sensitive to variations in crop yield, feedstock fuel used, and power plant efficiency, and that most NO_x emissions in the biopower life cycle (about 70 percent) are from combustion. Whether the feedstock is a fossil fuel or is biomass, the amount of NO_x produced during combustion depends on the nitrogen content of the fuel and the temperature of combustion. The higher the temperature, the more NO_x is produced. As a result, production of electricity from biopower produces NO_x at rates comparable to that of fossil fuels.

Emissions of Particulate Matter

Figure 5.7 illustrates the range of estimated life-cycle emissions of particulate matter (PM) from various renewable and non-renewable energy sources. PM emissions tend to be low (<100 mg/kWh) for all the energy technologies considered here, with the exception of coal, natural gas, and PV. However, many LCAs do not report on emissions of PM.

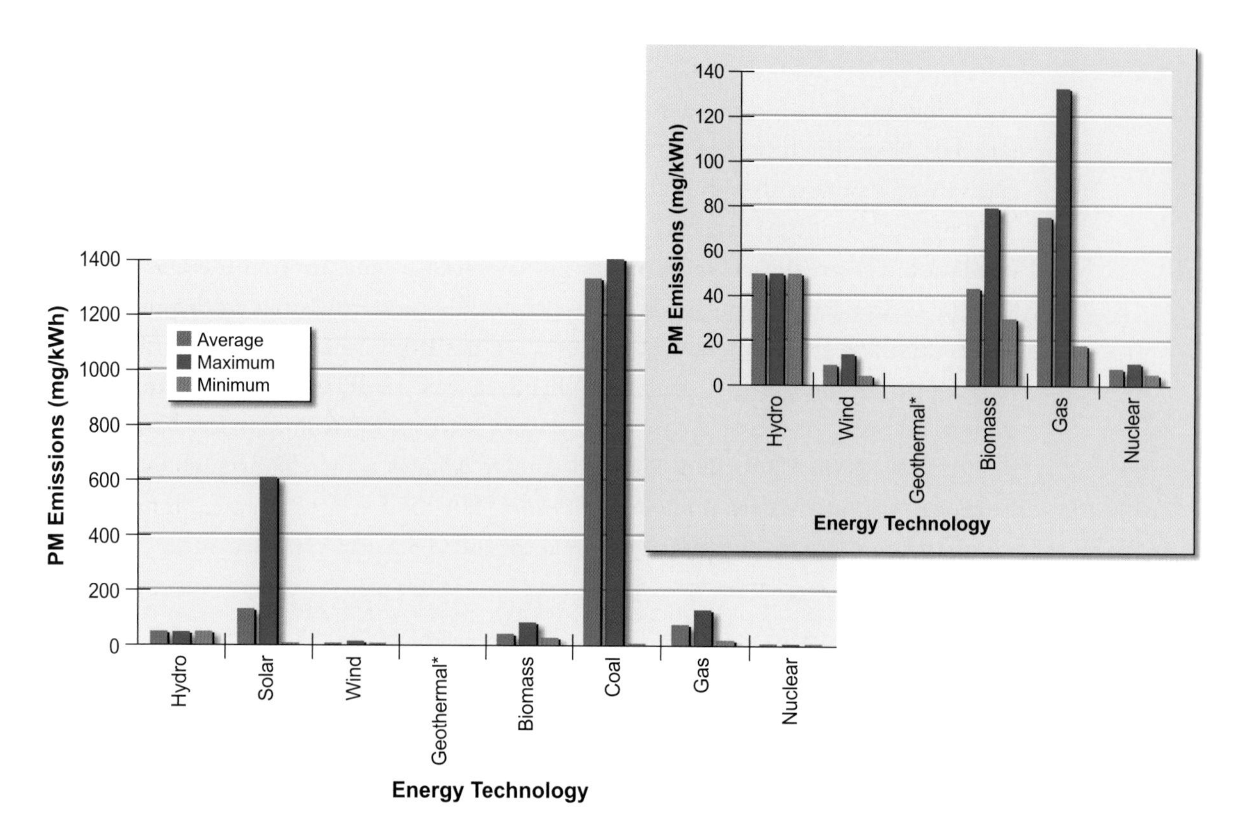

FIGURE 5.7 *Estimates of life-cycle particulate matter emissions for various electrical power generations technologies. No LCA data on emissions of particulate matter were found for geothermal, tidal, or energy storage technologies.*
Note: Asterisk indicates facility emissions only.
Source: Developed from data provided in Berry et al. (1998), Chataignere et al. (2003), Dones et al. (2005), European Commission (1997a,b,c,d), Frankl et al. (2004), Green and Nix (2006), Mann and Spath (1997), Odeh and Cockerill (2008), Spath et al. (1999), Spath and Mann (2000), and Spitzley and Keoleian (2005).

Solar Photovoltaic

Five LCAs for PM emissions from PV were found by the panel. Only one, a U.S. study, reported results higher than 100 mg/kWh: Spitzley and Keoleian (2005) reported a value for particulate matter of 610 mg/kWh for an older U.S. PV installation with lower insolation rates and a relatively large reliance on coal in electricity from the grid. The European data, on the other hand, showed emissions of PM ranging from 6 to 55 mg/kWh.

Land Use

Some have proposed that land use may be a limiting factor for the use of renewable energy technologies (Pimentel et al., 2002; Grant, 2003), supporting this argument with non-LCA land-use data based on calculations of power plant size and quantity of electricity generated. Other studies have focused on one aspect of an energy technology (e.g., reservoir size for hydropower) to derive a land-use estimate. These estimates of land use can be misleading because they fail to present an accurate understanding of the entire life-cycle land-use requirements of a technology. The LCA land-use data discussed here are from Spitzley and Keoleian (2005), whose land-use metric accounts for the total surface area occupied by the materials and products of an energy technology, including the time of land occupation over the total life-cycle energy generated. Figure 5.8 shows the results of this 2005

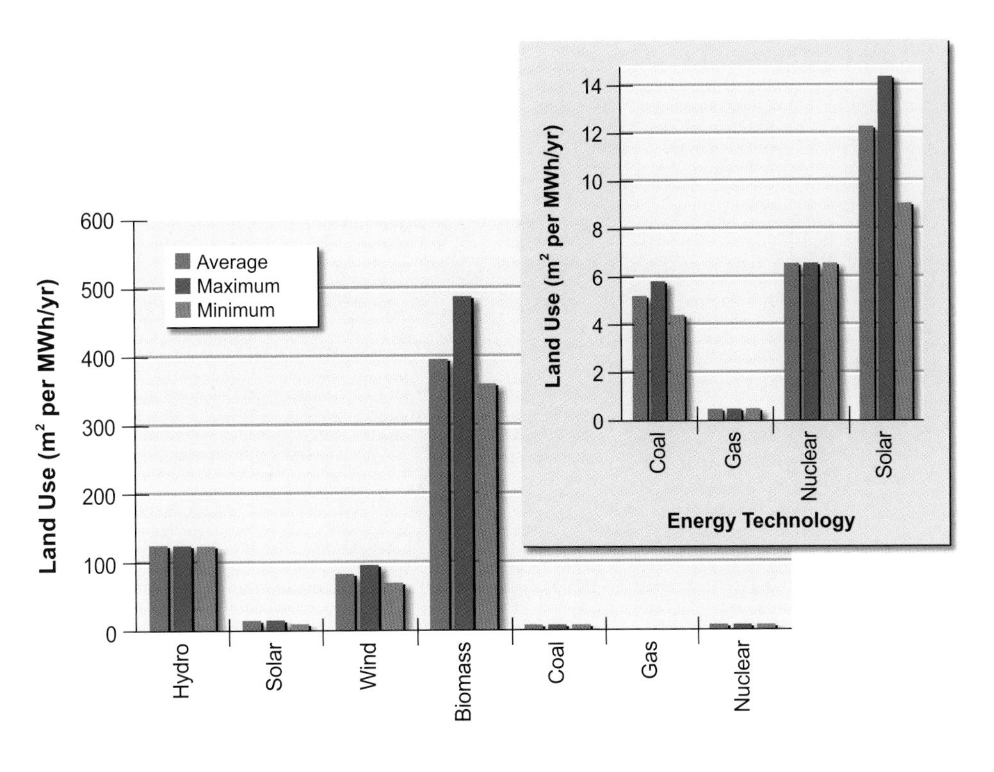

FIGURE 5.8 *Life-cycle cost assessment of land use for various renewable and non-renewable technologies in square meters per megawatt-hour per year. Note that the inset provides a smaller scale and more details for sources that are not distinguishable in the main figure.*
Source: Developed from data provided in Spitzley and Keoleian, 2005.

study's estimates of LCA land use for renewables and other electricity-generating technologies. Key assumptions in the Spitzley and Keoleian (2005) analysis are (1) exclusion of fuels and materials with insignificant land acquisition requirements compared to other life-cycle stages, and (2) inclusion of end-of-life land disposal requirements for nuclear fuel only. The Spitzley and Keoleian analysis does not allow for distinctions for intensity of land use.

A key factor affecting land use is the generating efficiency of the technology per unit area. By design, technologies using high energy density power sources use less land to produce more electricity at the point of generation than do the more diffuse renewable technologies. For this reason, analyses such as the ones cited here find that renewables have relatively large land-use requirements. To operate fossil-fuel and nuclear plants, however, the fuel must first be extracted or mined. Most LCAs, including those used in this study, do not account for that process in their assessment of land-use requirements. Moreover, the land used by some diffuse renewable electricity technologies usually allows for multiple uses, or the technology makes use of sites that also serve an alternate purpose (e.g., PV installations on roofs or sides of buildings, wind turbines on farms, and hydroelectric reservoirs that provide flood control, recreation, and water supply).

Figure 5.8 shows that studies found in the literature give natural gas, coal, and nuclear technologies low land-use values: 0.45 m^2/(MWh/yr), 4.4–5.8 m^2/(MWh/yr), and 6.5 m^2/(MWh/yr), respectively (without counting resource extraction). Of the renewable energy technologies, solar has the lowest land-use values, ranging from 9 to 14.3 m^2/(MWh/yr). The lowest estimated value for PV is for an installation in Phoenix where higher insolation rates yield more energy potential per unit area.

The two wind farms in the Spitzley and Keoleian (2005) study report land-use values of 69 and 94 m^2/(MWh/yr). The lower land-use value is from the wind farm with higher wind speed and reflects the greater power generation potential per unit area and per unit of equipment. Additionally, only about 1 percent of wind farm land is used by the turbines and associated facilities, thus allowing for multiple uses (e.g., grazing and agriculture).

Spitzley and Keoleian (2005) included one LCA example of hydroelectric power with a land-use value of 122 m^2/(MWh/yr) for a high-capacity, large-reservoir facility in the United States with a 50-year lifetime. The literature also contains a wide range of non-LCA land-use data for hydroelectric power. The range includes very small values for run-of-river hydroelectric facilities to very large values for low-capacity hydroelectric plants associated with very large reservoirs in developing nations.

Biopower is estimated to have the highest land-use requirements, with estimates ranging from 360 to 488 $m^2/(MWh/yr)$. The highest value is from a direct-fired boiler biopower facility with a small generating capacity. The other facilities all use IGCC and show very similar results of 360–375 $m^2/(MWh/yr)$. All four examples cited in the Spitzley and Keoleian (2005) study use cropping to supply biomass. Biopower from waste would be expected to have much lower land-use values.

Water Use

Water Use by Renewable Technologies

Although most renewable technologies use only a fraction of the water used by thermoelectric plants, some renewables, such as geothermal, hydroelectric, and solar thermal, can be water intensive. For example, flash geothermal plants consume reservoir water and require makeup water.[6] One plant in California uses 2,000 gal/MWh and requires 1,400 gal/MWh of makeup water (Table 5.1). This facility uses water recycled from a wastewater facility as an innovative makeup water source (DOE, 2006). Newer geothermal plant designs such as binary plants use little water. Hydroelectric power, which generates electricity from the kinetic energy of water itself, uses vast quantities of water. Evaporative loss from hydroelectric reservoirs has been estimated at 4,500 gal/MWh (DOE, 2006). CSP technologies can also be water intensive (see Table 5.1). Concentrated solar thermal power uses 770–920 gal/MWh, and solar power tower technologies use about 750 gal/MWh for evaporative cooling (DOE, 2006). Parabolic dish-engine solar technologies are air-cooled and use minimal water (DOE, 2006).

Energy technologies that withdraw and consume less water will have both public benefit and economic advantages in the marketplace moving forward. One option is to develop electricity from sources that use very little water, such as wind and PV. Other options include developing technologies that limit the use of water with fossil-fuel electricity sources or use alternate sources of water, such as reclaimed or saline water. For example, some power plants use mine water and/or

[6]Makeup water is the water added to the existing flow of cooling water to replace the water lost during passage through the cooling towers or other power plant processes.

TABLE 5.1 Water Use by Energy Technology

Technology		Consumption (Withdrawal) (gal/MWh)	
		DOE (2006)[a]	Feeley III et al. (2008)
Geothermal	Cooling tower	~1,400 (~2,000)	
Nuclear	Once through	~400 (25,000–60,000)	140 (31,500)
	Cooling tower	720 (800–1,100)	620 (1,100)
	Cooling pond	400–720 (500–1,100)	
Fossil/biopower	Once through	~300 (20,000–50,000)	60–140 (22,500–27,000)
	Cooling tower	480 (500–600)	460–520 (500–650)
	Cooling pond	300–480 (300–600)	4–800 (15,000–18,000)
NGCC	Once through	100 (7,500–20,000)	20 (9,000)
	Cooling tower	~180 (~230)	130 (150)
	Cooling pond		240 (6,000)
	Air cooled		4 (4)
IGCC, coal	Cooling tower	~200 (~380)	170 (230)
Concentrated solar power	Solar thermal	770–920 (770–920)	
	Power tower	760 (760)	
	Dish-engine	Minimal	
Hydroelectric		4,500 (reservoir evaporation)	

[a]DOE (2006) inaccurately reports water consumption for recirculating cooling system from EPRI (2002).

gray water from wastewater treatment plants. Other alternative sources of water include produced water from oil and gas operations and brackish groundwater aquifers. Developing alternate water sources requires careful consideration of economic and ecosystem impacts. For example, brackish groundwater requires additional conditioning to meet power plant water chemistry specifications. At the same time, groundwater withdrawals can affect freshwater aquifers and lead to saltwater intrusion. Relying on marine water has the same impacts for fish and other aquatic organisms as freshwater use.

Water Use by Thermoelectric Technologies

Generating electricity from energy technologies that rely on water to produce electricity from steam (thermoelectric generation) is very water-intensive. In recent years, concerns regarding water use have led to the denial of water permits for new power plant construction in various locations throughout the United States (Feeley III et al., 2008; DOE, 2006). In areas of the United States experiencing drought and population growth, thermoelectric power plants may be forced to reduce output as they compete with other users for a limited supply of water (Vinluan, 2007; MSNBC, 2008). The benefits of present and future water use by energy technologies must be carefully weighed in examining the implications of a finite water resource. Competition over water is intensifying, because water supports agricultural, industrial, and domestic needs, as well as the need for electricity.

Thermoelectric plants generate electricity using steam from a variety of fuel sources including fossil fuels, geothermal energy, concentrated solar power, and biopower. However, most thermoelectric power in the United States is generated from conventional fossil sources. Water use by thermoelectric power plants is categorized as water withdrawn or consumed. Thermoelectric power plants use large quantities in each category but withdraw more than is actually consumed. According to the U.S. Geological Survey (USGS, 2004), thermoelectric power plants withdrew about 136 billion gallons per day (billion gal/d) of freshwater for use and consumed about 3 billion gal/d of this amount. This accounted for about 40 percent of all freshwater withdrawals in the United States, and almost 15 percent of all non-agricultural consumption.

The USGS (2004) estimated that power plants withdrew an additional 59 billion gal/d of saline water, bringing the total daily water use in 2000 by power plants to 195 billion gal/d. However, Dziegielewski et al. (2006) argued that this was an underestimation. They reported that water-use data for thermoelectric power plants compiled by the USGS did not include water from public water supplies, nor was it clear whether water use by independent non-utility power plants was included. Independent non-utility power plants generated an additional 16 percent of electricity in 2000. On average, approximately 26 gallons of water is used to produce 1 kWh of electricity. Total per capita water withdrawals for electricity in 2000 amounted to 686 gallons per person per day, which is about four and a half times the direct per capita use (Table 5.2) (Dziegielewski et al., 2006).

Figure 5.9 illustrates the range of water withdrawal and consumption rates for a variety of technologies as compiled by DOE (2006). Power plants use water primarily for cooling, but significant quantities of water are also used in other plant activities. Because of this dependency, power plants have traditionally been sited near rivers, lakes, or oceans. Most of the water consumed by thermoelectric power plants is lost through evaporation. Cooling system options include once-through, recirculating, or air-cooled systems. Water use by thermoelectric technologies with these cooling options is shown in Table 5.2.

TABLE 5.2 Water Intensity of Thermoelectric Generators as Compared to Other Water Users

Category	Water Consumption (Withdrawal) (gal/d)
Average person U.S., total (indoor and outdoor), 2002	152
Average person Germany, total, 2002	51
Average person U.K., total, 2002	39
Washington, D.C., and area, 2005 (about 2 million people)	123.6 million
New York City, 2006 (about 8 million people)	1069 million
Average 500 MW coal plant, cooling tower	6.1 million (6.4–7.6 million)
Average 500 MW coal plant, once-through cooling	4 million (240–600 million)
Average 1 GW nuclear plant, cooling tower	17.3 million (19.9–27.1 million)
Average 1 GW nuclear plant, once-through cooling	10.3 million (600–1440 million)
Average 500 MW NGCC plant, cooling tower	2.2 million (2.8 million)

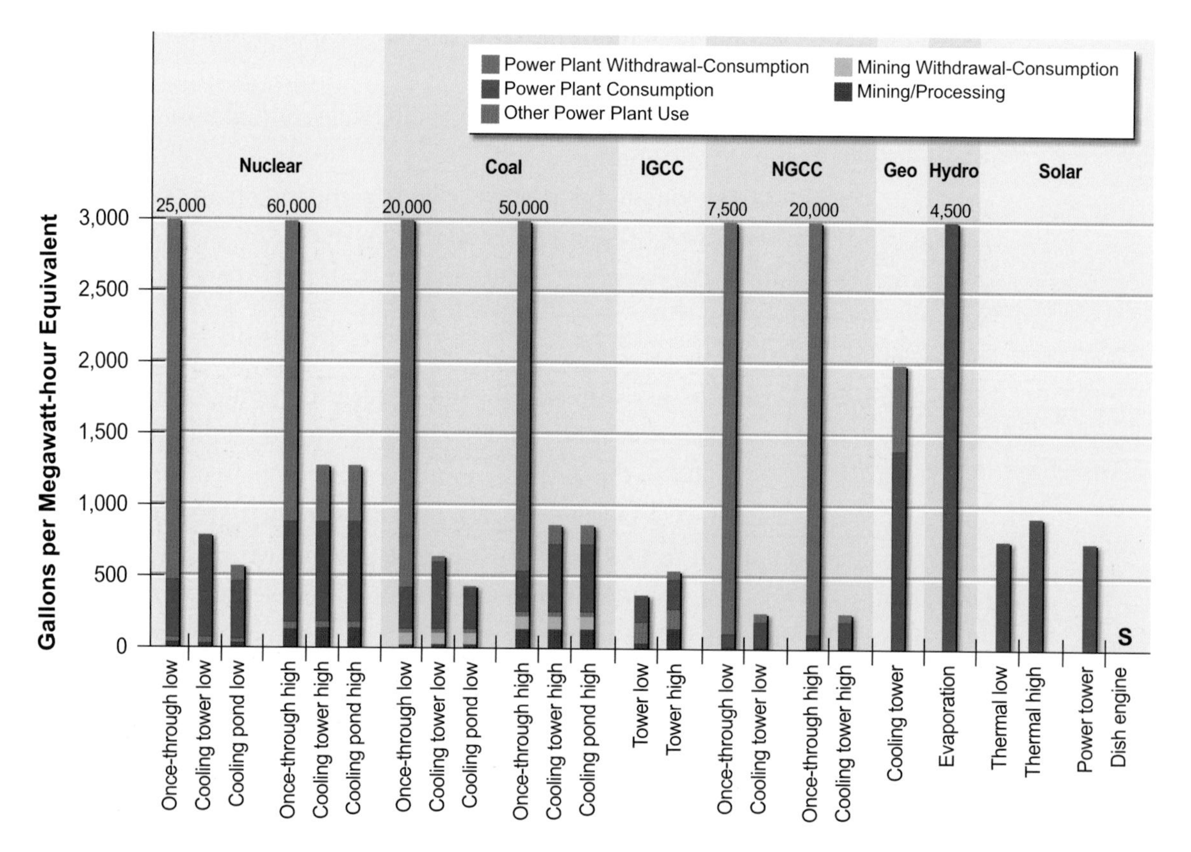

FIGURE 5.9 *Estimates of water withdrawal and consumption rates for various thermo-electricity generating technologies.*
Source: DOE, 2006.

LOCAL ENVIRONMENTAL IMPACTS—SITING AND PERMITTING

It is critical to consider siting and permitting issues. The permitting process addresses the wide range of localized impacts that might result from the construction of a renewables facility or related infrastructure, such as transmission lines. New renewable electric facilities can affect water supplies, ecosystems, and the natural landscape and hence can meet with local opposition. Though renewable facilities are obtaining permits and completing impact assessments, the knowledge of the full impacts of renewables and the guidance for permitting projects are nascent in comparison to those for fossil fuels. Further, renewable energy resources are generally more distributed than concentrated, especially those powering the technologies dominating the near term (wind power, solar PV, and CSP). As noted previously, renewables have relatively large land-use requirements. The process of siting and permitting these facilities has the potential to place burdens on local jurisdictions that regulate land use and create a hodgepodge of rules and requirements for renewable energy deployment.

Siting

Siting issues could be a significant concern with renewables. The NIMBY (not in my back yard) effect, which describes local opposition to a new development intended to distribute broad benefits, has delayed the construction of several major renewable energy projects in the United States. While proponents cite the environmental, economic, and energy security benefits to be gained from these projects, opponents cite the negative impacts, which often include potential damage to local ecosystems, loss of aesthetic value to the natural landscape, and the opportunity cost of land use. Biomass and biofuels, for example, require large amounts of land that could instead be used for agricultural purposes. Hydropower is becoming increasingly difficult to site; most major potential sites are already being used, and ecological considerations are preventing the exploitation of remaining ones. Siting renewable energy projects can also pit environmentalists against one another. In Cape Cod, Massachusetts, local residents who fear harm to aquatic life have fought the construction of 130 wind turbines; in southern California, advocates of solar power face resistance from environmental groups that fear potential disruption to the Mojave Desert ecosystem (Barringer, 2009). Local opposition has also stymied the development of new transmission lines (Silverstein, 2008). Review of siting issues occurs during the permitting process discussed below.

Permitting

Most if not all technologies for generating electricity will require multiple permits. These permits are intended to consider the local impacts on the land, water, and air that occur during the installation and operation of these technologies. Depending on the size and location of the generating facility, permits from local zoning boards, state agencies, and federal agencies may be required. In the case of traditional electricity-generating facilities, such as those that use coal and natural gas, there is a long and evolving permitting process that has been applied across the country. For most renewable technologies, the process is more in the developmental stage. As of January 2008, at least two states, California and Wisconsin, have enacted state laws preempting or limiting local siting jurisdiction for wind power (Green, 2008). Because wind power has been the most extensively deployed renewable electricity technology in recent times, guidance for permitting a wind power project is more advanced. A National Research Council (NRC) report contains a fairly extensive review of guidelines that have been developed for such projects (NRC, 2007). For biomass, geothermal, and solar, the guidance for permitting is less well developed. However, there are many current regulations that apply to all generating facilities. Table 5.3 summarizes some of the most important regulations that apply to all large electricity-generating facilities. Although an exhaustive review of local impacts and permitting issues is beyond the scope of this study, a short summary of permitting issues for wind, geothermal, and CSP is presented below.

Wind Power

Due to the increasing number of wind power projects, more information is being developed concerning the process for permitting them. The most prominent issues of concern are land use and the possible impacts on birds and bats. In addition, concerns have been raised about noise, aesthetics, and the use of herbicides to clear and maintain sites, particularly where endangered species are involved. Recent reports and references on permitting wind power projects include the American Wind Energy Association (AWEA) siting handbook, which presents information about regulatory and environmental issues associated with developing and siting wind energy projects in the United States (AWEA, 2008). The AWEA handbook covers the components of a typical wind power project: the stages of a wind power project; the federal, state, and local regulatory frameworks relevant for wind power; and the array of environmental and human impacts to consider

TABLE 5.3 Some of the Important Regulations That Apply to All Electricity Generating Facilities

Topic	Law: Statutory Citation	Regulation Name: Code of Federal Regulations Citation
Air Quality		
Clean Air Act conformity	Clean Air Act (CAA), Section 176(c)(1): P.L. 88-206, as amended; 42 USC 7401, *et seq.*	Determining Conformity of Federal Actions to State or Federal Implementation Plans: 40 CFR Part 51, Subpart W 40 CFR Part 93, Subpart B
Biota		
Eagles	Bald and Golden Eagle Protection Act: 16 USC 668a-668d, as amended	General Provisions (for taking, possession, etc. of wildlife and plants): 50 CFR Part 10 Eagle Permits: 50 CFR Part 22
Endangered and threatened species and critical habitats	Endangered Species Act (ESA), Section 7: P.L. 93-205, as amended; 16 USC 1536(a)-(d)	Interagency Cooperation: 50 CFR Part 402 Endangered Species Exemption Process: 50 CFR Parts 450–453
Essential fish habitat	Magnuson–Stevens Fishery Conservation and Management Act (also known as the Sustainable Fisheries Act): P.L. 104-297 (major amendments to P.L. 94-265); 16 USC 1801, *et seq.*	Magnuson–Stevens Act Provisions: 50 CFR Part 600 The following subparts deal with Essential Fish Habitat (EFH): Table of contents; Subpart A–General (purpose and scope, definitions, other acronyms); Subpart D–National Standards, section 600.305 General, particularly subsection 600.305(c); Subpart J–Essential Fish Habitat; Subpart K–EFH coordination, consultation, and recommendations
Fish and wildlife	Fish and Wildlife Coordination Act (FWCA): P.L. 85-624, as amended; 16 USC 661, *et seq.*	No implementing regulations

continued

TABLE 5.3 Continued

Topic	Law: Statutory Citation	Regulation Name: Code of Federal Regulations Citation
Marine mammals	Marine Mammal Protection Act (MMPA), Sections 103 and 104: P.L. 92-522; 16 USC 1361-1407	Fish and Wildlife Service General Permit Procedures: 50 CFR Part 13 Marine Mammals: 50 CFR Part 18 National Marine Fisheries Service Regulations Governing the Taking and Importing of Marine Mammals: 50 CFR Part 216
Migratory birds	Migratory Bird Treaty Act (MBTA): 16 USC 703-712	General Provisions (for taking, possession, etc. of wildlife and plants): 50 CFR Part 10 Migratory Bird Permits: 50 CFR Part 21
Cultural Resources		
Archaeological resources	Archaeological Resources Protection Act (ARPA): P.L. 96-95; 16 USC 470aa–mm	Protection of Archaeological Resources: 43 CFR Part 7
Historic resources	National Historic Preservation Act (NHPA), Sections 106 and 110: P.L. 89-665, as amended; 16 USC 470	Protection of Historic Properties: 36 CFR Part 800
Native American graves	Native American Graves Protection and Repatriation Act (NAGPRA): P.L. 101-601; 25 USC 3001, *et seq.*	Native American Graves Protection and Repatriation Regulations: 43 CFR Part 10
Native American religions	American Indian Religious Freedom Act (AIRFA): P.L. 95-341; 42 USC 1996 and 1996a	No implementing regulations
Land Use and Special Land and Water Designations		
Coastal zone areas	Coastal Zone Management Act (CZMA): P.L. 92-583, as amended; 16 USC 1451, *et seq.*	Federal Consistency with Approved Coastal Management Programs: 15 CFR Part 930
Farmland	Farmland Protection Policy Act (FPPA): P.L. 97-98, as amended; 7 USC 4201, *et seq.*	Prime and Unique Farmlands: 7 CFR Part 657 7 CFR Part 658

TABLE 5.3 Continued

Topic	Law: Statutory Citation	Regulation Name: Code of Federal Regulations Citation
National marine sanctuaries	Marine Protection, Research, and Sanctuaries Act (MPRSA), Title III: P.L. 92-532, as amended; 16 USC 1431-1445	National Marine Sanctuary Program Regulations: 15 CFR Part 922
National natural landmarks	Historic Sites Act: P.L. 74-292, as amended; 16 USC 461-467	National Natural Landmarks Program: 36 CFR Part 62
Wild and scenic rivers	Wild and Scenic Rivers Act (WSRA), Sections 5 and 7: P.L. 90-524, as amended; 16 USC 1271-1287	Wild and Scenic Rivers: 36 CFR Part 297
Wilderness areas	Wilderness Act: P.L. 88-577; 16 USC 1131-1133	Forest Service Regulations: Prohibitions: 36 CFR Part 261 Special Areas: 36 CFR Part 294

Note: Contents are listed by environmental topic.
Source: U.S. Department of Energy; see http://www.eh.doe.gov/nepa/tools/guidance/volume3/laws_regulations_table.html.

when siting wind power. An example of a state handbook on wind power permitting is the guidance developed by the Kansas Energy Council for siting wind power projects in that state (Kansas Energy Council, 2005). In terms of impacts on wildlife, the New York State Department of Environmental Conservation recently proposed guidelines on how to characterize bird and bat resources at onshore wind energy sites and how to estimate and document impacts (New York Department of Environmental Conservation, 2008). Although bird deaths are often characterized as one critical potential impact from wind turbines, the NRC (2007) study cited above concluded that, while the impacts on bat populations were unclear, there was no evidence that bird fatalities caused by wind turbines result in measurable demographic changes to bird populations in the United States (NRC, 2007).

One purpose of the NRC (2007) study was to develop an analytical framework for impact evaluation to inform siting decisions for wind-energy projects. The study organized impacts assessment into a three-dimensional action space that includes the relevant spatial jurisdictions (local, state/regional, and federal), project stage (pre-project, construction, operational, and post-operational), and environmental and human impacts (NRC, 2007). The NRC (2007) study found that because wind energy is new to many state and local governments, the quality of the permitting process is uneven, and it pointed out that a coordinated and consistent process would greatly aid planning and regulating wind-energy development at smaller scales. The report recommended that representatives of federal, state, and local governments work with wind developers and interested parties to develop guidance and permitting guidelines (NRC, 2007).

In order to better assess possible wildlife impacts of wind power, Secretary of the Interior Dirk Kempthorne in 2007 announced the creation of the Wind Turbine Guidelines Advisory Committee, which will function in accordance with the Federal Advisory Committee Act (FACA). The scope and objective of the committee, as outlined in its charter, is to provide advice and recommendations to the Interior secretary on developing effective measures to avoid or minimize impacts on wildlife and habitats related to land-based wind energy facilities. The committee members represent the varied interests associated with wind energy development and wildlife management.[7]

Another group that will address fauna issues is the recently formed American Wind Wildlife Institute, created through cooperation between members of the environmental community and the wind industry. The institute will focus on efforts to facilitate timely and responsible development of wind energy while protecting wildlife and wildlife habitat. It will do this through research, mapping, mitigation, and public education on best practices in wind farm siting and wildlife-habitat protection.

Geothermal

Because of the long history of geothermal (hydrothermal) projects in the western United States, there is a mature record of the permitting of these plants. Battocletti (2005) provides an overview of the geothermal permitting process. Most federal statutes listed in Table 5.3 that apply to geothermal development are similar to the statutes for fossil-fuel plants. Because much of the geothermal resources occur on lands managed by the U.S. Bureau of Land Management (BLM), the agency

---

[7]For additional information on the activities of the Wind Turbine Guidelines Advisory Committee, see http://www.fws.gov/habitatconservation/windpower/wind_turbine_advisory_committee.html.

has developed a set of reporting and permitting requirements that includes a notice of intent to conduct geothermal explorations, a geothermal drilling permit, and a monthly report of operations. California has its own geothermal permitting requirements, which are issued by the California Energy Commission (CEC, 2007).

Concentrating Solar Power

Permitting for CSP plants falls under general requirements for utility-scale solar projects. At present, CSP plants are the only utility-scale solar plants that have been built. Figure 2.3 shows that the U.S. resource base for CSP is located in the Southwest. As with geothermal energy, BLM manages much of this land and must issue permits for these plants. Because of concerns about the potentially large land resources needed for CSP projects, the BLM recently announced that it would produce a programmatic environmental impact statement to evaluate the environmental, social, and economic impacts associated with the 125 applications for solar energy development on BLM-managed public land (BLM, 2008). The announcement also called for a moratorium on the acceptance of any new applications for CSP development on BLM lands, but this policy was rescinded. In California, CSP plants greater than 50 MW in size require approvals from both the BLM and the CEC. To provide joint National Environmental Protection Act and California Environmental Quality Act review and a more efficient process, the BLM and the California Energy Commission have entered into a memorandum of understanding that contains projects of joint jurisdiction and provides a timeline for the joint review process.

Hydrokinetic (Tidal/Wave)

Hydrokinetic technologies are still very much in the pilot/demonstration stage. However, the Department of Energy, in conjunction with the Departments of Commerce and the Interior, is studying these issues at the request of the U.S. Congress[8] and will issue a report that is scheduled for publication in July 2009. The goal of that report is to address the potential effects of marine and hydrokinetic energy projects, options to prevent adverse impacts, and potential roles and components for environmental monitoring and adaptive management. For the pur-

---

[8]See http://www.ornl.gov/sci/eere/EISAReport/index.html.

poses of the report, the term "marine and hydrokinetic renewable energy" refers to electrical energy that comes from a wide range of sources, including waves, tides, and currents in oceans, estuaries, and tidal areas; free-flowing water in rivers, lakes, and streams; free-flowing water in man-made channels; and differentials in ocean temperature (ocean thermal energy conversion). The report will not address energy from impoundments or other diversionary structures. Given the scarcity of real operational data, the report will not constitute a definitive impact assessment but will highlight areas of potential concern and areas of research and monitoring necessary to gain needed data.

Other Localized Environmental Impacts

A wide spectrum of other environmental impacts are not addressed by LCA and are not discussed here. They are nevertheless of potential importance and, in some particular locations, can include the impact that raises the greatest concern on the part of local populations and regulators. For example, all large power plants and transmission corridors require large tracts of land that must be kept at least partially clear of unwanted vegetation to maintain security, operational performance, and access for maintenance. To the extent that herbicides are used to clear and maintain areas for such sites, localized impacts will occur. Other technology-specific impacts associated with the use of renewable sources of electricity include the following:

- *Hydroelectric*— Ecosystem changes including impacts on fish migration and mortality, habitat damage, degradation of water quality, and loss of sediment transport to delta systems (Goodwin et al., 2006; ORNL, 1993).
- *Solar (PV)*—Mobilization of trace metals (Fthenakis, 2004; EPRI, 2003).
- *Wind*—Potential climatic and meteorological perturbations, especially in the vicinity of large wind farms; noise pollution; aesthetic impacts; and bird and bat deaths (Keith et al., 2004; NRC, 2007; Morrison and Sinclair, 2004; GAO, 2005).
- *Biopower*—Ground and surface water pollution from fertilizers and depletion of water for irrigation (cultivated biomass); removal of organic material from soil (waste biomass) (Marland and Obersteiner, 2008).

- *Geothermal*—Metals (arsenic) and gas (H_2S) from power plant operations, groundwater and surface water pollution, and potential for land subsidence and induced seismicity (DiPippo, 2007).

FINDINGS

Shown below in bold text are the most critical elements of the panel's findings, based on its consideration of environmental impacts associated with generation of renewable electricity.

Energy is essential for modern life as we know it, and all energy use implies environmental impacts upstream of the point at which work is done. These impacts range widely in locus, intensity, and significance depending on the primary source of the energy and means used to deliver and convert it into useful work. Today's electric generation and delivery system already imposes significant impacts on the environment at the local, regional, and global levels.

Understanding of these impacts has improved with advancements in environmental sciences and in analytical processes used to assess present and future environmental impacts. These improvements in the ability to understand environment impacts, including advances made in life-cycle assessment, have advanced society's ability to improve the overall efficiency of energy resource decisions by improving the metrics that allow the comparison of impacts and potentially internalizing previously externalized costs.

Armed with better analytical tools and a greater appreciation of the systematic and long-term impacts of energy resource decisions, the basic question we face regarding environmental impacts, then, is the extent to which the continuation of impacts is acceptable to society, and more importantly, how the evaluation and consideration of potential environmental impacts should influence the policy that affects energy resource decisions. This panel's high-level assessment leads to a number of important conclusions when considering scenarios involving significant increases in the deployment of renewable energy.

Renewable electricity technologies have inherently low life-cycle CO_2 emissions as compared to fossil-fuel-based electricity production, with most emissions occurring during manufacturing and deployment. Renewable electricity generation also involves inherently low or zero direct emissions of other regulated atmospheric pollutants, such as sulfur dioxide, nitrogen oxides, and mercury. Biopower is an exception because it produces NO_x emissions at levels similar to those associated with fossil-fuel power plants. **Renewable electricity technologies (except biopower, high-temperature concentrated solar power, and some geothermal technologies) also consume significantly less water and have much smaller impacts on water quality than do nuclear, natural gas-, and coal-fired electricity generation technologies.**

Because of the diffuse nature of renewable resources, the systems needed to capture energy and generate electricity (i.e., wind turbines and solar panels and concentrating systems) must be installed over large collection areas. Land is also required for the transmission lines needed to connect this generated power to the electricity system. But because of low levels of direct atmospheric emissions and water use, land-use impacts tend to remain localized and do not spread beyond the land areas directly used for deployment, especially at low levels of renewable electricity penetration. Moreover, some land that is affected by renewable technologies can also be used for other purposes, such as the use of land between wind turbines for agriculture.

However, at a high level of renewable technologies deployment, land-use and other local impacts would become quite important. The land-use impacts have caused, and will in the future cause, instances of local opposition to the siting of renewable electricity-generating facilities and associated transmissions lines. State and local government entities typically have primary jurisdiction over the local deployment of electricity generation, transmission, and distribution facilities. Significant increases in the deployment of renewable electricity facilities will thus entail concomitant increases in the highly specific, administratively complex, environmental impact and siting review processes. While this situation is not unique to renewable electricity, nevertheless, a significant acceleration of its deployment will require some level of coordination and standardization of siting and impact assessment processes.

REFERENCES

AWEA (American Wind Energy Association). 2008. Wind Energy Siting Handbook. Washington, D.C.

Barringer, F. 2009. Environmentalists in a clash of goals. The New York Times, March 24.

Battocletti, L. 2005. An Introduction to Geothermal Permitting. Alexandria, Va.: Bob Lawrence and Associates, Inc.

Berry, J.E., M.R. Holland, P.R. Watkiss, R. Boyd, and W. Stephenson. 1998. Power Generation and the Environment—A U.K. Perspective. Brussels: European Commission. June.

Bertani, R., and I. Thain. 2002. Geothermal power generating plant CO_2 emission survey. IGA News 49:1-3.

BLM (U.S. Bureau of Land Management). 2008. BLM initiates environmental analysis of solar energy development. Press release, May 29, 2008. Washington, D.C.

Bloomfield, K.K., J.N. Moore, and R.M. Neilson, Jr. 2003. Geothermal energy reduces greenhouse gases. Geothermal Research Council Bulletin (March/April):77-79.

Carbon Trust. 2008. Life Cycle Energy and Emissions of Marine Energy Devices. Available at http://carbontrust.co.uk.

CEC (California Energy Commission). 2007. Geothermal Permitting Process. Sacramento, Calif.

Chataignere, A., and D. Le Boulch. 2003. Wind Turbine (WT) Systems. Final Report. ECLIPSE (Environmental and Ecological Life Cycle Inventories for present and future Power Systems in Europe). Brussels: European Commission. November.

Denholm, P.L. 2004. Environmental and Policy Analysis of Renewable Energy Enabling Technologies. Ph.D. dissertation. University of Wisconsin, Madison, Wis.

Denholm, P., and G. Kulcinski. 2003. Net Energy Balance and Greenhouse Gas Emissions from Renewable Energy Storage System. ECW Report Number 223-1. Madison, Wis.: Energy Center of Wisconsin. June.

DiPippo, R. 2007. Geothermal Power Plants: Principles, Applications, Case Studies and Environmental Impact. Edition 2. Oxford, U.K.: Butterworth-Heinemann.

DOE (U.S. Department of Energy). 2006. Energy Demands on Water Resources: Report to Congress on the Interdependency of Energy and Water. Washington, D.C. December.

DOE. 2007. Estimating Freshwater Needs to Meet Future Thermoelectric Generation Requirements: 2007 Update. DOE/NETL-400/207/1304. Washington, D.C. September 24.

Dones, R., T. Heck, M.F. Emmenegger, and N. Jungbluth. 2005. Life cycle inventories for the nuclear and natural gas energy systems, and examples of uncertainty analysis. International Journal of Life Cycle Assessments 10:10-23.

Dziegielewski, B., T. Bik, U. Alqalawi, S. Mubako, N. Eidem, and S. Bloom. 2006. Water Use Benchmarks for Thermoelectric Power Generation. Research Report of the Department of Geography and Environmental Resources. Carbondale, Ill.: Southern Illinois University. August 15.

Ecobilan. 2001. TEAM/DEAM. Sustainable Business Solutions. Bethesda, Md.: PricewaterhouseCoopers.

EPRI (Electric Power Research Institute). 2002. Water and Sustainability (Volume 3): Water Consumption for Power Production—The Next Half Century. Palo Alto, Calif.

EPRI. 2003. Potential Health and Environmental Impacts Associated with the Manufacture and Use of Photovoltaic Cells: Final Report. No.1000095. Palo Alto, Calif.

ETSO (European Transmission System Operators). 1999. Available at Carbon Trust website.

European Commission. 1997a. External Costs of Electricity Generation in Greece. Brussels.

European Commission. 1997b. ExternE National Implementation Denmark. Brussels.

European Commission. 1997c. ExternE National Implementation France. Brussels.

European Commission. 1997d. ExternE National Implementation Germany. Brussels.

Fargione, J., J. Hill, D. Tilman, S. Polasky, and P. Hawthorne. 2008. Land clearing and the biofuel carbon debt. Science 319:1235-1238.

Fearnside, P.M. 1995. Hydroelectric dams in the Brazilian Amazon as sources of "greenhouse" gases. Environmental Conservation 22:7-19.

Fearnside, P.M. 2002. Greenhouse gas emissions from a hydroelectric reservoir (Brazil's Tucurui Dam) and the energy policy implications. Water, Air, and Soil Pollution 133:69-96.

Feeley III, T.J., T.J. Skoneb, G.J. Stiegel, Jr., A. McNemar, M. Nemeth, B. Schimmoller, J.T. Murphy, and L. Manfredo. 2008. Water: A critical resource in the thermoelectric power industry. Energy 33:1-11.

Frankl, P., Corrado, A., and S. Lombardelli. 2004. Photovoltaic (PV) Systems. Final Report. ECLIPSE (Environmental and Ecological Life Cycle Inventories for present and future Power Systems in Europe). Brussels: European Commission. January.

Franklin Associates. Life cycle services. Available at http://www.fal.com/lifecycle.htm.

Fthenakis, V.M. 2004. Life cycle impact analysis of cadmium in CdTe PV production. Renewable and Sustainable Energy Reviews 8:303-334.

Fthenakis, V.M., and H.C. Kim. 2007. Greenhouse-gas emissions from solar-electric and nuclear power: A life cycle study. Energy Policy 35:2549-2557.

Fthenakis, V.M., H.C. Kim, and E. Alsema. 2008. Emissions from photovoltaic life cycles. Environmental Science and Technology 44:2168-2174.

Gagnon, L., and J. van de Vate. 1997. Greenhouse gas emissions from hydropower: The state of research in 1996. Energy Policy 25:7-13.

GAO (General Accountability Office). 2005. Wind Power: Impacts on Wildlife and Government Responsibilities for Regulating Development and Protecting Wildlife. GAO-05-906. Washington, D.C.

Goodwin, P., K. Jorde, C. Meier, and O. Parra. 2006. Minimizing environmental impacts of hydropower development: Transferring lessons from past projects to a proposed strategy for Chile. Journal of Hydroinformatics 8:253-270.

Grant, P.M. 2003. Hydrogen lifts off—with a heavy load. Nature 424:129-130.

Green, B.D., and R.G. Nix. 2006. Geothermal—The Energy Under Our Feet: Geothermal Resource Estimates for the United States. NREL/TP-840-40665. Golden, Colo.: National Renewable Energy Laboratory. November.

Green, J. 2008. Overview: Zoning for small wind turbines. Presentation for ASES Small Wind Division Webinar, January 17, 2008. Available at http://www.windpowering america.gov/pdfs/workshops/2008/sw_zoning_overview.pdf.

Heller, M.C., G.A. Keoleian, M.K. Mann, and T.A.Volk. 2004. Life cycle energy and environmental benefits of generating electricity from willow biomass. Renewable Energy 29:1023-1042.

Heller, M.C., G.A. Keoleian, and T.A. Volk. 2003. Life cycle assessment of a willow bioenergy cropping system. Biomass and Bioenergy 25:147-165.

Hendriks, C. 1994. Carbon Dioxide Removal from Coal-Fired Power Plants. Dordrecht, Netherlands: Kluwer Academic Publishers.

Hondo, H. 2005. Life cycle GHG emission analysis of power generation systems: Japanese case. Energy 30:2042-2056.

Kansas Energy Council. 2005. Wind Energy Siting Handbook: Guideline Options for Kansas Cities and Counties. Topeka, Kans.

Keith, D.W., J.F. DeCarolis, D.C. Denkenberger, D.H. Lenschow, S.L. Malyshev, S. Pacala, and P.J. Rasch. 2004. The influence of large-scale wind power on global climate. Proceedings of the National Academies of Sciences USA 101(46):16115-16120.

Keoleian, G.A., and G.M. Lewis. 2003. Modeling the life cycle energy and environmental performance of amorphous silicon BIPV roofing in the U.S. Renewable Energy 28:271-293.

Laws, E.A. 2000. Aquatic Pollution: An Introductory Text. New York: Wiley Interscience.

Lochbaum, D. 2007. Got water? Issue brief. Cambridge, Mass.: Union of Concerned Scientists. December 4.

Mann, M., and P. Spath. 1997. Life Cycle Assessment of a Biomass Gasification Combined-Cycle System. NREL/TP-430-23076. Golden, Colo.: National Renewable Energy Laboratory.

Marland, G., and M. Obersteiner. 2008. Large-scale biomass for energy, with considerations and cautions: an editorial comment. Climatic Change 87:335-342.

Meier, P. 2002. Life cycle Assessment of Electricity Generation Systems and Applications for Climate Change Policy Analysis. Ph.D. dissertation, University of Wisconsin, Madison, Wis.

Morrison, M.L., and K. Sinclair. 2004. Wind energy technology, environmental impacts of. Pp. 435-448 in Encyclopedia of Energy. Volume 6. St. Louis: Elsevier.

MSNBC. 2008. Drought could shut down nuclear power plants—Southeast water shortage a factor in huge cooling requirements. January 23, 2008. Available at http://www.msnbc.msn.com/id/22804065/.

Mudd, G.M., and M. Diesendorf. 2008. Sustainability of uranium mining and milling: Toward quantifying resources and eco-efficiency. Environmental Science and Technology 42:2624-2630.

New York Department of Environmental Conservation. 2008. Guidelines for Conduct of Bird and Bat Studies at Commercial Wind Energy Projects. Albany, N.Y.

NRC (National Research Council). 2007. Environmental Impacts of Wind-Related Projects. Washington D.C.: The National Academies Press.

Odeh, N.A., and T.T. Cockerill. 2008. Life cycle GHG assessment of fossil fuel power plants with carbon capture and storage. Energy Policy 38:367-380.

ORNL (Oak Ridge National Laboratory). 1993. Hydropower. ORNL Review 26(3&4). Oak Ridge, Tenn. Available at http://www.ornl.gov/info/ornlreview/rev26-34/text/hydmain.html.

Pacca, S., and A. Horvath. 2002. Greenhouse gas emissions from building and operating electric power plants in the Upper Colorado River Basin. Environmental Science and Technology 36:3194-3200.

Pacca, S., D. Sivaraman, and G.A. Keoleian. 2007. Parameters affecting the life cycle performance of PV technologies and systems. Energy Policy 35:3316-3326.

Pimentel, D., M. Herz, M. Glickstein, M. Zimmerman, R. Allen, K. Becker, J. Evans, B. Hussain, R. Sarsfeld, A. Grosfeld, and T. Seidel. 2002. Renewable energy: Current and potential issues. BioScience 52:1111-1120.

Schleisner, L. 2000. Life cycle assessment of a wind farm and related externalities. Renewable Energy 20:279-288.

Searchinger, T., R. Heimlich, R.A. Houghton, F. Dong, A. Elobeid, J. Fabiosa, S. Tokgoz, D. Hayes, and T. Yu. 2008. Use of U.S. croplands for biofuels increases greenhouse gases through emissions from land-use change. Science 319:1238-1240.

Serchuk, A. 2000. The Environmental Imperative for Renewable Energy: An Update. Renewable Energy Policy Project. Washington, D.C.

Silverstein, K. 2008. Transmission developers jolted. EnergyCentral.com, January 14, 2008. Available at http://www.energycentral.com/centers/energybiz/ebi_detail.cfm?id=447.

Spath, P., and M. Mann. 2000. Life Cycle Assessment of a Natural Gas Combined-Cycle Power Generation System. NREL/TP-570-27715. Golden, Colo.: National Renewable Energy Laboratory. September.

Spath, P, and M. Mann. 2004. Biomass Power and Conventional Fossil Systems with and without CO_2 Sequestration—Comparing the Energy Balance, Greenhouse Gas Emissions and Economics. NREL/TP-510-32575. Golden, Colo.: National Renewable Energy Laboratory. January.

Spath, P., M. Mann, and D. Kerr. 1999. Life Cycle Assessment of Coal-fired Power Production. NREL/TP-570-25119. Golden, Colo.: National Renewable Energy Laboratory. June.

Spitzley D., and G.A. Keoleian. 2005. Life Cycle Environmental and Economic Assessment of Willow Biomass Electricity: A Comparison with Other Renewable and Non-Renewable Sources. Report CSS04-05R (March 2004, revised February 10, 2005). Center for Sustainable Systems, University of Michigan, Ann Arbor, Mich.

Storm van Leeuwen, J.W. 2008. Nuclear power—The energy balance, energy insecurity, and greenhouse gases. An updated version of "Nuclear power—The energy balance" by J.W. Storm van Leeuwen and P. Smith, published in 2002. Available at http://www.stormsmith.nl/.

USGS (United States Geological Survey). 2004. Estimated Use of Water in the United States in 2000. USGS Circular 1268. Available at http://pubs.usgs.gov/circ/2004/circ1268/pdf/circular1268.pdf.

Vattenfall AB. 2004. Certified Environmental Product Declaration of Electricity from Vattenfall's Nordic Hydropower. Registration No. S-P-00088. Vattenfall AB Generation Nordic. Stockholm. February. Available at http://www.environdec.com/reg/088/.

Vattenfall AB. 2005. Certified Environmental Product Declaration of Electricity from Forsmark Nuclear Power Plant. Registration No. S-P-00021. Vattenfall AB Generation Nordic. Stockholm. June. Available at http://www.environdec.com/reg/021/.

Viel, J.A. 2007. Use of Reclaimed Water for Power Plant Cooling. Report ANL/EVS/R-07/3. Environmental Science Division. Argonne, Ill.: Argonne National Laboratory.

Vinluan, F. 2007. Drought could force shutdown of nuclear, coal plants. Triangle Business Journal, November 23.

White, S. 1998. Net Energy Payback and CO_2 Emissions from Helium-3 Fusion and Wind Electrical Power Plants. Ph.D. dissertation. UWFDM-1093. Fusion Technology Institute, University of Wisconsin, Madison, Wis.

White, S. 2006. Net energy payback and CO_2 emissions from three midwestern wind farms: An Update. Natural Resources Research 15:271-281.

ANNEX

TABLE 5.A.1 Estimates of Life-Cycle Greenhouse Gas Emissions in CO_2 Equivalent (g/kWh) for Electricity Generation Technologies

Technology	CO_2	Notes
Solar		
	39	Meier 2002. 8 kW, a-Si, 20% capacity, 30 yr lifetime. 157 m^2. Colorado.
	70 w/BES	Denholm 2004. Storage (10-50% capacity, 20 yr lifetime) added to Meier (2002) PV system with T&D.
	53	Hondo 2005. 15% capacity, 30 yr lifetime. Rooftop 3 kW, p-Si, 10 MW/yr, system efficiency 10%.
	44 or 26	Future scenarios Hondo 2005. 1% capacity, 30 yr lifetime. Case 1, p-Si w/ production rate 1 GW/yr, 10% system efficiency. Case 2, a-Si, 1 GW/yr, 8.6% system efficiency.
	55	European Commission 1997d. ExternE. Germany. 4.8 kW, mc-Si (technology from 1990), 25 yr lifetime.
	51	European Commission 1997d. ExternE. Germany. 13 kW, mc-Si (technology from 1993), 25 yr lifetime.
	43	Frankl et al. 2004. ECLIPSE. Italy. 1 kW, sc-Si, 25 yr lifetime, 13% conversion efficiency. Insolation 1740 kWh/m^2/yr.
	51	Frankl et al. 2004. ECLIPSE. Italy. 1 kW, mc-Si, 25 yr lifetime, 10.7% conversion efficiency. Insolation 1740 kWh/m^2/yr.
	44	Frankl et al. 2004. ECLIPSE. Italy. 1 kW, a-Si, 20 yr lifetime, 6% conversion efficiency. Insolation 1740 kWh/m^2/yr.
	45	Frankl et al. 2004. ECLIPSE. Italy. 1 kW, CIGS, 20 yr lifetime, 9% conversion efficiency. Insolation 1740 kWh/m^2/yr.
	66	Spitzley and Keoleian 2004. Data from Keoleian and Lewis 2003. 2 kW, a-Si. 20 yr lifetime. Detroit. 6% conversion efficiency. Insolation 1380 kWh/m^2/yr (technology from 1900s).
	44	Spitzley and Keoleian 2005. 2 kW, a-Si, 20 yr lifetime. Phoenix, Arizona.
	71	Spitzley and Keoleian 2005. 2 kW, a-Si, 20 yr lifetime. Portland, Oregon.
	35	Fthenakis et al. 2008. UTCE, ribbon Si, 11.5% conversion efficiency. (This case and the next seven all have the same assumptions for the following parameters: solar insolation of 1700 kWh/m^2 per yr, performance ratio of .8, 30 yr lifetime. Did not include a case with crystal-clear project energy mix (natural gas and hydroelectric).)
	43	Fthenakis et al. 2008. UTCE, mc-Si, 13.2% conversion efficiency.
	44	Fthenakis et al. 2008. UTCE, s-Si, 14% conversion efficiency.
	21	Fthenakis et al. 2008. UTCE, CdTe, 9% conversion efficiency.
	44	Fthenakis et al. 2008. U.S., ribbon Si, 11.5% conversion efficiency.
	52	Fthenakis et al. 2008. U.S., mc-Si, 13.2% conversion efficiency.
	54	Fthenakis et al. 2008. U.S., s-Si, 14% conversion efficiency.
	26	Fthenakis et al. 2008. U.S., CdTe, 9% conversion efficiency.

TABLE 5.A.1 Continued

Technology	CO_2	Notes
Wind	15	White 1998. 25 yr lifetime. Capacity 24% actual. Class 2 to 4 wind. Includes replacement of all blades. Note: White (2006) updated LCA on actual performance and found similar results—14. Wind results specific to site; hard to generalize and dependent on energy used to produce materials—in the United States, coal. Wind dismantling assumed to be same as construction. No recycling of metals taken into account.
	24 w/PHS	Denholm 2004. PHS (10-50% capacity).
	105 w/CAES	Denholm 2004. CAES (70-85% capacity, 30 yr lifetime).
	29 (20 future)	Hondo 2005. 20% capacity both. 300 kW (future case 400 kW).
	7	European Commission 1997d. ExternE. Germany. 0.25 MW, 20 yr lifetime. Recycle metals.
	7	Chataignere et al. 2003. ECLIPSE. Europe. 0.6 MW, 20 yr lifetime. 1995-1998 technology, onshore.
	12	Chataignere et al. 2003. ECLIPSE. Europe. 1.5 MW, 20 yr, onshore.
	9	Chataignere et al. 2003. ECLIPSE. Europe. 2.5 MW, 20 yr, offshore.
	14.5	European Commission 1997b. ExternE. Denmark. 0.5 MW turbine, onshore.
	22	European Commission 1997b. ExternE. Denmark. 0.5 MW turbine, offshore.
	8	European Commission 1997a. ExternE. Greece. Onshore.
	9	Berry et al. 1998. 0.3 MW, onshore.
	1.7	Spitzley and Keoleian 2005. Turbine data from Schleisner 2000. 30 yr lifetime, 25 MW, Class 6 wind, 36% capacity.
	2.5	Spitzley and Keoleian 2005. Turbine data from Schleisner 2000. 30 yr lifetime, 25 MW, Class 4 wind, 24% capacity.
Biopower	49	Mann and Spath 1997. IGCC with 80% capacity, 30 yr lifetime. Assumes 95% carbon closure. Biopower from energy crops. 600 MW via several small plants.
	–667	Spath and Mann 2004. Added CO_2 capture and storage (CCS) to Mann and Spath (1997) example from above.
	–410	Spath and Mann 2004. 0.6 GW direct-fire boiler with biomass from waste streams.
	–1368	Spath and Mann 2004. 0.6 GW direct-fire boiler with biomass from waste streams with CCS.
	18	European Commission 1997c. ExternE. France. Cropping.
	15	Berry et al. 1998. Biopower source mainly willow and poplar. Lp IGCC.
	49	Spitzley and Keoleian 2005. 30 yr lifetime. 113 MW, hybrid poplar based on Mann and Spath 1997. Lp IGCC.
	40	Spitzley and Keoleian 2005. 20 yr lifetime. 75 MW, willow based on Heller et al. 2003. Hp IGGC. EPRI model.
	39	Spitzley and Keoleian 2005. 20 yr lifetime. 113 MW, willow based on Heller et al. 2003. Lp IGGC. NREL model.
	52	Spitzley and Keoleian 2005. 20 yr lifetime. 50 MW, willow based on Heller et al. 2003. Direct fire. EPRI model.

continued

TABLE 5.A.1 Continued

Technology	CO_2	Notes
Geothermal		
	15	Hondo 2005. 60% capacity, 30 yr lifetime. Double flash type.
	47–97*	Serchuk 2000. Only includes reservoirs emissions, not LCA.
	91*	Bloomfield et al. 2003. A weighted average of all geothermal capacity (including binary plants with no CO_2 emissions) per unit of electricity produced (not LCA).
	122*	Bertani and Thain 2002. A weighted average of existing plant operation per unit of electricity produced not LCA. Actual range 4–740 g CO_2 e/kWh from 85 plants in 11 countries.
Hydroelectric		
	20	Gagnon et al. (1997) present summary of a hydropower LCA survey using data from Finland, Canada, China, Japan, and Switzerland. Range in data 15 to 165 g CO_2e/kWh; average 20 g CO_2e/kWh. 100 yr lifetime. Includes data from river run and reservoir systems, alpine and prairie, small and large plants. Emissions very dependent on climate, topography, size of reservoir, construction materials, type of ecosystem flooded. Lowest case: 15 CO_2e from large reservoir in cold climate where emissions from flooded biomass drop to 0 at year 50. Worst case was in Finland where peat land flooded. LCA includes plant construction and decaying biomass from reservoir. A Brazilian reservoir is mentioned that due to very large size and low generation capacity has an estimated CO_2e of 237 (Fearnside's estimate is even higher).
	11	Hondo 2005. 45% capacity, 30 yr lifetime. Assumed river run w/small reservoir and did not include CO_2 from flooded biomass.
	26	Spitzley and Keoleian 2005. 50 yr lifetime. 1296 MW. Large reservoir type. Used data from Pacca and Horvath (2002).
Tidal		
	25–50	Preliminary, not rigorous. NOTE: production of steel for turbine is 25 g/kWh of CO_2. ETSO (1999) from Carbon Trust website.

TABLE 5.A.1 Continued

Technology	CO_2	Notes
Coal		
	974	White 1998. 75% capacity, 40 yr lifetime. Average U.S. plant with SO_2 control.
	1050	Denholm 2004. With T&D based on White 1998.
	975	Hondo 2005. 70% capacity, 30 yr lifetime. Average Japanese plant with SCR and FGD.
	1042	Spath et al. 1999. Average, 360 MW, 60% capacity, 1995, 30 yr lifetime. FGC and ESP (same as baghouse?).
	960	Spath et al. 1999. NSPS, 425 MW, 60% capacity, 1995, 30 yr lifetime. Same as average but with low NO_x burners or staged combustion for increased removal of airborne pollutants.
	757	Spath et al. 1999. Future LEBS, 404 MW, 60% capacity, 30 yr lifetime, 1995. Unspecified technologies used to decrease emissions.
	681	Spath and Mann 2004. Biomass residue co-fired w/coal.
	43	Spath and Mann 2004. Biomass residue co-fired w/coal w/CCS.
	847	Spath and Mann 2004. Coal based on Hendriks 1994.
	247	Spath and Mann 2004. Coal w/CCS.
	861	Odeh and Cockerill 2008. IGCC.
	167	Odeh and Cockerill 2008. IGCC w/CCS via selexol.
	984	Odeh and Cockerill 2008. Subcritical pulverized coal with SRC, ESP, FGD.
	879	Odeh and Cockerill 2008. Supercritical pulverized coal with SRC, ESP, FGD.
	255	Odeh and Cockerill 2008. Supercritical pulverized coal (same as above) w/CCS via MEA.
Gas		
	469	Meier 2002. 75% capacity over 30 yr lifetime. Average 620 MW, NGCC. Assumed CH_4 release rate of 1.4% (can range from 1 to 11%). Missouri plant.
	500	Denholm 2004. NGCC w/T&D based on Meier 2002.
	608	Hondo 2005. 70% capacity, 30 yr lifetime, LNG-fired.
	518	Hondo 2005. 70% capacity, 30 yr lifetime, LNGCC.
	499	Spath and Mann 2000. Average case NGCC with SCR.
	245	Spath and Mann 2004. Added CCS to Spath and Mann 2000.
	488	Odeh and Cockerill 2008. NGCC.
	200	Odeh and Cockerill 2008. NGCC w/CCS via MEA.

continued

TABLE 5.A.1 Continued

Technology	CO_2	Notes
Nuclear		
	15	White 1998. PWR. 75% capacity, 40 yr lifetime. Enrichment by gas centrifuge (not normally used in United States). Data for construction, operations, decommissioning, and waste disposal from others. Only fuel considered in land reclamation. Spent fuel disposal data 30 yrs old.
	25	White (2006) updates value to reflect 100% enrichment by gas diffusion—25.
	16	Denholm 2004. With T&D based on White 1998.
	24 (22)	Hondo 2005. Disposal costs not included, only 50 yr dry storage for spent fuel. Assumes 67% enrichment in United States. Analysis very sensitive to enrichment conditions, e.g., values range from 30 to 10 g CO_2/kWh if all U.S. versus all Japan enrichment. 70% capacity, 30 yr lifetime. Accounted for CH_4 leakage during resource extraction. Did not include decommissioning land for mining and milling, just electricity to mine and mill. LLW stored w/o maintenance in near-surface waste disposal sites. Note: Future case 22 w/recycling includes HLW disposal but not disposal transport. Lower due to enrichment savings. Includes one-time MOX reprocessing of spent fuel.
	20	European Commission 1997d. ExternE. Germany. Capacity 1375 MW. PWR.
	3	Vattenfall 2004. Sweden. Industry EDP. PWR and BWR. Two sites. 85% capacity, 40 yr lifetime.
	108	Storm van Leeuwen and Smith 2008. Average lifetime baseline case. 30 years at 82% capacity. Very detailed LCA.
	24	Fthenakis and Kim 2007. Baseline case represents average United States.
	55	Fthenakis and Kim 2007. Worst case, poor ores typical of Australia (0.05% U), most energy for enrichment from coal requiring 3000 kWh/SWU of energy, EIO method for construction.
	16	Fthenakis and Kim 2007. Best case, rich Canadian ores (12.7% U), 20% energy for enrichment from coal, rest U.S. grid mix requiring 2400 kWh/SWU of energy, process analysis for construction.
Storage		
PHS		
	3	Denholm and Kulcinski 2003. 74% efficient (?=capacity), 60 yr lifetime. Assumes dams and reservoirs permanent.
	5.6	Denholm 2004. With T&D. 74% efficient, 60 yr lifetime. Assumes dams and reservoirs permanent.
CAES		
	291	Denholm and Kulcinski 2003. 40 yr lifetime.
	292	Denholm 2004. With T&D. 65% efficient, 40 yr lifetime. Excludes primary electricity generation. Based on a 2.7 GW proposed facility in Ohio. Assumes negligible leaks, no energy intensive maintenance on cavern. Uses natural gas to compress air.

TABLE 5.A.1 Continued

Technology	CO_2	Notes
BES		
	80.5 Pb-acid	Denholm and Kulcinski 2003. 20 yr lifetime.
	64.9 V redox	
	50.4 Pb-acid	Denholm 2004. With T&D. 20 yr lifetime, excludes the stored electricity. Assumes large system w/energy:power ratio of 8 hr. Pb-acid oversized 30% due to limited depth of discharge. VRB 75%, PSB 63%, Pb-acid 66% efficient.
	32.6 PSB	
	40.2 V redox	

Note: a-Si, amorphous silicon; BES, battery energy storage; CAES, compressed air energy storage; CCS, carbon capture and storage; CIGS, copper indium gallium selenide; FGC, flue gas clean-up; FGD, flue gas desulphurization; LEBS, low emission boiler system; mc-Si, multicrystalline silicon; MEA, monoethanolamine; PB-acid, lead acid; pc-Si, polycrystalline silicon; PHS, pumped hydro storage; PSB, sodium-bromide/sodium-polysulfide battery; sc-Si, single-crystalline silicon; SCR, selective catalytic reduction; T&D, transmission and distribution; V redox, vanadium acid; VRB, vanadium redox battery.
All studies listed use LCA method. Not all studies are comparable. Denholm (2004) includes all life-cycle costs plus T&D emissions in LCA (most LCAs do not include T&D).

6 Deployment of Renewable Electric Energy

Renewable energy technologies are poised to become an important component of the electricity supply mix. However, it is not a foregone conclusion that the United States will achieve and maintain a high rate of deployment of renewable electricity. The current financial situation (as of 2009) is impacting the renewables market at many levels, but the issues discussed in this chapter are nonetheless important for expanding the market for renewable electricity technologies. Renewables face challenges involving the deployment and commercialization of innovative technologies—the stages that follow technological innovation and development. These challenges include the risk of introducing new technologies into competitive markets; the investment in the long-term, market-enabling research and development activities needed to help move technologies along the learning curve; and the impact of policy measures that share the risk of product innovation and market transformation. The proverbial investment valley of death[1] can prevent new technologies from advancing past the demonstration phase due to a lack of capital. Manufacturing capacity, policy, business and market innovation, and access to financing must coincide with technology innovations for the continued successful deployment of renewable sources of electricity.

As noted in Chapter 3, in many ways new renewable electricity technologies, and the thinking that will enable them, represent disruptive rather than

[1]A stage after product development but before commercialization when the financial investment required to move a new technology to commercialization may exceed the ability of a new business to raise capital.

incremental changes in long-established industry sectors. Disruptive technologies have two important characteristics. First, they typically present different performance attributes—such as providing a carbon-free source of electricity—that, at least at the outset, are not valued by a majority of customers. Second, the performance attributes (e.g., costs) for disruptive technologies that customers do value can improve at such a rapid rate that the new technology can overtake established markets. Figure 6.1 shows how the performance of a disruptive technology that was once lagging that of an earlier established technology can improve at a faster rate. However, such performance improvements are speculative and are not preordained. In the case of renewable electricity technologies, on a conventional cost-of-energy basis traditional sources of electricity generation initially outperform non-hydropower renewables. The attraction of technologies that use renewable resources, together with government incentives, has been responsible for much of their market presence. However, owing to improvements in renewable technologies and cost increases for fossil fuels and nuclear power, renewables are gaining the ability to match the cost performance of traditional generating sources both in the wholesale power market and on the customer side of the meter.

This chapter explores the logistical and market barriers to commercial-scale deployment of renewable electricity. Although individual renewable energy technologies have unique developmental and economic characteristics, there are common, non-technical challenges as well, including (1) constraints on capacity for larger-scale manufacturing and installation and limitations on the availability of trained employees for manufacturing, installation, and maintenance; (2) integration of intermittent resources into the existing electricity infrastructure and market; (3) market requirements such as capacity for competing in price and performance with conventional lower-cost coal, nuclear, and natural-gas-fired power plants; and (4) risk and related issues, including business risk, cost issues, and unpredictability of and inconsistency in regulatory policies.

Because of the robust regulatory and business activities related to wind and solar energy industries, many examples discussed in this chapter come from these sources. However, they are used to indicate deployment issues associated to some degree with other renewable sources of electricity.

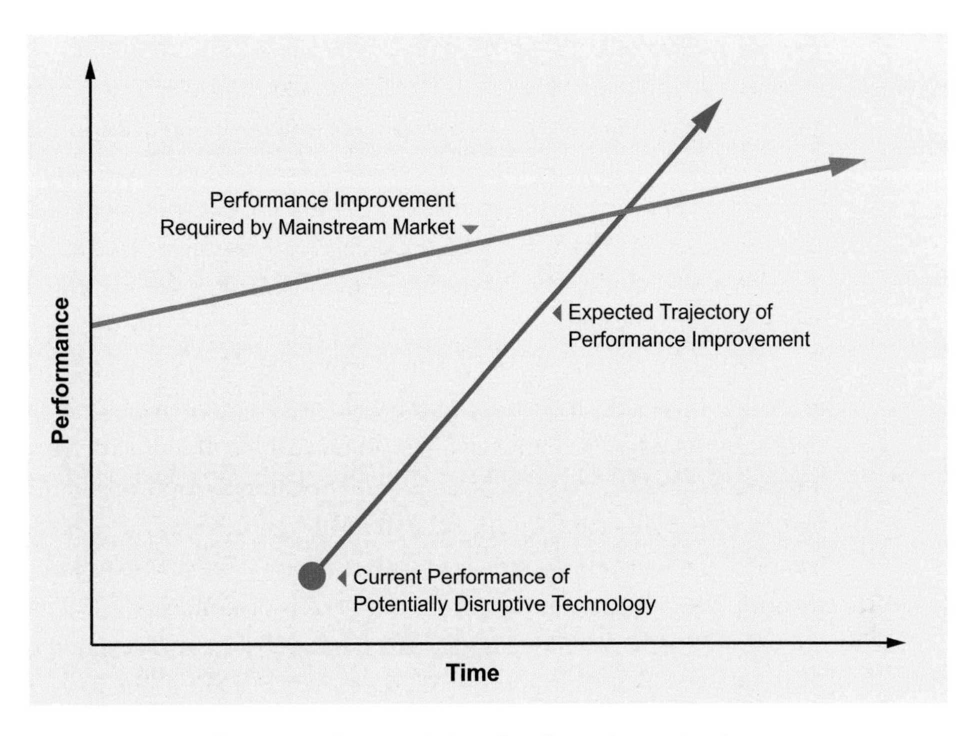

FIGURE 6.1 *Performance characteristics of a disruptive technology.*
Source: Bowen and Christensen, 1995.

DEPLOYMENT CAPACITY CONSIDERATIONS

Capacity constraints, such as restricted supplies of basic raw material inputs, limitations on manufacturing capacity, competition for larger construction project management and equipment, and limited trained workforce, have the potential to derail large-scale deployment and integration of renewable electricity resources. Thus, to grow the renewable electricity market, which is increasingly driven by the private sector, will require continued and ramped up investment in order to deploy, operate, and maintain these technologies.

Materials, Manufacturing, and Development Considerations

Raw and Basic Materials

Renewable energy technologies potentially can be restricted by a scarcity of key raw materials. A common example is solar photovoltaics (PV). Recent shortages of polycrystalline silicon have increased prices for PV modules, though these shortages were expected to ease by 2009 (Bradford, 2008). In addition, while silicon is relatively abundant, a scarcity of silver could limit use of traditional crystalline and polycrystalline silicon, as well as nano-silicon-based cells, in the long term. Likewise, limited availability of naturally occurring indium could restrict more efficient thin-film solar cell technologies using copper indium gallium selenide (CIGS). Solar cell raw material components and limiting material are summarized in Table 6.1, and global reserves for key materials are shown in Figure 6.2 (Feltrin and Freundlich, 2008).

There are also issues related to global competition for such basic materials as steel and cement that hinder large-scale deployment of renewables and increase renewable energy development costs. Wind turbine manufacturers are particularly affected by these material shortages. Global competition for essential elements has, in recent years, driven up the costs of commodities and limited the materials available for wind energy projects. Table 6.2 projects the raw materials needed through 2030 to support the 20 percent wind scenario (DOE, 2008a), and Figure 6.3 shows the predicted near-term U.S. and global raw material usage for wind turbines. Global competition for these resources is not limited to renewables. It applies to all types of generation and to the construction sector generally. Longer-term goals are achievable, but the broader use of renewables will require a well-defined strategy for deployment.

TABLE 6.1 Critical Limiting Raw Materials Needed for Fabrication of Solar Cells

Solar Cell	Limiting Material	Usage
Poly/c-Si	Silver (Ag)	n-electrode
a-Si	Indium (In)	TCO substrate
CdTe	Tellurium (Te)	Cell material
CIGS	Indium (In)	Cell material
Dye-sensitized	Indium (In)	TCO
	Tin (Sn)	TCO
	Platinum (Pt)	TCO
Conductive MJC III-V	Germanium (Ge)	Substrate
	Gallium (Ga)	GaAs substrate
Conductive MJC III-V, lift-off	Indium (In)	Cell material
	Gold (Au)	Electrode

Note: CIGS, copper-indium-gallium-arsenide; TCO, transparent conductive oxide.
Source: Adapted from material in Feltrin and Freundlich, 2008.

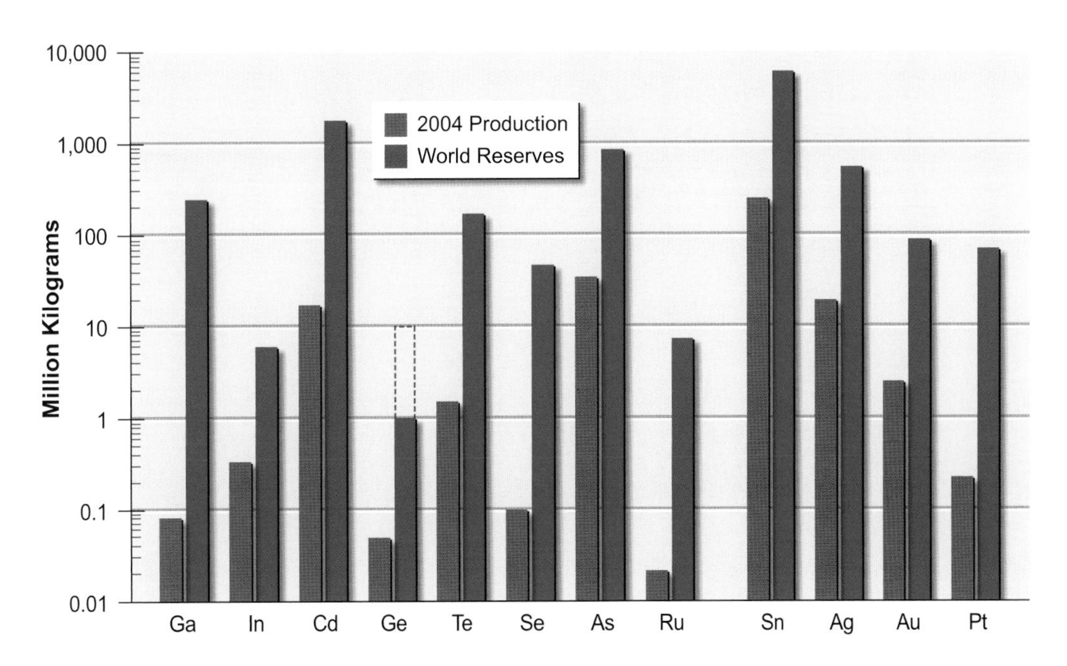

FIGURE 6.2 *Estimated (2004) annual production levels and world material reserves of raw materials used in PV cell manufacturing. Note that because data are not available on world reserves of germanium (Ge), the solid bar represents U.S. reserves and the dashed lines represent a best guess about world reserves.*
Source: Feltrin and Freundlich, 2008.

Manufacturing and Development

Wind Power Industry

Developers face shortages of wind turbines due to continuing strong demand for wind power both in the United States and globally (AWEA, 2008). Wind turbine manufacturers are still in the process of making the capital investments necessary to increase their capacity to catch up with the growing demand. Projections have suggested that the mismatch between turbine supplies and wind developer demands would level out as soon as 2009 (EER, 2007). Meanwhile, manufacturers continue to play catch-up, with typical delays of 6 months or more from turbine order to delivery. Though lead times have lengthened due to the rapid growth in wind turbine installations, wind and solar PV projects have an advantage over traditional power plants because of their shorter time between purchase of the equipment and placing it on line (Bierden, 2007).

Overall wind power project costs have increased due to recent increases in

TABLE 6.2 Yearly Raw Materials Required in 2030 to Meet Wind Turbine Demand in 20 Percent Wind Scenario (in units of thousands of metric tons)

Year	kWh/kg[a]	Permanent Magnet	Concrete	Steel
2006	65	0.03	1,614	110
2010	70	0.07	6,798	464
2015	75	0.96	16,150	1,188
2020	80	2.20	37,468	2,644
2025	85	2.10	35,180	2,544
2030	90	2.00	33,800	2,308

[a] Proposed scenario for energy density improvement for wind turbine growth during the 2006–2030 period.
Source: Adapted from material in Wiley, 2007.

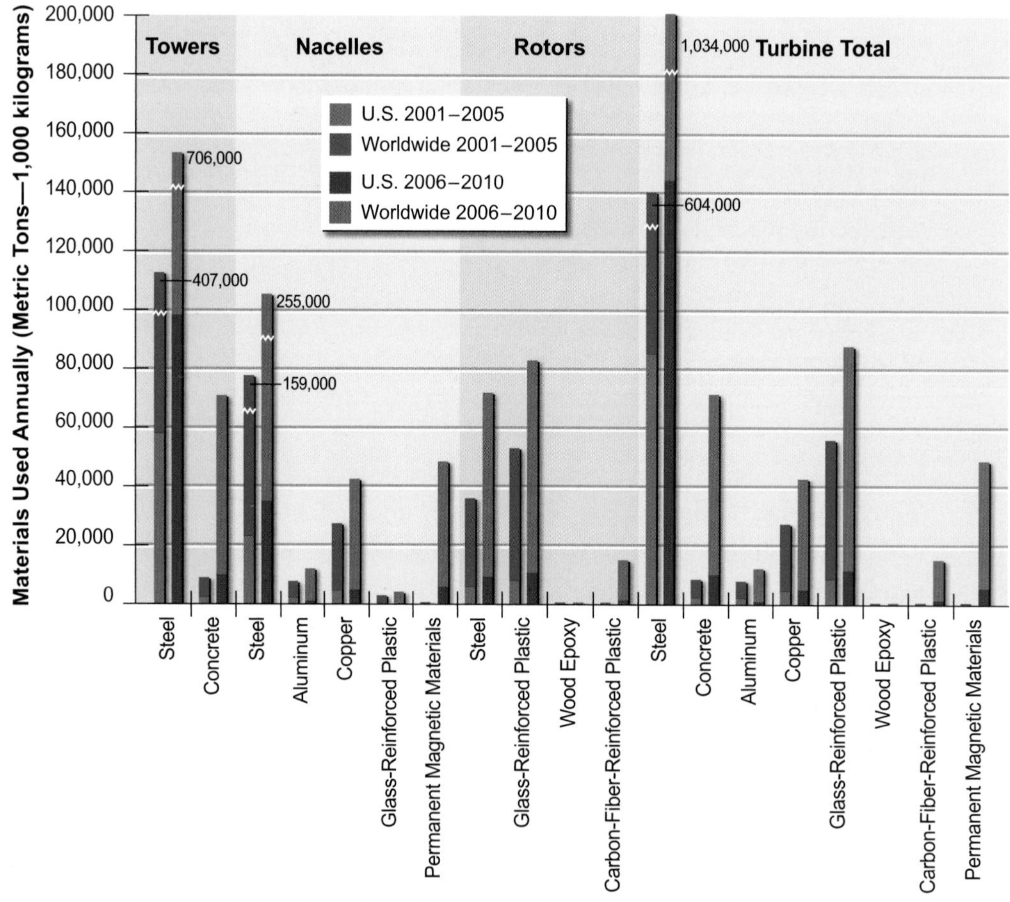

FIGURE 6.3 *U.S. and worldwide wind turbine material usage. Source: Ancona and McVeigh, 2001.*

Aluminum	Copper	Glass-Reinforced Plastic	Carbon Fiber Composite	Adhesive	Core
1.2	1.6	7.1	0.2	1.4	0.4
4.6	7.4	29.8	2.2	5.6	1.6
15.4	10.2	73.8	9.0	15.0	5.0
29.6	20.2	162.2	20.4	33.6	11.2
27.8	19.4	156.2	19.2	31.4	10.4
26.4	18.4	152.4	18.4	30.2	9.6

wind turbine prices (DOE, 2008b). Figure 6.4 shows the recent trend in turbine costs. These prices have increased due to increased costs for materials and energy inputs; component shortages; upscaling of turbine size and improvements in turbine design; declining value of the U.S. dollar; and attempts to increase profitability in the wind turbine manufacturing industry (DOE, 2008b). The increase in project costs as of year 2000 reversed the long-term decline in project costs, which includes the turbine as well as other balance of system components (Figure 6.5). The upturn in the price of turbines might, however, be partially offset by an increase in the kilowatt-hour output per kilowatt turbine capacity with the use of power electronics, variable-speed drives, and more stringent requirements of ride-through faults in utility system operation.

The increased demand for wind turbines worldwide has expanded wind turbine manufacturing facilities in the United States. Though General Electric (GE) remains the dominant turbine manufacturer, other domestic and foreign manufacturers have entered the market or expanded their operations (DOE, 2008b). Component manufacturers of blades, gearboxes, and other elements are spread across the United States (Sterzinger and Svrcek, 2004). However, lower wages have caused many manufacturers to locate factories overseas (DOE, 2008b). In general, the strong growth nationally and internationally has resulted in an expansion of all segments of the wind industry, including manufacturers, as well as parts of the industry related to installation and operations and maintenance.

There have been changes in the wind power development sector of the industry (EER, 2007). Independent power producers (IPPs) have shown increased interest in wind power projects; IPPs develop a variety of electricity generation facili-

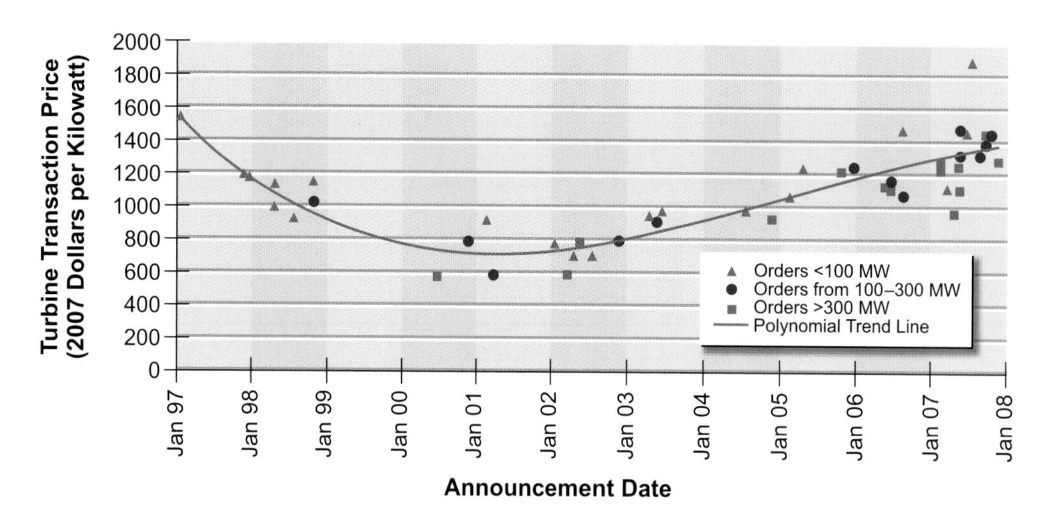

FIGURE 6.4 *Wind turbine prices over time.*
Source: DOE, 2008b.

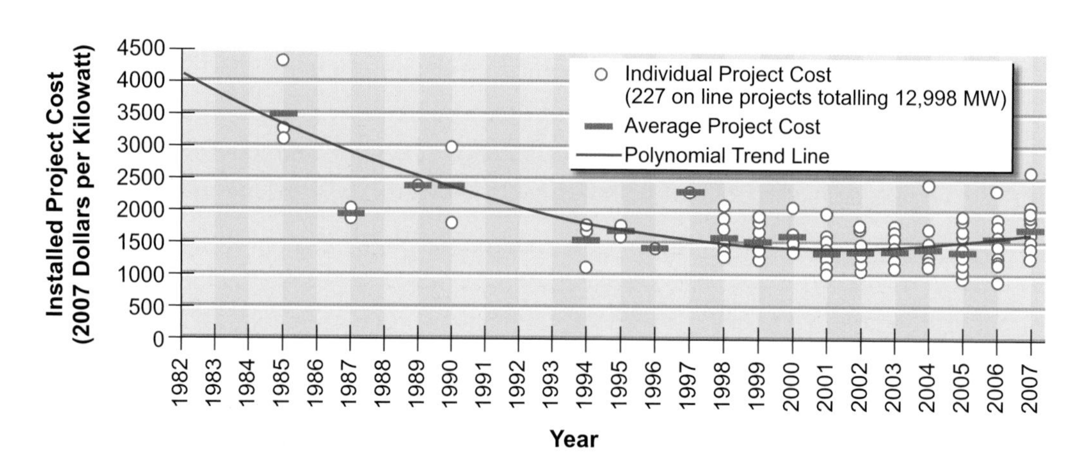

FIGURE 6.5 *Installed wind project costs over time.*
Source: DOE, 2008b.

ties for the wholesale electricity market. IPPs have to compete against developers whose sole focus is the development of wind power projects (termed the *pure play wind developers*). Further, globalization has become a factor in the U.S. market, with developers from Europe initiating projects in the United States. Most of these European developers provide wind through long-term contracted sales to utilities, though they also sell to power markets. A variant is the purchase of Energy

East, a New York state utility, by Iberdrola S.A., a Spanish energy company that develops wind power projects worldwide. As noted in Chapter 4, there also is a market for renewable energy credits (RECs) that can be sold separately from electric power. Finally, some utilities are beginning to develop their own wind power projects instead of purchasing wind power through long-term contracts with wind developers.

Solar PV Industry

Like wind power, the large growth rate for solar PV, both within the United States and globally, has caused shortages in manufacturing capacity and raw materials. As with wind power, it has also resulted in increasing prices and changes within the industry. As noted in the section on raw materials, the primary cause for shortages in PV is a shortage in polycrystalline silicon. Originally, the primary use of polycrystalline silicon was for semiconductors in the electronic industry, with solar PV manufacturers using a small fraction of silicon production and even using silicon recycled from the electronics industry. Recently, the solar PV industry has become the largest consumer of polycrystalline silicon, bringing new entrants into the industry that include producers specifically oriented to the solar PV industry, and even solar PV manufacturers looking to become more integrated along the supply chain (Prometheus Institute, 2007). Despite these new entrants, there was still a shortage of polycrystalline silicon, which had driven up the price for solar silicon PV modules (Figure 6.6), though this shortage was expected to subside by 2009. Recent articles project 2009 to see this decrease in costs for solar PV, though the decline in price has been attributed to both increasing supplies and decreasing demands due to the global economic slowdown (Patel, 2009).

Solar companies that are expected to perform well in the current solar PV market are generally those with stable silicon supplies (EIA, 2007). Conversely, companies that are thought to have insufficient or inflated silicon supplies have not done well in the market (Greentech Media, 2007). Another current positive market characteristic is less reliance on polycrystalline silicon. There is more competition among distinctively different technologies in the solar PV industry compared to the wind turbine market. As shown in Figure 6.6, shortages of polycrystalline silicon have spurred increases in the thin-film solar PV technologies that do not require as much or any silicon. Figure 6.7 shows the impacts on shipments by U.S. manufacturers of this shift toward thin-film PV.

The rapid growth and projected demand for solar PV have spurred increases

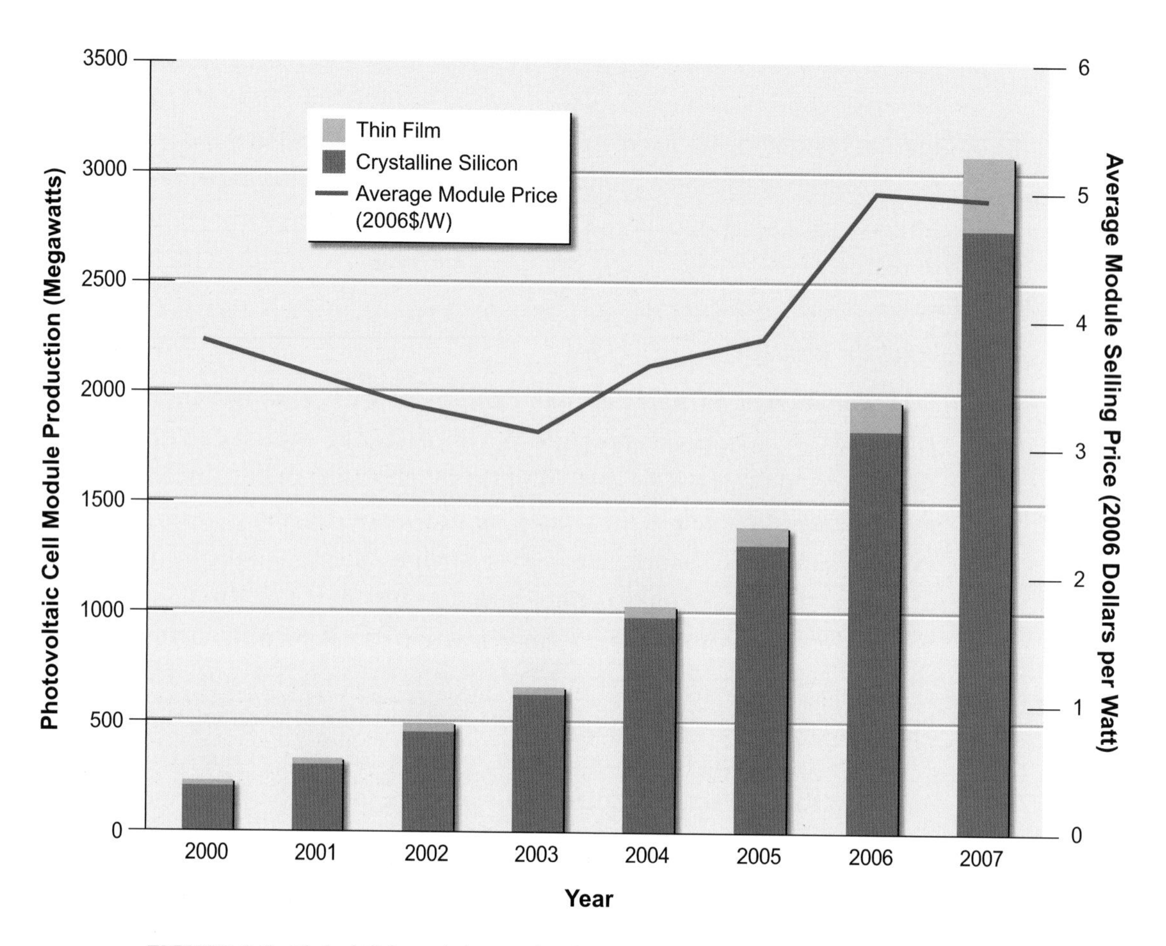

FIGURE 6.6 *Global PV module production 2000–2007 and average module price during the same timeframe.*
Source: Courtesy of Paula Mints, Principal Analyst, Navigant Consulting PV Services Program.

in both PV prices and demand for manufacturers to increase their manufacturing capacities. PV manufacturing in the United States is dominated by First Solar of Arizona, which has responded to market demand by expanding manufacturing capacity in Ohio and Germany, and it has announced additional capacity expansion in Malaysia (Prometheus Institute, 2007). Together, this expanded capacity is expected to bring First Solar's total manufacturing capacity to more than 1 GW/yr by the end of 2009. This capacity expansion substantially increased income for this company in 2008 (Greentech Media, 2008). By 2010, SunPower and SolarWorld are expected to add an additional 984 MW of capacity.

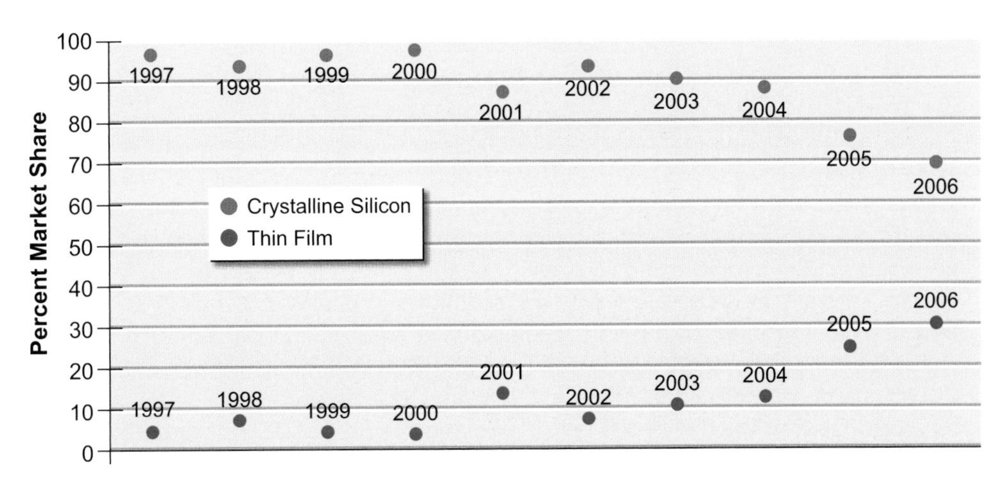

FIGURE 6.7 *Crystalline silicon shipment and thin-film shipment market shares in the United States, 1997–2006.*
Source: EIA, 2007.

The largest customer category for PV modules/cells has shifted from wholesale distributors to installers (EIA, 2007), reflecting the recent trend toward large commercial PV installations, such as those at Wal-Mart and the Google headquarters in California. The commercial sector was the largest market for PV in 2006 and grew more than 100 percent from 2005 (EIA, 2007). Additionally, some PV manufacturers have begun to enter the installation business to become more fully integrated along the PV supply chain (Greentech Media, 2007). Box 6.1 provides some background on the history and characteristics of the market for solar PV.

Workforce Requirements

Direct Requirements

Another limiting variable to the large-scale manufacturing and deployment of new renewable electricity systems is the need for a trained and capable workforce that grows as market demand grows. Educating this workforce requires the development of high-quality training infrastructures that include accredited institutions, skill testing, and certification. Table 6.3 shows the direct jobs and economic activity in the renewable electricity industry for 2006 (ASES, 2007).

The renewable energy industry in the United States opened 450,000 jobs in 2006 (ASES, 2007). Meeting a renewable energy portfolio standard of 20 percent by 2020 is projected to require an additional 185,000 jobs related to renewable

BOX 6.1 **Evolution of the Market for Solar Photovoltaics**

The market for solar photovoltaics (PV) has evolved from niche, off-grid applications to a wide array of applications that provide power to the grid. For years, the primary market for PV cells and modules was in remote, stand-alone power for communication and navigation systems, cathodic protection, and village power, and in consumer products such as calculators, watches, and portable lighting products. Recently the grid-connected market has become the prominent user of PV modules and systems. The solar PV market has segments, distinguished by system size, such as residential (<10 kW), small commercial (10 kW to 100 kW), large industrial and public (100 kW to 1 MW), and utility scale (>1 MW). The economics of PV installations is directly related to the size of the installation and the degree of integration for the installation company across the PV supply value chain. Generally, the larger systems with greater degrees of integration into the grid will realize greater cost-reductions through economies of scale.

In the United States, a bifurcated market for PV systems has developed, depending on whether the system is installed on a customer's premises (behind the meter) or as a utility-scale generation resource. Behind-the-meter systems compete by displacing customer-purchased electricity at retail rates, while utility-scale plants must compete against wholesale electricity prices. Thus, behind-the-meter systems can often absorb a higher overall system cost structure. Much of the development of solar has occurred in this behind-the-meter market.

Residential systems, one type of behind-the-meter systems, tend to be custom-designed based on roof space, pitch, and orientation. System dealers need to stock a variety of products and components and to manage product inventory; installers incur costs in project permitting and contracting for utility interconnection. Residential system installers have begun to address some of these issues. Some are

energy (UCS, 2007). However, the renewable energy sector faces a challenge in meeting an increasing demand for educated and skilled workers. In fact, these workforce needs apply across the entire energy sector, which is faced with an aging workforce and a shortage of technically skilled people. Companies developing wind, solar, biomass, and geothermal recoverable resources will require an influx of skilled employees for sales marketing, customer services, and business support services in order to support the wide-scale deployment of these energy sources. Already, the shortage of skilled workers in the solar industry is partially blamed for upward cost pressures (EIA, 2007). A variety of sector-specific technical jobs, outlined in Table 6.4, will be drawn upon (Council on Competitiveness, 2007). To give a better picture of the variety of renewable energy positions, customizing PV module systems that integrate racking hardware, grounding wires, wiring connections, and connections between panels. These systems can be factory-produced, reducing on-site installation costs. A homeowner will invest the needed capital to pay for the system purchase and installation, with cost recovery occurring over some period of time from displaced electricity savings. Economic payback periods can be quite long, but early-adopter residential investors are less sensitive to overall system economics because of other purchase motivations.

In another version of the behind-the-meter systems, a commercial business or government agency might install and run a system itself, or have a system installed through a third-party ownership or solar services model. Under the solar services model, third-party companies install and own a PV system on behalf of a host business or public agency. The system is located behind the meter on a utility customer's premises. The third-party company acts as technology integrator, project developer, and system operator, and secures the project financing as well. The solar electricity is sold to the host customer at a rate below the prevailing utility retail rate.

Businesses and public agencies generally adhere to strict economic payback criteria. For example, businesses have an internal rate of return (IRR) hurdle (often >15 percent) that must be met for any corporate investment to be undertaken. At today's costs, PV system investments may not meet the IRR hurdle. The success of this model often relies on key factors, including (1) net metering, which allows valuation of displaced grid electricity at the prevailing retail rate, and (2) the ability of the third-party entity to raise capital at rates well below the IRR hurdle of the private companies. Other factors include the availability of federal and state incentives; the existence of a time-of-use utility tariff in which the utility's high-price rate tiers match well with the solar electricity output; and an existing market for solar renewable energy credits (RECs), the sale of which provides additional value to the solar generation.

Table 6.5 delineates selected occupations at a typical wind turbine manufacturing plant in Ohio, illustrating that the renewable energy sector employs a wide range of people at all levels of skills and education.

Indirect Requirements

In addition to the basic manufacturing and operation workforce needs, there is an equally pressing need in the related electric utility infrastructure, where the turnover of an aging traditional electric utility employee base is outpacing the supply of skilled replacements. The shortfall may be as high as 10,000 by 2010 (DOE, 2006). According to the Center for Energy Workforce Development, at least half of the electric utilities' technical workforce, including power line workers,

TABLE 6.3 Direct Jobs in the Renewable Energy Sector in 2006

Industry Segment	Revenues/Budget (billion $)	Direct Jobs
Wind	3.0	16,000
Photovoltaics	1.0	6,800
Solar thermal	0.1	800
Hydroelectric power	4.0	8,000
Geothermal	2.0	9,000
Biopower	17.0	66,000
Federal government (including direct-support contractors)	0.5	800
DOE laboratories (including direct-support contractors)	1.8	3,600
State and local governments	0.9	2,500

Source: Based on data from ASES, 2007.

TABLE 6.4 Breakdown of Renewable-Energy-Specific Positions

Wind	Solar	Biomass	Geothermal
Electrical and mechanical engineers and technicians	Electrical, mechanical, and chemical engineers and technicians	Chemists and biochemists	Geologists, geochemists, and geophysicists
Aeronautical engineers	Material scientists	Agricultural specialists	Hydrologists
Construction workers	Physicists	Microbiologists	Hydraulic engineers
Meteorologists	Construction workers, architects, and builders	Electrical, mechanical, and chemical engineers and technicians	HVAC contractors

Source: Adapted from material in Council on Competitiveness, 2007.

TABLE 6.5 Selected Occupations of Employees at a 250-Person Wind Turbine Manufacturing Company in Ohio in 2006

Occupation	Employees	Earnings
Engine and Other Machine Assemblers	31	$36,300
Machinists	27	40,500
Team Assemblers	16	30,100
Computer-Controlled Machine Tool Operators	12	40,600
Mechanical Engineers	10	71,600
First-Line Supervisors/Managers of Production/Operating	10	59,600
Inspectors, Testers, Sorters, Samplers, and Weighers	8	40,400
Lathe and Turning Machine Tool Setters/Operators/Tenders	6	40,000
Drilling and Boring Machine Tool Setters/Operators/Tenders	4	39,800
Welders, Cutters, Solderers, and Brazers	4	39,900
Laborers and Freight, Stock, and Material Movers	4	29,800
Maintenance and Repair Workers	4	44,100
Tool and Die Makers	4	43,600
Grinding/Lapping/Polishing/Buffing Machine Tool Operators	4	34,800
Multiple Machine Tool Setters/Operators/Tenders	4	40,800
Industrial Engineers	3	70,400
Industrial Machinery Mechanics	3	46,000
Engineering Managers	3	108,300
Shipping, Receiving, and Traffic Clerks	3	32,100
General and Operations Managers	3	120,600
Industrial Production Managers	3	93,100
Industrial Truck and Tractor Operators	3	34,200
Purchasing Agents	3	56,200
Cutting/Punching/Press Machine Setters/Operators/Tenders	3	31,400
Production, Planning, and Expediting Clerks	3	45,200
Milling and Planning Machine Setters/Operators/Tenders	3	40,600
Mechanical Drafters	2	39,900
Customer Service Representatives	2	39,100
Bookkeeping, Accounting, and Auditing Clerks	2	35,600
Office Clerks, General	2	29,400
Sales Representatives, Wholesale and Manufacturing	2	55,300
Janitors and Cleaners	2	29,800
Sales Engineers	2	72,500
Accountants and Auditors	2	59,800
Tool Grinders, Filers, and Sharpeners	2	44,000
Executive Secretaries and Administrative Assistants	2	43,200
Mechanical Engineering Technicians	2	50,900
Electricians	2	49,600
Other Employees	48	49,700
Employee Total (126 occupations in the industry)	250	$46,400

Source: ASES, 2007. Used with permission of the American Solar Energy Society. Copyright 2007 ASES.

mechanics, installers, repairers, and first- and second-line supervisors, may retire in five to ten years (CEWD, 2007). These traditional electric utility roles are essential to the large-scale deployment and integration of renewable energy sources.

Training and Certification

To meet the growing demand for skilled workers, a variety of workforce development strategies are needed (Great Valley Center, 2003). Recent initiatives attempt to address the insufficient supply of skilled workers by instituting renewable energy-specific training, certification, and licensing programs. Some leading examples include:

- *New York State Energy Research and Development Authority.* NYSERDA supports the development of an in-state network of training programs to provide accessible and quality instructional opportunities for those already in the renewable energy trades or those planning on entering the profession. NYSERDA has invested in developing seven accredited solar training centers and continuing education programs across the state through partnerships with community colleges, trade schools, universities, and trade unions (NYSERDA, 2005).
- *Florida Solar Energy Center.* The center is a state-supported research and training institute in the area of renewable energy, with courses in photovoltaics and energy efficiency programs. In addition, the center develops curricula for national and international training on renewable energy, in partnership with other organizations, and offers these programs through distance learning.
- *Green Energy Ohio.* A partnership of the Great Lakes Renewable Energy Association and Florida Solar Energy Center, Green Energy Ohio has a 5-day Photovoltaic (PV) Installer Apprentice Program. It is designed for individuals beginning a career as a PV system integrator, combining classroom sessions with field experience to introduce students to distributed generation technologies and interconnection issues, with a focus on solar energy.
- *Sonoma State University Energy Management and Design.* The Energy Management and Design (EMD) Program provides either a B.A. or a B.S. degree in environmental studies. It provides management and design training in the application of a wide variety of energy efficiency

and renewable energy technologies. All EMD students must complete an internship, which provides experience in a professional setting. The program has several external relationships, including the California Energy Commission, Lawrence Berkeley National Laboratory, Pacific Gas and Electric Company, Sacramento Municipal Utility District, California Association of Building Energy Consultants, and the Northern California Solar Energy Association.

- *Midwest Renewable Energy Association.* The MREA hosts a series of educational and hands-on workshops throughout the year, instructed by experienced renewable energy experts in small classroom settings and on-site installation locations. Workshop participants come from varied backgrounds, including homeowners, builders, educators, architects, engineers, and others. Participants can receive Continuing Education Units for attending workshops.
- *Central Carolina Community College.* The college offers a course on introduction to PV system design, related to the properties and installation of solar panels that produce electricity. This course has been approved by the North American Board of Certified Energy Practitioners.

The federal government also is increasing its role in training. The Energy Independence and Security Act of 2007 authorizes a Department of Labor energy efficiency and renewable energy worker-training program. This legislation also establishes a grant program within the DOE's Office of Solar Energy Technologies to create and strengthen solar-industry workforce training and internship programs for installation, operation, and maintenance of solar-energy devices.

To improve the workforce, there is a critical need to develop quality training programs that test and certify skill acquisition and capability. The Interstate Renewable Energy Council recommends criteria necessary for the design and implementation of workforce training programs (Weissman and Laflin, 2006). They include the need for the training institution to offer programs under the auspices of recognized third-party or government accreditation standards and development of curriculum based on industry-approved task analyses. The North American Board of Certified Energy Practitioners, an industry-based, non-profit credentialing organization that assesses competency and certifies solar installers, will be adding categories of certificates over time.

RENEWABLE ELECTRICITY INTEGRATION

The electric system balances and delivers power generation from a portfolio of resources to power demand centers that vary in scale and location. A modern electric grid is essential for overall reliability. Large-scale integration of renewables into the electricity system may require improved technologies to expand and upgrade the transmission and distribution system capabilities, and changes to utility and grid operations that can occur during system upgrades. This section discusses the potential impacts on utility and grid operations associated with renewable electricity deployment. Other aspects of renewables integration are considered in this report in Chapter 3, which discusses the technologies themselves; Chapter 4, which provides estimates of integration costs; and Chapter 7, which discusses scenarios that involve increased deployment of renewables and that include grid capacity needs.

System operators seek to ensure that generation and transmission resources meet the on-peak load within the entire control area with sufficient generating reserves and transmission capability to cover contingencies, in order to meet mandatory federal system reliability criteria. In most cases, a generation reserve margin[2] of 15 percent is deemed adequate. If a given resource, such as wind, cannot be counted on to be available on peak with some degree of certainty, the grid operator cannot count on it to meet the resource adequacy requirement. In California, the wind resource available on peak is on average less than 7 percent of the nameplate rating[3] of the aggregated machines and frequently is of the order of 1 percent (CALISO, 2007). In New York, GE estimated an average onshore capacity value of 9 percent and an offshore value of 36 percent (GE Energy, 2005).

A grid can support some intermittent resources without electricity storage if sufficient excess capacity is available to maintain resource adequacy. As described below and in Chapter 7, in many cases the amount of intermittent renewable resources that can be supported is approximately 20 percent, particularly for utilities that rely primarily on hydropower or natural-gas-fired generation. Hydropower and natural-gas-fired plants can ramp levels of generation up or down fairly rapidly, and are able to incorporate a higher fraction of renewables than

---

[2]Percentage by which available capacity is expected to exceed forecasted peak demand across the region.

[3]"Nameplate rating" is the 100 percent rated capacity of the device (nominally in kilowatts or megawatts).

utilities that rely on nuclear and coal-fired generation, which cannot ramp up or down quickly. Increases in ancillary services and additions of storage or other dispatchable resources may be necessary to maintain system reliability as the system mix is diversified. For example, a New York State study demonstrated that wind generation could be introduced up to about 10 percent of total capacity of the network without adding storage (GE Energy, 2005), although the location of the wind resources might require additional transmission due to wind capacity generated constraints. Dispatchable resources, typically natural-gas-fired units and sometimes hydropower, and interruptible demand are used to compensate for the lack of dispatchable resources when scheduled wind or other capacity diminishes.

Transmission Considerations

The prospect of large-scale development of renewable energy resources raises issues of electricity transmission grid adequacy, planning, and expansion. Because they are manufactured technologies, renewable energy generating systems can be constructed much more rapidly than can baseload fossil-fuel or nuclear plants. Dramatic expansion of the renewable electricity base in the United States implicates the need for improvement and expansion of the transmission grid. Of course, transmission grid development would be required as part of a significant expansion of the conventional electricity resource base as well. The primary difference is in the timing of this development.

Large conventional power plants, such as nuclear and coal units, require much lead time and many years of planning and construction, and therefore, transmission to serve these units can be addressed later in the construction process. This is not so with wind or solar farms which can be constructed in a matter of months or a few years (Bierden, 2007; Sheedan and Hetznecker, 2008). A systems perspective will be required in order to dramatically increase the contribution of large-scale renewables to electricity supply, one which undertakes consideration of transmission and other infrastructure needs for the whole utility or electricity control area[4] well in advance of generation plant construction. Though there is scant experience with this approach to electricity system development, there would be potential efficiency benefits in planning and constructing the grid as part of

[4]An electricity control area is a portion of the grid over which a single entity has responsibility for maintaining the balance of supply and demand and for ensuring reliability. The control area includes the service area for multiple utilities.

an overall resource development process. These benefits could include least-cost route planning, sensitivity to environmental and cultural resources, planning for maximum beneficial use by all generation resources, and proactive design for system reliability and security. These benefits might be sufficient to offset some of the costs of building large-scale transmission systems in advance of generation interconnection demand, but further study is required.

In Texas, rapid growth of the development of west Texas wind resources without a coordinated transmission development plan has led to curtailment of renewable electricity generation and system congestion. The Texas Legislature created the Competitive Renewable Energy Zones model to rationalize renewable electricity development and necessary transmission support. After 3 years of regulatory proceedings, the Public Utility Commission of Texas adopted a plan. The challenge that Texas faces now is whether it can efficiently and rapidly extend the transmission system to support its world-leading pace of renewable electricity development. The Texas experience highlights the importance of transmission system planning and development for the rest of the nation. Outside of the Electric Reliability Council of Texas (ERCOT), these planning and development functions would be even more challenging due to the involvement of multiple states and regulatory jurisdictions.

Reaching out for outstanding resources capable of sustaining large installations is typical of early stage power option development—whether it is fuel or some other key supporting resource. Mine-mouth coal plants, rail-dependent coal plants, large hydropower, geothermal (ring of fire), large wind, large solar, coastal/riverine nuclear (for cooling), and carbon capture and storage (CCS)-ready coal are all examples. Two important points bear consideration. First, investment in infrastructure to reach these resources supports capturing economies of manufacturing scale and thereby driving down prices, while also improving technological performance to support deployment in less optimal (and less remote) locations. Second, the cost of building infrastructure to capture remote wind, for example, can be mitigated by planning lines to deliver system-wide reliability and security benefits. Conventional plants can generally no longer locate near load centers due to air quality non-attainment designations associated with most large urban areas, so expanding capacity to meet increasing demand will require building *all* power plants in relatively remote locations. Large "interstate" transmission highways can support not just renewables but also conventional energy, reliability, and system security. Finally, in many cases the energy and fuel savings associated with capturing large amounts of renewables can help offset the incremental remaining costs of new transmission infrastructure.

The Committee on America's Energy Future (AEF), through its subgroup on electric power transmission and distribution (T&D subgroup) is considering the needs of the grid in its upcoming report (NAS-NAE-NRC, 2009). The T&D subgroup is focusing on the need to add more transmission lines, modernize vintage equipment, and introduce new technologies, which includes introducing advanced equipment; measurements, communications, and control technologies; and improved decision support tools. The benefits from an improved grid as discussed by the AEF T&D subgroup include economic, security, and environmental benefits, extending beyond additional needs for increased renewable electricity generation.

Studies on the Integration of Renewables

Numerous studies of the integration of renewables into the electric system have looked almost exclusively at integration of wind power, because wind is the intermittent renewable with generation capacity approaching levels where integration becomes important. Several reviews analyze, summarize, and cross-compare these various state and national studies (see, for example, Parsons et al., 2006; Holttinen et al., 2007; DOE, 2008a).[5] Although the individual studies use different assumptions and modeling techniques, these reviews provide some synthesis and offer general observations. First, it appears that large, diverse balancing areas with robust transmission are a key factor to reducing wind's impacts as its market penetration increases. Second, although wind speeds vary continuously (e.g., from second to second), wind power fluctuation is comparatively lower for very short periods of time, such that appropriate attention to the ramp rates for a given system can overcome these fluctuations (Wan, 2004). As a result, regulation impacts (variations measured in seconds or minutes) are small, but load-following and unit commitment can entail much higher costs. Third, improved wind forecasting can play an important role in reducing integration challenges and costs, particularly for unit commitment. Finally, based on techniques and methods used to analyze wind's impacts, Parsons et al. (2006) concluded that it is important to focus on

[5]Many additional reviews are available on the Utility Wind Integration Group website (http://www.uwig.org) and two large studies (the eastern wind integration and transmission study, and the western wind and solar integration study) are expected in 2009.

balancing the system as opposed to individual parts; in other words, not every movement needs to be matched on a one-to-one basis in order to maintain a stable grid.

The individual studies tend to focus on associated costs (and in some cases economic benefits), with less attention to the more technical aspects and limitations of integration. Most U.S. studies model potential impacts of meeting a state's renewables portfolio standard and do not necessarily seek to identify upper limits for market penetration. Current systems can handle wind market penetrations of up to 10–20 percent based on capacity, and costs tend to increase with penetration (Parsons et al., 2006). New York State conducted perhaps the most comprehensive domestic study to date to examine the potential impacts of integrating wind power equal to 10 percent of the state's estimated peak load for 2008 (NYSERDA, 2005). It concluded that the state could accommodate at least that amount of market penetration of wind generation with only minor adjustments to its existing system and practices (NYSERDA, 2005). Minnesota conducted a similar study that estimated that the Midwest Independent System Operator market could accommodate wind penetration of at least 25 percent of Minnesota's retail electricity sales, at an added integration operating cost of ~$4.41/MWh (MPUC, 2006). A Colorado utility commissioned studies of 10, 15, and 20 percent wind penetration (nameplate capacity relative to peak load) and the associated cost implications, which showed that integration costs would rise as market penetration increased, but that improved forecasting would help reduce these costs, estimated to be ~$5.13/MWh (EnerNex, 2008).

The California Intermittency Assessment Project (Davis and Quach, 2007) considered a portfolio of renewables that could be used as the state pursues a goal of obtaining 33 percent of its electricity from renewables by 2020. The study concludes that the 2020 goal is feasible (assuming some needed transmission line upgrades), and that most of this would be provided by wind and solar, two intermittent resources that together would represent 31 percent penetration of California ISO's market. Another large regional assessment, the Northwest Wind Integration Action Plan (NPCC, 2007), synthesizes several studies from the Pacific Northwest and finds no technical barriers to achieving a regional goal of 6000 MW of wind, although cost may become a limiting factor as some of the states reach 30 percent wind penetration. A recent international assessment (Holttinen et al., 2007) also indicates that the barriers to substantial wind penetration are not technical, but social and economic. This review draws on some of the only practical experience[6] to date (e.g., in Germany and Denmark); it suggests that in the regions where wind already meets more than 20 percent of gross demand, interconnection capacity has allowed grid operators to maintain balance within the grid, and curtailment of wind power is rare.

Another approach to improving wind power integration is to have an interconnected, geographically dispersed resource base (NERC, 2009). Archer and Jacobson (2007) considered the impacts of connecting wind farms with transmission throughout a given geographical area and found that such an interconnected system increases the capacity factor associated with the wind power. Using hourly wind data, this study simulated 19 sites in the Midwest with wind speeds greater than 6.9 meters/second and found that an average of 33 percent and a maximum of 47 percent of yearly averaged wind power from the pooled resources could be used as baseload electric power. Similarly, Hawkins and Rothleder (2006) found that having wind generation spread into five separate geographical areas with different weather patterns and power production patterns improved the management of the wind power production in California.

Co-Siting of Electricity Generators

Co-siting of generators has the potential to smooth temporal variations of electricity generation associated with intermittent renewable resources. For example, as shown in the wind resources map in Chapter 2 (Figure 2.1), some of the best wind resources are located away from demand centers and existing transmission capacity. If an intermittent resource is located at some distance from any load center and if there is insufficient transmission in the vicinity of the resource, transmission lines have been and will continue to be built or upgraded. The dilemma is how to size the transmission line. If size is based on the nameplate rating of the aggregated generation, the line will not be fully used. If the line is sized based on a criterion related to average capacity, then there will be times when some of the remote generation will be curtailed. Though no transmission line is loaded to capacity all of the time, increasing the usage through co-location could improve the economics of additional transmission capacity by smoothing temporal variations in electricity generation. It should be noted that in the concept of co-location

---

[6]Because of low existing wind power generating capacity in the United States, most studies are based on modeling.

of resources for pooling transmission resources, the effect is the same if the plants are close but not actually on the same site.

Conventional and Renewable Electricity Generators

Co-location of a conventional fossil-fired power generator in close proximity to the renewable resource can address renewable intermittency. For example, Mesa Energy has recently announced plans to build 500–600 MW of natural gas base-load capacity and a separate 300 MW of peaking capacity in close proximity to its West Texas wind farm development (Stern, 2008). The approach taken by Mesa and others is intended to allow the transmission line not only to be dedicated to the wind resource to increase its utilization, but also to be amortized as part of the greater generation/transmission system in that area. Consideration must be given, of course, to fuel transportation costs, since generators would not necessarily be located near fossil resources or existing distribution networks.

Two (or More) Renewable Electricity Generators

In certain locations within the United States, two (or more) renewable resources may be co-located to take advantage of temporal synergies, including both daily and seasonal fluctuations. Wind and solar are intermittent resources that can interact synergistically in locations where solar energy peaks during daylight hours and wind energy peaks during late-night hours. Meteorological conditions may also create synergies between solar and wind power, such as in areas of the country where low barometric pressure fronts create more windy and cloudy conditions, and stable, high-pressure conditions create sunny, stagnant conditions.

Co-location might also help renewable generation located in remote regions. The California Renewable Energy Transmission Initiative has looked at the use of Competitive Renewable Energy Zones (CREZs) to aggregate projects based on their physical location and shared transmission needs (Black & Veatch, 2008). Significant progress has been made in Texas's CREZs, and recently specific projects have been purposed to take advantage of the new transmission structure.[7] The issue of co-locating renewables is also the subject of the NREL Western Wind and Solar study looking at the costs and operating impacts due to the variability and uncertainty of wind, PV and concentrating solar power (CSP) on the grid (Lew,

---

[7]See http://www.seco.cpa.state.tx.us.

2008). Hydropower has been used in Europe to balance wind power production. In particular, Denmark is able to balance its large penetration of wind (almost 20 percent of generation in 2007) in part due to its interconnection to hydropower production in Norway and Sweden and its interconnection to Germany (Sharman, 2005).

Renewable Generators and Storage

Greater use of storage technologies as discussed in Chapters 3 and 4 is an important consideration with respect to renewables' contribution to electricity generation for the later time periods (post-2020) if the level of generation by non-hydropower renewables reaches 20 percent and above. In the mid-term (2020–2035), storage capacity could help relieve the sizing of transmission lines from remote renewable resources as well as increase the flexibility of those resources. Storage tied to intermittent renewables can have three distinct purposes: (1) to increase the flexibility of the resources; (2) to increase the use of transmission line(s) connecting the resource to the grid; and (3) to increase the on-peak availability of renewable electricity. The last is particularly relevant if the resource is remotely located from the load centers, though it would have similar value for local wind resources. Several types of electricity storage used or under consideration for supplementing renewable electricity include pumped hydropower, compressed air energy storage (CAES), and advanced stationary batteries. In the renewable electricity context, chemical energy storage that uses the electricity generated from distributed technologies and the customer side of the meter, such as solar, wind, or other renewable resources, could run an electrolysis process that creates hydrogen or another fuel. As noted in Chapter 3, storage could increase the interconnection between the generation and transportation sectors. Because large-scale storage is not necessary until later timeframes (post-2020), it is not necessary at this time to identify what approach to storage would be the most functional with renewables.

RENEWABLE ENERGY MARKETS

Deploying new technologies requires a concerted effort in overcoming market barriers (Box 6.2) and in meeting investment requirements. Businesses generally adhere to a strict economic payback criterion: an internal rate of return (IRR), often >15 percent, that must be met for any corporate investment to be under-

BOX 6.2 Key Barriers, Opportunities, and Stakeholders Affecting the Wide-Scale Deployment and Integration of Renewable Energy Sources

Barriers

Relatively High Costs
- High capital and operating cost uncertainty
- Cost of financing
- Competing demands for funds, zero-sum game economics in a flatter demand
- Increased energy-intensity phase
- Economies of scale and technology learning

Lack of Knowledge
- Inadequate workforce
- Complex decision making
- Capability of the product must be understood by the market

Market Inertia and Risk Aversion
- Inertia
- Perception of risk
- Sunk investments
- Small number of market actors
- Structure of energy industry

Infrastructure Limitations
- Electric transmission and distribution system
- Insufficient supply and distribution channels
- Limited complementary technology (energy storage)

Lack of Performance Validation and Experience
- Performance uncertainty
- Lack of maturity—20 year (wind) versus 100 year (coal) sectors

Opportunities
- Economic growth
- Industry sector leadership (local and export markets)
- Energy security
- Climate spin-off technologies
- Collateral benefits (climate security, reduced health care costs, enabling green building sector, economic development, education, and so on)

Stakeholders
- Customers
- Manufacturers
- Regulators (power, environmental)
- Policy makers

taken. Surmounting market barriers to new renewable energy commercialization requires both alleviating consumer hesitancy to try new products and mitigating business risks associated with adopting technologies prior to a predictable, measurable market demand (Brown et al., 2007). Significant difficulties are associated with attempting to enter and successfully compete in well-established, highly regulated, structured markets characterized by large, well-capitalized incumbents. As a general rule, financial markets demand higher returns from investments by new market entrants than from incumbents. This fact reflects the risk associated with being a market challenger and constitutes a market barrier for the challenger.

Basic Market Structure

Renewable energy markets have some distinctive attributes compared to traditional electricity markets. Markets for renewable electricity can be viewed from two perspectives: distributed power production, where the electric power is generated and used on-site; and wholesale power production,[8] where the electricity is sold and distributed to customers through the transmission and distribution grid. In each of these renewable electricity market structures, the energy generated and its associated renewable attributes can be, and frequently are, sold separately.[9] Rather than being lumped together with all the generation on the local electrical grid, renewable attributes can follow a contract path to a customer, giving the end purchaser a legitimate claim of a specific percentage of renewable electricity in that electricity purchase. The math is fairly simple: 1 kWh of conventional system power, plus 1 kWh worth of wind energy attributes (uniquely used and sold to a single ultimate customer), equals one unit of renewable electricity. The ability to track and sell the electricity and the renewable attributes associated with that energy substantially increases renewable energy market opportunities by increasing the number of ways that renewable energy sales can occur. Box 6.3 indicates the wide array of markets that can be available to renewable electricity for one of these two attributes. Market flexibility opens more opportunities for renewable

---

[8]The aim of wholesale renewable electricity markets is to increase sales of renewable electricity to retail customers.

[9]One customer purchases the kilowatt-hours of energy, while another can buy the attributes (e.g., avoided air emissions) associated with renewable electricity generation. Renewable attributes sold as separate products are variously called a renewable energy certificate, renewable energy credit, tradable renewable energy credit, or tradable renewable energy certificate.

BOX 6.3 Types of Renewable Electricity Ownership and Markets

Customer Ownership

With customer-owned renewable energy generation equipment, the self-generation of electricity displaces an existing or potential purchase of electricity from the grid or other source of generation. Solar photovoltaics and small-scale wind are examples of distributed generation (customer-owned generation). In this case, a homeowner will invest the needed capital to pay for the system purchase and installation, with cost recovery occurring over some period of time from displaced electricity savings.

Third-Party Ownership

Under the third-party ownership model, third-party companies install and own renewable energy systems on behalf of a host utility customer. Such systems are located behind the meter on a utility customer's premises. The third-party company acts as technology integrator, project developer, and system operator, and often secures the project financing. The electricity generated is sold to the host customer at a rate at or below the prevailing utility retail rate.

The success of this model often relies on several key factors, including (1) net metering, which allows valuation of displaced grid electricity at the prevailing retail rate; and (2) the ability of the third-party entity to raise capital at rates well below the IRR hurdle of the private companies. Other factors include the availability of federal and state incentives; the existence of a time-of-use utility tariff in which the utility's high-price rate tiers match well with the solar electricity output; and an existing market for renewable energy credits, which provides additional value.

Wholesale Energy

Wholesale, or utility-scale, projects generally compete against wholesale power prices. Larger projects can usually achieve lower costs from economies of scale throughout the value chain. In the case of PV, price competition can also encourage a greater degree of vertical integration throughout the PV value chain to achieve available cost savings and margin compression. In the most extreme example, a project developer with PV module manufacturing capabilities might have a distinct advantage in offering the lowest-priced solar projects.

Utility-scale projects can take the form of either utility ownership or third-party ownership using traditional power purchase agreements (PPAs). To date, most utilities have used the PPA model for solar project development, primarily because of technology risk considerations[1] and the fact that utility property is not currently eligible for the federal ITC.

Utility Ownership

Some key benefits of utility ownership include long-term amortization of capital investments (often 30 years), the ability to manage electricity transmission requirements, and the ability to manage grid impacts. Utilities can also earn a rate of return on owned assets unlike with PPAs, which are a cost pass-through to customers. However, utilities typically do not vertically integrate into equipment manufacture, and thus are subject to paying market prices for the renewable technologies.

Utility-Sponsored Green Pricing

In many jurisdictions, customers can buy electricity from a renewable energy facility through the grid via a utility-sponsored green pricing program. These programs tend to have a cost premium for subscribing to renewable electricity, though customers may also enjoy the system benefits associated with renewable energy purchases, including the value of saved fuel costs or costs associated with pollution control equipment required by fossil generation.

Renewable Energy Credits or Certificates

Customers can directly buy renewable energy credits to match with their conventional electricity purchases in order to green their electricity use. The U.S. EPA Green Power Partners Program calls attention to voluntary renewable energy credits purchases by large electricity users.

Retail Electricity Choice

In relatively few jurisdictions, customers may choose their retail electricity supplier. Some of these suppliers offer renewable electricity options.[2] Given the mechanics of electricity markets, these suppliers are typically selling a retail product comprised of generic system power combined with renewable energy credits.

---

[1]Typically, electric utility companies are not willing to take on the cost and operational risk of new technology deployment. For example, only now are some utilities willing to own wind energy projects after several years of successful third-party project development. Likewise, utilities are shunning investment in commercially unproven coal-fired technology such as integrated gasification combined cycle (IGCC).

[2]During the height of California's competitive markets, some 250,000 customers bought renewable energy-based retail electricity supply products.

continued

BOX 6.3 Continued

Renewables Portfolio Standards

Many customers are indirect purchasers of renewable energy through their utilities. In states where utilities or retail suppliers are required, typically by legislation, to procure a minimum amount of renewable energy for their total portfolio (under a renewables portfolio standard), all customers contribute indirectly to increasing the renewable electricity in the grid. The notion behind the RPS concept is that customers acting together can procure more renewable energy at a lower individual cost than they would as individual retail purchasers. Most RPSs also incorporate tradable renewable energy certificate programs designed to use market forces to reduce the total cost of procuring the renewable energy.

Utility Resource Planning or Integrated Resource Planning

In recent years, public utility regulatory authorities have been revisiting the process of comprehensive resource planning review for regulated utilities. Utilities submitting resource or integrated resource plans are often required to procure increasing amounts of renewable energy generation as a way to diversify supply, reduce costs, reduce emissions, and support economic development. Acquisition costs are rate-based and spread to all utility customers according to traditional regulatory approaches.

electricity to become part of the supply mix and increases the number of potential customers.

Market Infrastructure and Inertia

Mature industries have established, complex, and interdependent industry structures, where specialization in supporting sectors, such as maintenance and service, law, regulation, financial support, and insurance, develop and become more efficient as a business sector expands. In well-established markets, supply chain actors integrate effectively; parts and tools become standardized; and operating standards are adopted. Emerging industries are in a transition between novel and mature status; product demand, growth potential, and market infrastructure needs are unknown. Emerging renewable energy technologies are speculative, as both the individual companies and the industry itself have yet to be established in the larger electricity sector. Expanding the use of renewable energy technologies will require an appropriate distribution chain to move the product from the manufacturer through to the consumer (OECD/IEA, 2003). Some technologies may not easily fit into an existing market infrastructure. Moreover, the availability, quality, and location of resources vary by region and thus will have an important impact on how regional infrastructure will develop. Failure of the emerging renewable energy industry to mature in a timely fashion poses a significant impediment to substantially increasing renewable energy availability.

Established markets and incumbent technologies are better insulated from policy and regulatory variations than are emergent ones. Markets and industries organize around and within their regulatory and policy frameworks and become conservative, maintaining stability and reducing risk. The large, interconnected nature of established industries, especially one as large as the electric service sector, also distributes risk broadly and efficiently through a well-developed value chain. Renewable electricity is only beginning to realize this kind of stability, and only in parts of the United States.

Market inertia is an initial challenge to renewable energy deployment. Inertia (or resistance to change) is found in well-established conventional energy markets. Likewise, consumers are resistant to change. A study on the differing views of the target consumer and the manufacturer when making a value assessment during new product introduction and adoption concluded that, when sizing up a new product, consumers tend to assign its product value based on a perceived value (using experiences with similar products as a point of reference) as opposed to the actual product value (Gourville and Sellers, 2006). The consumer also practices a high level of loss aversion and assigns greater weight to potential shortcomings than to potential benefits. Manufacturers marketing a new product, on the other hand, tend to overvalue the new product and overestimate the probability that the consumer will see the same value. Firms may also tend to invest in improvements to the cost-competitiveness of products they already manufacture or service. This type of technology lock-in helps explain why disruptive innovations typically result from new businesses and not existing firms (Brown et al., 2007).

DEPLOYMENT RISK AND RELATED ISSUES

Risk in its simplest form is the likelihood that things will not turn out the way we expect. For financial and economic performance, risk is the degree of likelihood that an investment will not yield sufficient returns. The impacts of this risk can be found in the cost of capital and rates of returns required in the market. Investments in renewable energy markets are made all along the value chain, and all assume a certain ultimate range of sales of generated electricity and related products to the markets for renewable energy. In theory, the internal rate of return demanded by investors and the interest charged by lenders reflect the risk associated with a particular investment.

Economic and financial risk issues for renewable technologies involve a host of uncertainties associated with the electricity market. The growth in renewable energy as a share of U.S. electricity generation comes at the expense of incumbent conventional supply sources, and their potential growth in those sectors. While far from purely competitive due to an embedded system of rules, regulations, subsidies, habits, and customer expectations, the U.S. electricity system is still highly competitive, with mature energy companies directly competing with emerging renewable energy companies. Ultimately, renewable electricity will have to compete in this market environment, and investors and policy makers will assess the risks associated with the possible profit outcomes for renewable electricity business ventures against risks in other electricity sources, or in other investment opportunities in general.

Perspectives of Risk by Public and Private Sector Investors

Bringing increased renewable energy technologies to market requires public and private sector investment and commitment to overcome the various market barriers. Public and private sector investors will approach risk and related issues associated with renewable energy technologies from different perspectives, as outlined in Table 6.6. Private sector investment is typically based on the ability to achieve an acceptable return on investment and on projected business growth. The private sector's decisions on whether to move ahead with a renewable electricity project usually begin with an evaluation of the business case for such an investment. On the other hand, public sector investments incorporate different criteria, such as whether to attempt to spur the development of technologies that meet public sector goals by taking on some of the early development and deployment risks that the private sector may not assume. Government investment in high-risk research and development provides the information and impetus for the private sector to pursue public sector goals, and it also reduces deployment risks to private sector investors by developing technology certifications and standards. Decades of research funded by DOE in the areas of energy efficiency and fossil energy have

TABLE 6.6 Key Characteristics and Perspectives of Public and Private Sector Investors

	Public Sector Investors	Private Sector Investors
Key goals	• Develop promising technology options that meet public sector needs by reducing early technology risks that private sector investors would otherwise not assume • Private sector will subsequently exercise its option to invest	• Profitable investments in technology-based businesses that address real market needs —Investments that are technology neutral within the context of meeting customer needs
Investment focus	• *Technology-focused* development of high-quality innovations —Early, high-risk RD&D —Technology performance and cost reduction —Technology certification and performance verification	• Early, prudent investments in *market-focused businesses* that emphasize: —Strong management teams —Products—not technologies —Market development and access to these markets; customer driven
Biggest concern	• Technical showstoppers	• Customer and market showstoppers
Other key contributing investor insights/ expertise/ strengths	• *Technology*-based perspectives on: —Capabilities, benefits, and applications —Technical competition (possible) • *Macro market perspectives* on energy needs and trends • Perspectives on *public policy and public good needs* and trends, as well as the potential for impact • Standards development	• *Business and financial* perspectives on: —Market-driven, customer benefits —Broader (beyond energy) sets of industry applications —Market competition • Specific market perspectives and trends for energy and other applications including market beachhead, and entry strategies • Ability to factor public policy impacts into investment and business formation decisions effectively
Key constraints on collaborations	• Investment collaborations must abide by governmental regulations including those for fairness of opportunities, and not competing with the private sector • Commercialization viewed as responsibility of private sector	• Investment collaborations should reduce the risk and improve the profitability of investments
Key enablers needed	• Collaborations that *accelerate the deployment* and use of the technology in which the public sector invests	• Access to the information, people, knowledge, and data necessary for sound investments • Entrepreneurs that are predisposed to, and/or already focused on, market/customer product and business development issues

TABLE 6.6 Continued

	Public Sector Investors	Private Sector Investors
Differences in funding process	• Competitive written proposals judged mainly by a technology-focused review team; decisions sometimes appealed • Non-disclosure agreements (NDAs) not unusual	• Final decisions based in large part on presentations by management team; supported by extensive due diligence; decisions seldom reconsidered and not subject to review by higher authority • NDAs very rarely used
Pay off	• Technology is commercialized and public-good goals are met, including energy diversity, security, and environmental protection • Public sector has no direct ownership	• Profit through capital appreciation, i.e., increase in value of ownership stake. Profits are often realized at later investment stages through an exit strategy

Source: Murphy and Edwards, 2003.

yielded significant economic, environmental, and security benefits; new technical options with applications in other fields; and general contributions to the stock of scientific and technical knowledge (NRC, 2001).

Product Cost Evolution

New products, including renewable energy technologies, experience high initial cost in the market. Inexperience in the manufacturing and deployment process is one reason for higher costs and contributes to the greater risks for fiscal, regulatory, and other market participants in adopting a new technology. As outlined below, costs can come down as market participants gain experience with a new technology. However, using this market experience to reduce costs is greatly aided by the participation of early adopters and niche markets. Their feedback to other market participants, such as the technology manufacturers, installers, and regulators, can be critical for reducing costs.

Learning Curve

Technology learning based on increasing economies of scale in production, additional research and development, learning-by-doing, learning-by-using, and improvements in product distribution and service can result in cost reductions (IEA, 2008). Learning curves measure and illustrate the reduction in unit cost of

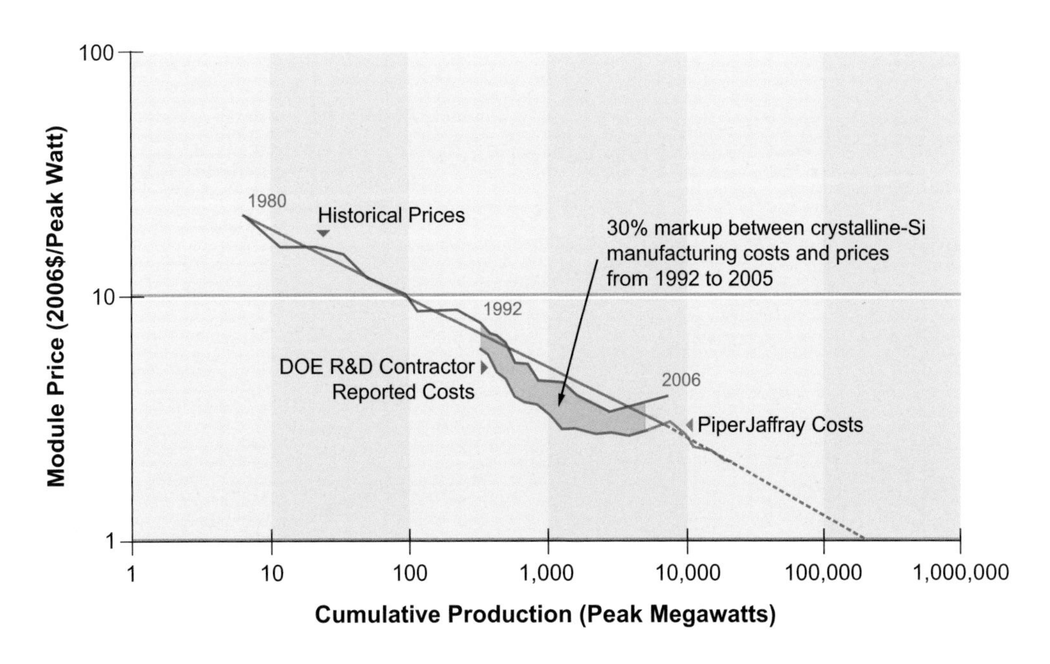

FIGURE 6.8 *Learning curve cost reductions for crystalline silicon PV modules. Source: Department of Energy; presented in Cornelius, 2007.*

a technology as a function of the cumulative increase in sales and deployment of the technology (Jamasb, 2006). As demand increases, the manufacturer of the technology develops mechanisms to overcome barriers not directly related to the cost or performance of a technology. Subsequently, other market actors learn how to use the technology more efficiently. Figures 6.8 and 6.9 illustrate the historical decrease in cost of solar PV systems with a cumulative increase in installed capacity. This trend can be an indicator of future decreases; however, the panel recognizes the difficulty of achieving this goal. Other energy technologies have experienced similar cost decreases, as shown in the slope of the cost reduction curve for the natural gas combustion turbine in Figure 6.9. Other examples of estimated cost reductions associated with learning include large automobile manufacturers that show a cost reduction of 25 percent to 30 percent for a 10-fold increase in production volumes (personal communication, K.G. Duleep, February 26, 2008); the developer of the Tesla Roadster, a limited production electric sports car, contends that the cost for the vehicle could be reduced by a factor of two for a 10-fold increase in production volume (Newsweek, 2008); and an assessment of cost reductions needed for solar PV to reach 10 percent of electricity generation that

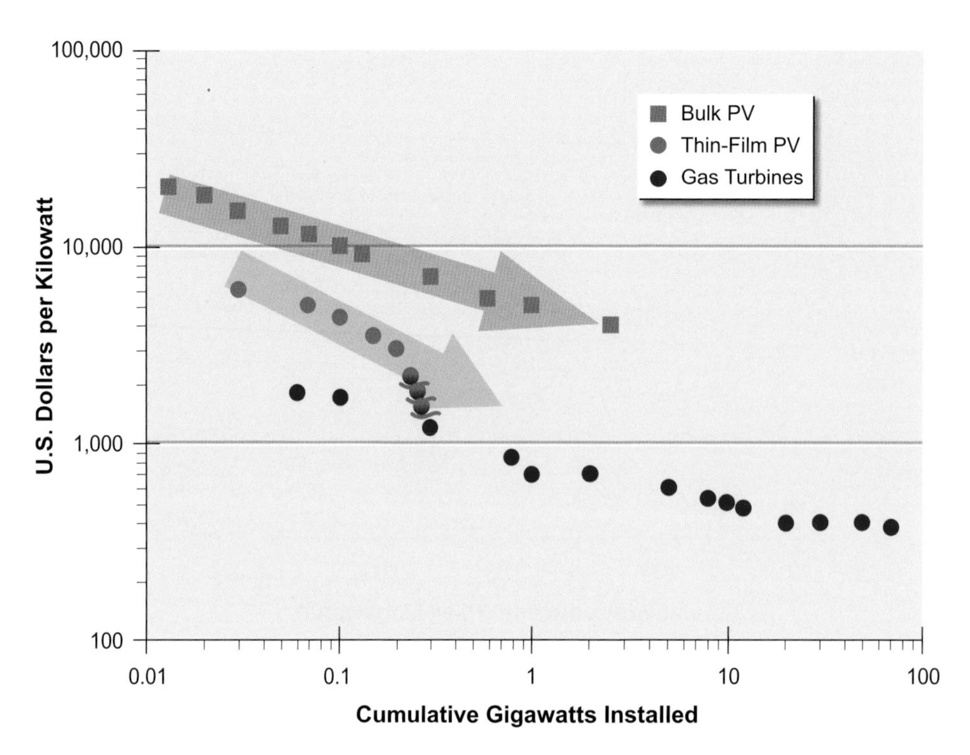

FIGURE 6.9 *Learning curve cost reductions for crystalline silicon solar PV, thin-film solar PV, and natural gas combustion turbines.*
Source: Courtesy of Charles Gay and Applied Materials.

assume an 18 percent reduction in installed PV cost (per MW) for every doubling of the market size (Pernick and Wilder, 2008).

Diffusion Curve

Technology deployment is a combination of market pull, where early stage consumers create a demand for the technology, and scientific push, where the technology developer actively promotes the technology to the market. Experience is gained in the market when innovators and early adopters willingly take new technology risks and encourage more risk adverse consumers that the technology is worth the investment (Mathur et al., 2007). The adage is that everyone wants to be the first to be the second to adopt a new technology. Market growth can be illustrated as in Figure 6.10 through use of the diffusion curve showing demand for a product increasing as early adopters start building on the experience of innovators. Market mechanisms, such as learning-by-doing and learning-by-using, bring the product to the inflection point on the curve where demand increases rapidly. Prior to the inflection point, the demand for the technology may not be self-sustaining and could benefit from public/private partnerships to share early market risk and provide a feedback mechanism for integrating market experience with research and development activities. In addition, the shape of the diffusion curve varies by the characteristics of the technology and the local market, as shown in the diffusion curve in Figure 6.10.

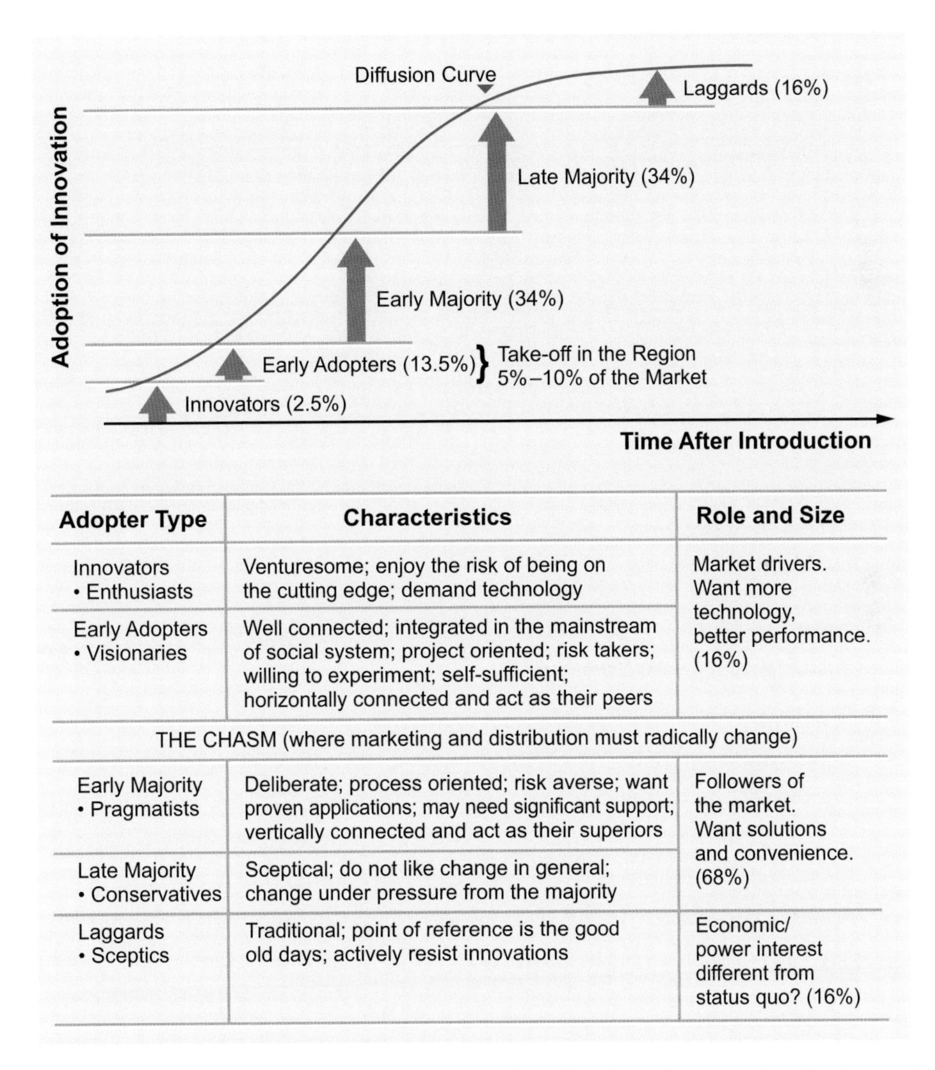

Adopter Type	Characteristics	Role and Size
Innovators • Enthusiasts	Venturesome; enjoy the risk of being on the cutting edge; demand technology	Market drivers. Want more technology, better performance. (16%)
Early Adopters • Visionaries	Well connected; integrated in the mainstream of social system; project oriented; risk takers; willing to experiment; self-sufficient; horizontally connected and act as their peers	
THE CHASM (where marketing and distribution must radically change)		
Early Majority • Pragmatists	Deliberate; process oriented; risk averse; want proven applications; may need significant support; vertically connected and act as their superiors	Followers of the market. Want solutions and convenience. (68%)
Late Majority • Conservatives	Sceptical; do not like change in general; change under pressure from the majority	
Laggards • Sceptics	Traditional; point of reference is the good old days; actively resist innovations	Economic/ power interest different from status quo? (16%)

FIGURE 6.10 *Generalized diffusion curve for adoption of new technologies and key characteristics of the various adopters.*
Source: IEA, 2008.

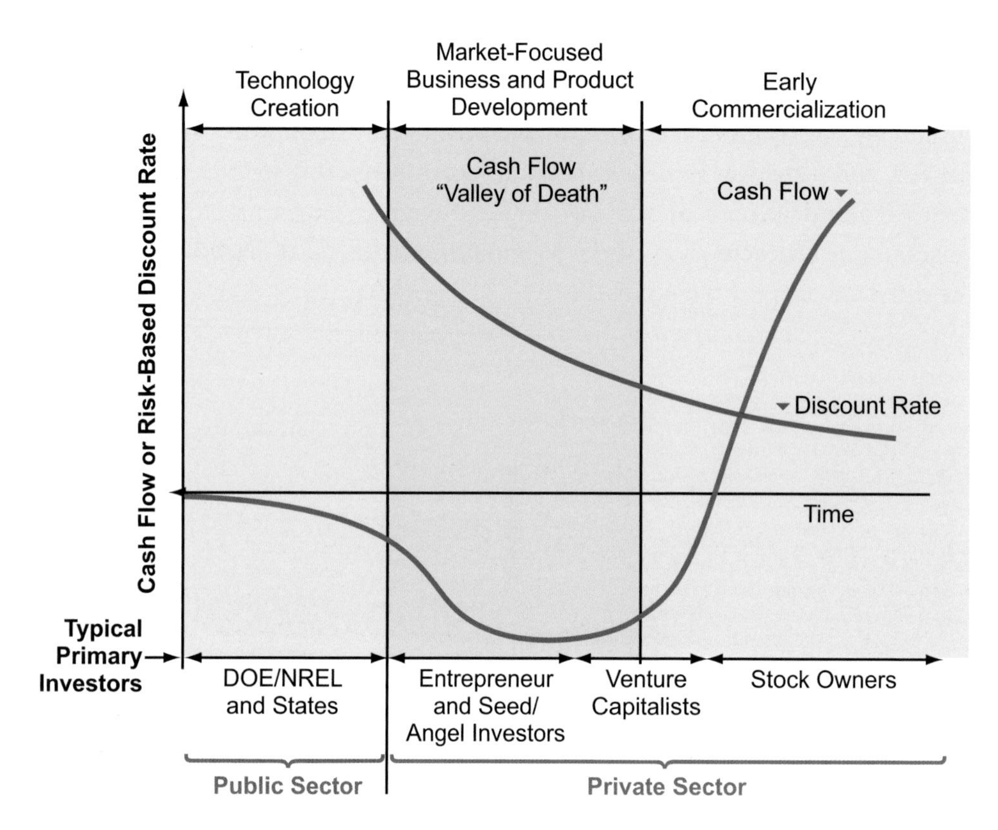

FIGURE 6.11 *The cash flow valley of death for the process from product development to commercialization.*
Source: Murphy and Edwards, 2008.

Commercialization Risk

Investment is necessary to move the new technology to the point of commercialization. Similar to the diffusion curve that illustrated the different market participants as demand for the product increases, Figure 6.11 illustrates the relative financial investment necessary to bring a new technology to the point of commercialization. Figure 6.11 depicts three broad stages of development where investment is needed: (1) the technology creation stage, when the public sector focuses its investment; (2) the cash flow valley of death[10] stage, after product development

[10]The cash flow valley of death occurs at the point when the financial investment required to commercialize a new technology may exceed the ability of a new business to raise capital. For clean energy technologies, this occurs during the transition from public sector financing to private sector funding.

but before commercialization, when public sector financing may not be available and there is typically a dearth of private capital; and (3) the early commercialization stage, when a company has an improved position with respect to obtaining private sector investment (Murphy and Edwards, 2003). The risks associated with the introduction and successful deployment of a new technology are directly tied to whether the developers of the technology can successfully navigate through these various stages, especially the stage between technology innovation and commercial introduction.

Issues Related to Electricity Rates

Several issues related to the basic approach to electricity services rate regulation have a significant impact on the renewable electricity deployment risks. These issues particularly arise in three areas: (1) the treatment of intermittent resources; (2) the development of supporting infrastructure (hardware and policy) for bulk power (transmission); and (3) the development of supporting infrastructure (hardware and policy) for distributed energy resources. Numerous regulatory and policy initiatives have been launched to address these issues in recent years. The most significant risk facing the large-scale deployment of renewable electricity in this regard is whether policy makers and regulators will move to address these issues in an orderly, predictable, and sustainable fashion.

The relationships between rates and market behavior by suppliers and consumers are complex and vary by location; indeed, rate is a term that describes a whole host of tariffs,[11] rates, and charges that affect those that interact with the electricity system. A renewable electricity facility, for example, faces a wholesale rate at which it can sell electricity and another rate for sales of capacity. There is one rate for transmission service, another for interconnection, yet another for standby service, and there may be ancillary services charges as well. As delivered to the end-user customers, the final bill for the electricity may include congestion charges, and ultimately will be bundled with distribution, metering, and billing-rate elements. Uncertainties about the application and charges in any such rate elements may slow progress toward greater deployment of renewable electricity.

A few examples of where growth in renewable electricity market penetration may conflict with the current rate structure and where regulatory risk associated with rates may be significant include:

[11] A government-approved contract rate.

- The volume-based, average rate per customer class model for consumption favors baseload generation capacity and fails to create incentives for resources like photovoltaics that generate electricity on or near peak.
- Net metering schemes that do not assign full retail value to generation occurring behind-the-meter may not encourage distributed generation.
- Transmission capacity reservation and shortfall charges that drive high availability for dispatchable resources (such as natural gas turbines) can effectively preclude cost-effective deployment of intermittent resources.
- Rate structures driven by efforts to encourage all-requirements loads and customers in order to build demand for capital investments often penalize partial-requirements loads coupled with self-generation. Renewed interest in demand-response and interruptible loads may require reexamination of rate-making fundamentals.

There is a chicken-and-egg problem associated with rates. Most often in the United States, rates are calculated based on extrapolation from a historical test year of experience, and adjudicated in contested rate cases. While the general constructs of rate making are well understood, there are variations in all the jurisdictions with authority to impose them. These jurisdictions are primarily states and the federal government, but also include municipal governments, electric cooperative boards, and multistate electric reliability and transmission authorities. Because there has been relatively little experience in the United States with large-scale deployment of renewable electricity (above the scale where significant impacts are experienced), there is relatively little actual data on which to construct fair and non-discriminatory rates. Any period of expansion in the amount of renewable electricity will therefore be accompanied by risk related to how the rate structure treats renewables.

Policy and Regulatory Risk

The relationships among markets to policy and regulation can be contributory, supportive, symbiotic, and parasitic. This is true for the electricity market as well as all sectors of the economy. All participants in the electricity market seem to agree that policy and regulation can have a profound impact on energy markets and that predictability and sustainability are highly valued. Electricity markets operate within a web of interlocking, overlapping, and sometimes conflicting policy prescriptions and legal and regulatory structures.[12] The key risks engendered by this pervasive regime relate to the degree to which one can expect that future policies will conform to reasonable expectations. For example, uncertainty surrounding the renewal of federal production tax credit policy for renewables carries a potential impact for the renewables industry in the billions of dollars. Regulation is the tool for implementing policy in the electric industry, even when that implementation involves relaxation of regulation. As the United States Supreme Court has held, when business is "affected with the public interest," such regulation is proper (Munn v. State of Illinois, 94 US 113 [1876]). There are few industries so affected with the public interest as that of electricity.

Renewable electricity will always be fundamentally affected by wider regulatory and policy conditions existing in electricity markets for several reasons. First, of course, is the ubiquity of electric service in the United States. Second, the dominant industry model is one based on spreading of costs through franchised service via regulated utilities. Even when some degree of competitive market structure exists as it does in much of the electricity sector today, the industry remains highly regulated. Third, the most significant environmental attributes of electricity are also spread broadly through energy security and reducing greenhouse gases. Indeed, greenhouse gas emissions are part of a global budget of atmospheric gases. Finally, the technologies and businesses of renewable electricity are young and relatively immature. Development of renewables depends on research and development, as well as special subsidies and manipulation of the existing markets, for renewables to succeed against well-established incumbents that enjoy embedded subsidies of their own.

Electricity Sector Regulation

Regulators and policy makers in the electricity sector are often uncertain about how to deal with new market entrants, new technologies, and new product and service models. Charged with protecting the general public interest, regulators, and policy makers often approach innovation with caution, and on an ad hoc basis. Regulation and policy designed for incumbent industries may not be well suited to emerging technologies and businesses, but efficient alternatives are not often apparent. New market entrants often face risk due to lack of clarity and

---

[12]The various incentives for renewable energy are catalogued by the Database of State Incentives for Renewables and Efficiency (DSIRE), available at www.dsireusa.org.

specificity; newcomers must spend proportionally more time and money to engage with the regulatory systems than well-known incumbents.

Large-scale deployment of renewable electricity will add a new dimension to this uncertainty. For example, relatively simple and clear regulatory and management solutions exist for wind penetration rates of 1 percent or 2 percent, but the need for potentially expensive regulatory changes and ancillary services may occur as system penetration rates reach 10 percent to 20 percent and higher. Moreover, effective response to system-scale issues requires comprehensive reviews and solutions. Regulatory processes, such as integrated resource planning, rate cases, and broad revisions of transmission system pricing regimes, place heavy demands on scarce regulatory resources.

Climate Regulation

Climate change regulation and policy are emerging in many local and regional jurisdictions around the United States. Many other countries have also implemented climate regulations. Indeed, increasing attention and concern about the potential for global climate change is having impacts on business decision making and risk evaluation, especially companies operating in the power sector and energy-intensive industries. Renewable energy industries should benefit greatly from comprehensive and effective regulation to reduce or avoid greenhouse gas emissions. Greenhouse gas regulation will likely affect the relative costs of renewable electricity and non-renewable fossil-fuel and nuclear power options and spur more rapid technology improvement in renewables. However, there are risks. Greenhouse gas regulation is itself a new thing, and changes and inconsistencies are inevitable. Because this regulation will have a direct impact on the costs and market opportunities for both incumbent and emerging technologies, the degree of orderliness and predictability of changes in regulations constitutes a significant risk factor for large-scale deployment of renewables.

Lash and Wellington (2007) categorize business risks associated with the public and regulatory climate change concerns as follows:

- *Regulatory risk*. Rates and direct regulation of emissions.
- *Supply chain risk*. Higher component and energy costs as suppliers pass along increasing carbon-related costs to their customers.
- *Product and technology risk*. Ability to identify new market opportunities for climate-friendly products and services.

- *Litigation risk*. Threat of lawsuits against companies that generate significant carbon.
- *Reputational risk*. Public, or consumer, perception on the role of the company as a steward of the environment.
- *Physical risk*. Risk posed by climate change as droughts, floods, and storms become more frequent and more severe.

These risks and benefits are summarized in Box 6.4. Deployment of renewable energy technologies can help electricity generators mitigate climate-change-related risks through reduced risk exposure, direct reductions in greenhouse gas emissions, improved ability to take advantage of climate policy incentives, reduced resource use, and improved perception of corporate social responsibility (Pater, 2006).

Environmental Policy

As discussed in Chapter 5, renewable electricity deployment is particularly site specific, whether for resource availability or access to infrastructure. The permitting process is intended to consider the local impacts on the land, water, and air that occur during the installation and operation of these technologies. As a result, local, state, and national governmental policies and regulations affecting the siting of generation and associated facilities will have a major impact on renewable energy deployment. The range of local, state, and national regulations confronting development also grows, and the risk of variability and inconsistency likewise increases as the scale of renewable energy deployment grows.

FINDINGS

Shown in bold below are the most critical elements of the panel's findings, based on its examination of issues related to the deployment of renewable electricity into the U.S. electricity supply.

Policy, technology, and capital are all critical for the deployment of renewable electricity. In addition to enhanced technological capabilities, adequate manufacturing capacity, predictable policy conditions, acceptable financial risks, and access to capital are all needed to greatly accelerate the deployment of renewable electricity. Improvements in the relative position of renewable electricity will

BOX 6.4 Risks and Benefits for Renewable Electricity Generation Under Climate Regulation

Risk management	• Hedge against fuel-price volatility • Hedge against grid outages • Get ahead in the futures markets • Prepare for regulatory change • Reduce insurance premiums • Reduce future risks of climate change
Emissions reduction	• Generate emissions reduction credits/offsets • Reduce fees for emissions • Avoid remediation costs
Policy initiatives	• Production tax credit, accelerated depreciation, property tax break • Preferential loan treatment • Renewables portfolio standard • Renewable energy certificates • System benefit funds • Rebates, feed-in tariffs, net metering • Sales-tax exemption • Local R&D incentives • Other financial incentives
Reduced resource use	• Reduce water use and consumption • Reduce energy use
Corporate social responsibility	• Improve stakeholder relations • Satisfy socially responsible investing portfolio criteria
Societal economic benefits	• Rural revitalization, jobs, economic development • Avoided environmental costs of fuel extraction/transport • Avoided costs of transmission and distribution infrastructure expansion

require consistent and long-term commitments from policy makers and the public. Investments and market-facing research that focuses on market needs as opposed to technology needs are also required to enable business growth and market transformations.

Successful technology deployment in emerging energy sectors such as renewable electricity depends on sustained government policies, both at the project and at the program level, and continued progress requires stable and orderly government participation. Uncertainty created when policies cycle on and off, as has been the case with the federal production tax credit, can hamper the development of new projects and reduce the number of market participants. Significant increases in renewable electricity generation will also be contingent on concomitant improvements in several areas, including the size and training of the workforce; the capabilities of the transmission and distribution grids; and the framework and regulations under which the systems are operated. As with other energy resources, the material deployment of renewable electricity will necessitate large and ongoing infusions of capital. However, renewable energy requires a greater allocation of capital than do the conventional fossil-based energy technologies to manufacturing and infrastructure requirements.

Integration of the intermittent characteristics of wind and solar power into the electricity system is critical for large-scale deployment of renewable electricity. Advanced storage technologies will play an important role in supporting the widespread deployment of intermittent renewable electric power above approximately 20 percent of electricity generation, although electricity storage is not necessary below 20 percent. Storage tied to renewable resources has three distinct purposes: (1) to increase the flexibility of the resources in providing power when the sun is not shining or the wind is not blowing, (2) to allow the use of energy on peak when its value is greatest, and (3) to facilitate increased use of the transmission line(s) that connect the resource to the grid. The last is particularly relevant if the resource is located far from the load centers or if the system output does not match peak load times well, as is often the case with wind power. However, wind power's development is occurring long before widespread storage will be economical. Although storage is not required for continued expansion of wind power, the inability to maximize the use of transmission corridors built to move wind resources to load centers represents an inefficient deployment of resources. Several parties are currently exploring the co-location of natural-gas-fired generation and other types of electricity generation with wind power generation to bridge this gap between storage technology and asset utilization. The co-siting of conventional dispatchable generation sources (such as natural-gas-fired combustion turbines or combined cycle plants) with renewable resources could serve as an interim mechanism to increase the value of renewable electric power until advanced storage technologies are technically feasible and economically attractive. The location of such natural-gas-fired generation could be at or near the wind resource, or at an appropriate site within the control area. Another possibility is the co-siting of two (or more) renewable resources, such as wind and solar resources, which might on average interact synergistically with respect to their temporal patterns of power generation and needs for transmission capacity.

Finally, it is important to note that the deployment needs and impacts from renewable electricity deployment are not evenly distributed regionally. Development of solar and wind power resources has been growing at an average annual rate of 20 percent and higher over the past decade. Overall electricity demand is forecasted to continue to grow at just under 1 percent annually until 2030, with the southeastern and southwestern regions of the United States expected to see most of this growth. Although some of this growth may correspond to areas where renewable resources are available, some of it will not, indicating the possible need for increases in electricity transmission capacity.

REFERENCES

Ancona, D., and J. McVeigh. 2001. Wind Turbine—Materials and Manufacturing Fact Sheet. Princeton Energy Resources International. Prepared for Office of Industrial Technologies, U.S. Department of Energy. Washington, D.C.

Archer, C.L., and M.Z. Jacobson. 2007. Supplying baseload power and reducing transmission requirements by interconnecting wind farms. Journal of Applied Meteorology and Climatology 46:1701-1717.

ASES (American Solar Energy Society). 2007. Renewable Energy and Energy Efficiency: Economic Drivers for the 21st Century. R. Bezdek, principal investigator, Management Information Services, Inc. Washington, D.C.

AWEA (American Wind Energy Association). 2008. Wind Power Outlook 2008. Washington, D.C.

Bierden, P. 2007. The process of developing wind power generators. Presentation at the second meeting of the Panel on Electricity from Renewable Resources, December 6, 2007. Washington, D.C.

Black & Veatch. 2008. Renewable Energy Transmission Initiative RETI Phase 1B—Resource Report. RETI-1000-2008-003-F. Prepared for RETI Stakeholder Steering Committee, Renewable Energy Transmission Initiative (RETI), Sacramento, Calif.

Bowen, J.L., and C.M. Christensen. 1995. Disruptive technologies, catching the wave. Harvard Business Review 73(1):43-53.

Bradford, T. 2008. Solar energy market update 2008—PV and CSP. Presentation at Solar Market Outlook: A Day of Data, February 19, 2008. New York.

Brown, M., J. Chandler, M.V. Lapsa, and B.K. Sovacol. 2007. Carbon Lock-in: Barriers to Deploying Climate Change Mitigation Technologies. ORNL/TM-2007/124. Oak Ridge, Tenn.: Oak Ridge National Laboratory. November.

CALISO (California Independent System Operator Corporation). 2007. Annual Report. Folsom, Calif.

CEWD (Center for Energy Workforce Development). 2007. An Action Plan for Workforce Development. Washington, D.C.

Cornelius, C. 2007. DOE Solar Energy Technologies Program. Presentation at the first meeting of the Panel on Electricity from Renewable Resources, September 18, 2007. Washington, D.C.

Council on Competitiveness. 2007. Energy Security, Innovation, and Sustainability. Washington, D.C.

Davis, R.E., and W. Quach. 2007. Intermittency Analysis Project: Appendix A. Intermittency Impacts of Wind and Solar Resources on Transmission Reliability. CEC-500-2007-081-APA. PIER Renewable Energy Technologies Program. Sacramento, Calif.: California Energy Commission.

DOE (U.S. Department of Energy). 2006. Workforce Trends in the Electric Utility Industry. Prepared for the United States Congress. Washington, D.C.

DOE. 2008a. 20% Wind Energy by 2030—Increasing Wind Energy's Contribution to U.S. Electricity Supply. Office of Energy Efficiency and Renewable Energy. Washington, D.C.

DOE. 2008b. Annual Report on U.S. Wind Power Installation, Cost, and Performance Trends. Office of Energy Efficiency and Renewable Energy. Washington, D.C.

EER (Emerging Energy Research). 2007. U.S. Wind Power Markets and Strategies, 2007-2015. Cambridge, Mass.

EIA (Energy Information Agency). 2007. Solar Thermal and Photovoltaic Collector Manufacturing Activities, Renewable Energy Annual 2006. Washington, D.C.: U.S. Department of Energy, EIA.

EnerNex. 2008. Detailed Analysis of 20% Wind Penetration. Addendum to Wind Integration Study for Public Service of Colorado. Prepared by EnerNex Corporation for Xcel Energy. Denver, Colo. December 1.

Feltrin, A., and A. Freundlich. 2008. Material considerations for terawatt level of deployment of photovoltaics. Renewable Energy 33:180-185.

GE Energy. 2005. The Effects of Integrating Wind Power on Transmission System Planning. Prepared for the New York State Energy Research and Development Authority. New York.

Gourville, J.T., and E. Sellers. 2006. Eager sellers and stony buyers: Understanding the psychology of new-product adoption. Harvard Business Review 84(6):98-106.

Great Valley Center. 2003. Workforce Implications in Renewable Energy. Modesto, Calif.

Greentech Media. 2007. Silicon Shortages Slows Solar Industry—The Green Year in Review. Cambridge, Mass. December 31.

Greentech Media. 2008. First Solar Posts Blockbuster 2Q. Cambridge, Mass. July 30.

Hawkins, D., and M. Rothleder. 2006. Evolving role of wind forecasting in market operation. Presented at the CAISO in IEEE Power Systems Conference and Exposition, Atlanta, Ga. Available at http://ieeexplore.ieee.org.

Holttinen, H., B. Lemström, P. Meibom, H. Bindner, A. Orths, F. van Hulle, C. Ensslin, A. Tiedemann, L. Hofmann, W. Winter, A. Tuohy, M. O'Malley, P. Smith, J. Pierik, J.O. Tande, A. Estanqueiro, J. Ricardo, E. Gomez, L. Söder, G. Strbac, A. Shakoor, J.C. Smith, B. Parsons, M. Milligan, and Y. Wan. 2007. Design and operation of power systems with large amounts of wind power: State-of-the-art report. VTT Working Papers 82. VTT Technical Research Centre of Finland. October. Available at http://www.vtt.fi/inf/pdf/workingpapers/2007/W82.pdf.

IEA (International Energy Agency). 2008. ETP 2008: Technology Learning and Deployment—A Workshop in the Framework of the G8 Dialogue on Climate Change, Clean Energy and Sustainable Development—Final Version 29. Paris. June.

Jamasb, T. 2006. Technical change theory and learning curves: Patterns of progress in energy technologies. Cambridge Working Papers in Economics, University of Cambridge, Cambridge, U.K. March 20. Available at http://www.dspace.cam.ac.uk/handle/1810/131682.

Lash, J., and F. Wellington. 2007. Competitive advantage on a warming planet. Harvard Business Review 85(3):95-102.

Lew, D. 2008. Western wind and solar integration study. Presentation at Stakeholders Meeting, August 14, 2008. Denver, Colo.

Mathur, A., A.P. Chikkatur, and A.D. Sagar. 2007. Past as prologue: An innovation diffusion approach to additionality. Climate Policy 7:230-239.

MPUC (Minnesota Public Utilities Commission). 2006. Final Report: 2006 Minnesota Wind Integration Study. Volume 1. Saint Paul, Minn.

Murphy, L.M., and P.L. Edwards. 2003. Bridging the Valley of Death: Transitioning from Public to Private Sector Financing. NREL/MP-720-34036. Golden, Colo.: National Renewable Energy Laboratory.

NAS-NAE-NRC (National Academy of Sciences-National Academy of Engineering-National Research Council). 2009. America's Energy Future: Technology and Transformation. Washington, D.C.: The National Academies Press.

Navigant Consulting. 2008. Photovoltaic Shipments and Competitive Analysis 2007/2008. Report # NPS-Supply3. Washington, D.C.

NERC (North American Electric Reliability Corporation). 2009. Accommodating High Levels of Variable Generation: Special Report. Princeton, N.J.

Newsweek. 2008. PayPal's cofounder hopes to produce a practical $30,000 all-electric car in four years.

NPCC (Northwest Power and Conservation Council). 2007. The Northwest Wind Integration Action Plan. Portland, Ore. Available at http://www.nwcouncil.org/energy/wind/library/2007-1.pdf.

NRC (National Research Council). 2001. Energy Research at DOE: Was It Worth It? Energy Efficiency and Fossil Energy Research 1978 to 2000. National Academy Press: Washington, D.C.

NYSERDA (New York State Energy Research and Development Authority). 2005. The Effects of Integrating Wind Power on Transmission System Planning, Reliability, and Operations: Report on Phase 2—System Performance Evaluation. Albany, N.Y. March.

OECD (Organisation for Economic Co-Operation and Development)/IEA. 2003. Creating Markets for Energy Technologies. Paris: OECD Publishing. Available at http://www.iea.org/textbase/nppdf/free/2000/creating_markets2003.pdf.

Parsons, B., M. Milligan, J.C. Smith, E. DeMeo, B. Oakleaf, K. Wolf, M. Schuerger, R. Zavadil, M. Ahlstrom, and D.Y. Nakafuji. 2006. Grid impacts of wind power variability: Recent assessments from a variety of utilities in the United States. European Wind Energy Conference Paper. NREL/CP 500-39955. National Renewable Energy Laboratory. Washington, D.C.: U.S. Department of Energy. July.

Patel, S. 2009. PV sales in the U.S. soar as solar panel prices plummet. POWER Magazine. March 1.

Pater, J.E. 2006. Framework for Evaluating the Total Value Proposition of Clean Energy Technologies. Technical Report NREL/TP-620-38597. National Renewable Energy Laboratory. Washington, D.C.: U.S. Department of Energy. February.

Pernick, R., and C. Wilder. 2008. Utility Solar Assessment (USA) Study: Reaching Ten Percent by 2025. Washington, D.C.: Clean Edge, Inc.

Prometheus Institute: For Sustainable Development. 2007. U.S. Solar Industry: The Year in Review 2006. SEIA/Prometheus Institute Joint Report. Cambridge, Mass.

Roy, B.S., S.W. Pacala, and R.L. Walko. 2004. Can large wind farms affect local meteorology? Journal of Geophysical Research 109:D19101.

Sharman, H. 2005. Why wind power works for Denmark. Civil Engineering 158:66-72.

Sheehan, G., and S. Hetznecker. 2008. Utility scale solar power. Next Generation Power & Energy (5; October).

Sherwood, L. 2007. U.S. market trends: Solar and distributed wind. Presented at Interstate Renewable Energy Council, September 24, 2007. Long Beach, Calif.

Stern, G. 2008. Lassoing panhandle wind: Oilman plans huge complex. EnergyBiz Magazine (January/February):12-13.

Sterzinger, G., and M. Svrcek. 2004. Wind Turbine Development: Location of Manufacturing Activity. Renewable Energy Policy Project. Washington, D.C.

UCS (Union of Concerned Scientists). 2007. Cashing in on Clean Energy National Analysis. Cambridge, Mass.

Wan, Y.H. 2004. Wind Power Plant Behaviors: Analyses of Long-Term Wind Power Data. NREL Technical Report NREL/TP-500-36551. Golden, Colo.: National Renewable Energy Laboratory.

Weissman, J., and K. Laflin. 2006. Trends in practitioner training for the renewable energy trades. Solar 2006: Renewable Energy—Key to Climate Recovery. Proceedings of 35th ASES Annual Conference, July 9-13, 2006, Denver, Colo. Boulder, Colo.: American Solar Energy Society.

Wiley, L. 2007. Utility scale wind turbine manufacturing requirements. Presented at WPA/NWCC Wind Energy and Economic Development Forum, April 24, 2007. East Lansing, Mich.

7 Scenarios

This chapter considers the extent to which renewable technologies might contribute to the future U.S. electric power supply. To come to conclusions about the level that renewables might contribute to electricity generation, we focus on scenarios of the technologic, economic, environmental, and implementation-related characteristics that may enable a greater fraction of renewable electricity. How much these factors might affect the market penetration of any individual renewable resource would depend on the rate at which generation from additional renewables is introduced. Under business as usual conditions without major policy initiatives to speed deployment, the introduction of renewables into electricity markets can continue at a moderate pace, with the growth rate and technology learning following a conventional S curve. But if policy makers or external conditions were to bring a sense of urgency to addressing concerns such as energy security or climate change, the question would become how to accelerate the market penetration of renewables while minimizing impacts on electricity's price, the environment, the reliability of electricity service, and the ability of industry to manufacture and deploy relevant technologies. The scenarios selected by the panel allow exploration of such issues.

The scenarios discussed below in this chapter were chosen to represent aggressive but achievable rates of renewables deployment in the U.S. electricity sector, provided that significant policy and financial resources are devoted to the effort. Scenarios do not represent a simple extrapolation of historical growth rates; instead, they reflect a more integrated perspective on the conditions required to scale up renewables deployment. The panel's criteria in choosing the particular scenarios it presents were whether the scenario was developed with input from multiple stakeholder groups and whether it underwent peer review. The panel also considered the degree to which each scenario assessed not simply deployment rates and cumulative levels of generation but also economic, financial, human, and environmental facets. Many of the scenarios described here have been released over the past few years, which helps ensure that inputs to the scenarios reflect recent conditions.

OBJECTIVES FOR SCENARIOS

Scenarios provide conceptual and quantitative frameworks to describe and assess how renewable resources' contribution to electricity supply might be significantly increased. Such scenarios are a primary way to quantify materials and manufacturing requirements, human and financial resource needs, and environmental impacts that come with greatly expanding electricity generation from renewable electricity sources. These scenarios typically use qualitative analysis, quantitative assumptions, and computational models of the energy, economic, and/or electricity systems. They attempt to integrate the environmental, technologic, economic, and deployment-related elements into an internally consistent analytical framework. The panel considered two types of scenarios. The first type analyzes increased market penetration of a single resource, such as solar or wind. A prominent example is the 20 percent wind study (DOE, 2008) described in more detail in the following section. Examples for solar energy include the *Solar America Initiative* (DOE, 2007b), the *U.S. Photovoltaic Industry Roadmap* (SEIA, 2001, 2004), and the 10 percent solar study (Pernick and Wilder, 2008). The scenarios described here are used to assess issues such as:

- Land-use impacts, manufacturing and employment requirements, and economic costs associated with an assumed market penetration of a single renewable resource (e.g., 20 percent electricity generation from wind power or more than 50 percent electricity generation from solar);
- The additional transmission, distribution, and other technologies needed to incorporate or enhance the use of intermittent renewable resources in the electricity market; and
- The cost-reduction trajectories needed to make solar electricity widely competitive with other electricity sources.

A second type of scenario examines how renewables interact with other sources of electricity, other sources of energy, and end-use energy demands (CCSP, 2007; EIA, 2008a). Through the use of long-term energy–economic models, these scenarios enable assessment of the potential impacts of demographic, economic, and regulatory factors on renewable electricity within a framework that considers the whole energy sector. The scenarios described here are used to explore issues such as:

- How wider energy–economic interactions and the electricity market could affect market penetration by renewables;
- The impacts of environmental, economic, and/or energy policies on end-use demand and electricity generation from renewables and other sources.

These scenarios, as with the reference case scenario presented in Chapter 1, are not predictors of the future, and the results of scenarios are not forecasts. Rather, they are descriptions of one set of conditions that could result in significantly increased market penetration by one or several renewables over what is estimated based on present-day conditions and a business-as-usual future. They demonstrate the costs, benefits, and scale of the challenges associated with increasing the integration of renewables into the electricity sector.

EXAMPLES OF HIGH-PENETRATION SCENARIOS

20 Percent National Wind Penetration Scenario

The American Wind Energy Association and DOE's National Renewable Energy Laboratory (NREL) developed a scenario assuming that 20 percent of electricity generation would come from wind power by 2030 (DOE, 2008). The scenario included assessments of the wind resource base, materials and manufacturing requirements, environmental and siting issues, transmission and system integration, costs, and public policy drivers (Smith and Parsons, 2007). The scenario estimated that more than 300 GW of new wind power capacity would be needed to meet a goal of 20 percent market penetration by wind, of which about 250 GW would be installed onshore and 50 GW installed offshore. Under this scenario, in 2030 wind power would produce about 1.2 million GWh out of a total U.S.

electricity generation of 5.8 million GWh. All impacts for the 20 percent wind scenario (such as costs and impacts on CO_2 emissions) were estimated through a comparison to a base case that assumed no new wind capacity additions after 2006, which is a more pessimistic base case in terms of wind power than both the AEO 2007 and AEO 2008 versions (EIA, 2008b,c). Because the 2008 DOE report contained "influential scientific information" as defined by the Office of Management and Budget's (OMB's) Information Quality Bulletin for Peer Review, it was subjected to interagency peer review.

Manufacturing, Materials, and Resources

Manufacturing and other requirements to implement a 20 percent wind scenario are significant. Figure 7.1 shows the amount of annual installed capacity needed to increase to 300 GW by 2030 from approximately 12 GW in 2006. Though the scenario limited the annual capacity increase to 20 percent, it assumed an

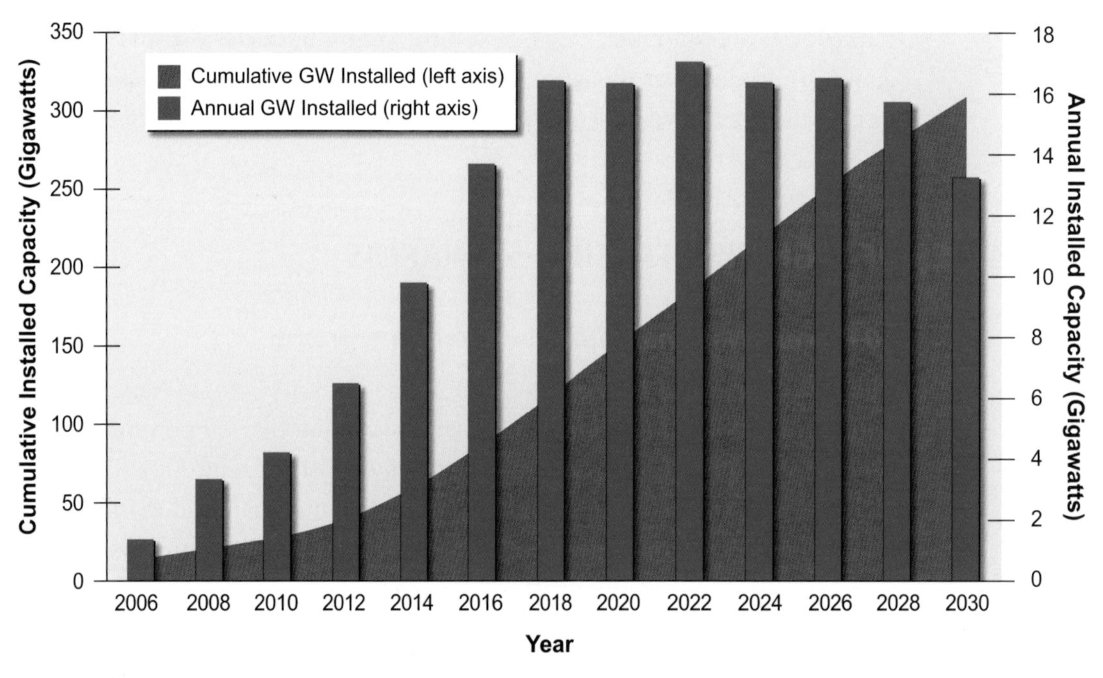

FIGURE 7.1 *Annual and cumulative generation needed to achieve 20 percent wind generation of electricity by 2030.*
Source: Lawrence Berkeley National Laboratory; presented in Wiser, 2008.

extremely large expansion of manufacturing, materials, and installation capacities. It projected that by 2018 the amount of annual installed capacity in the United States would be more than 16 GW, compared to a global wind turbine manufacturing output of about 15 GW in 2007 (DOE, 2007a). As discussed in Chapter 1 of the present report, an additional 5 GW of capacity was added in the United States in 2007 and more than 8 GW in 2008, both exceeding the trajectory for the 20 percent wind scenario. Even assuming that growth outside the United States would be more modest, this scenario would require a continued large expansion of the manufacturing base. Global growth in wind power is likely to continue to be strong. For example, the Commission of the European Communities' roadmap for renewables proposes that the European Union establish a mandatory target of 20 percent for renewable energy's share of energy consumption in the EU by 2020, much of which would be met with wind power (Commission of the European Communities, 2007).

The 20 percent wind scenario also contains critical challenges to fulfill materials, capital, and employment requirements. Table 7.1 shows the level of raw materials needed to meet this scenario. While some quantities would be small relative to global production, Smith and Parsons (2007) concluded that supplying fiberglass, core materials (balsa and foam), and resins could be difficult, as would supplying a sufficient number of wind turbine gearboxes. Assuming that the average-sized wind turbine would be in the 1–3 MW range, with modest introduction of large 4- to 6-MW turbines, there could be a total of almost 100,000 wind turbines installed (Wiley, 2007; DOE, 2008). The average number of turbines

TABLE 7.1 Raw Materials Requirements for 20 Percent Wind Scenario (thousands of tons per year)

Year	Concrete	Steel	Aluminum	Copper	Glass-Reinforced Plastic	Carbon Fiber Composite	Adhesive	Core
2010	6,800	460	4.6	7.4	30	2.2	5.6	1.8
2015	16,200	1,200	15	10	74	9	15	5
2020	37,000	2,600	30	20	162	20	34	11
2025	35,000	2,500	28	19	156	19	31	10
2030	34,000	2,300	26	18	152	18	30	10

Source: Adapted from material in Wiley, 2007.

TABLE 7.2 Net Present Value Direct Electricity Sector Costs for 20 Percent Wind Scenario and No-New-Wind Scenario

	NPV Direct Costs for 20 Percent Wind Scenario (billion U.S. 2006$)	NPV Direct Costs for No-New-Wind-After-2006 Scenario (billion U.S. 2006$)
Wind technology O&M costs	51	3
Wind technology capital costs	236	0
Transmission costs	23	2
Fuel costs	813	968
Conventional generation O&M	464	488
Conventional generation capital costs	822	905
Total	2,409	2,366

Note: NPV, net present value; O&M, operation and maintenance.
Source: DOE, 2008.

installed would have to increase from its present level of 2,000 per year to 7,000 per year by 2017 (DOE, 2008).

The NREL Wind Development System (WinDS) model, which simulates generation capacity expansion in the U.S. electricity sector for wind and other technologies through 2030, estimates that the 20 percent wind scenario would result in a direct increased cost for the total electricity sector of $43 billion (U.S. 2006$) in net present value (NPV) over the no-new-wind case. Table 7.2 shows the breakdown of direct electricity sector costs for the 20 percent wind scenario and the no-new-wind scenario. Overall, increases in wind power generation costs (capital and operation and maintenance [O&M] expenses) would be partially offset by lower capital, O&M, and fuel costs for other electricity sources. The total capital costs for wind under this scenario would be $236 billion NPV, and O&M cost would be $51 billion NPV. These cost estimates do not consider the total capital required for potential investments in manufacturing capacity, expanded employment training, or other needs, and do not represent the indirect costs to the economy. According to the scenario, in 2030, 20 percent market penetration by wind would provide well over 140,000 direct manufacturing, construction, and operations jobs, as indicated by DOE's Job and Economic Development (JEDI) model (Goldberg et al., 2004; Wiley, 2007; DOE, 2008). This projection would include more than 20,000 jobs in manufacturing, almost 50,000 jobs in construction, and more than 75,000 jobs in operations (DOE, 2008).

Integration of Wind Power into the Electricity System

Under this high-market-penetration scenario, integrating 20 percent wind power into the electricity system would require investment in the electricity grid and other parts of the electricity system. Transmission could be the biggest obstacle to seeing levels of wind power rise to 20 percent. Studies of wind integration at the utility and state level show that incorporating significant amounts of wind power into the electricity grid, while feasible, would require improvements in the transmission grid, wind forecasting, and other modifications to the electricity system, which would impose additional costs (Zavadil et al., 2004; GE Energy, 2005; DeMeo et al., 2005; UWIG, 2006; Parsons, 2006). The 20 percent wind integration study included a conceptual framework of the regional transmission system upgrades needed to move electricity from high-resource to high-demand areas (Figure 7.2). The study estimated the cost of expanded transmission at $23 billion,

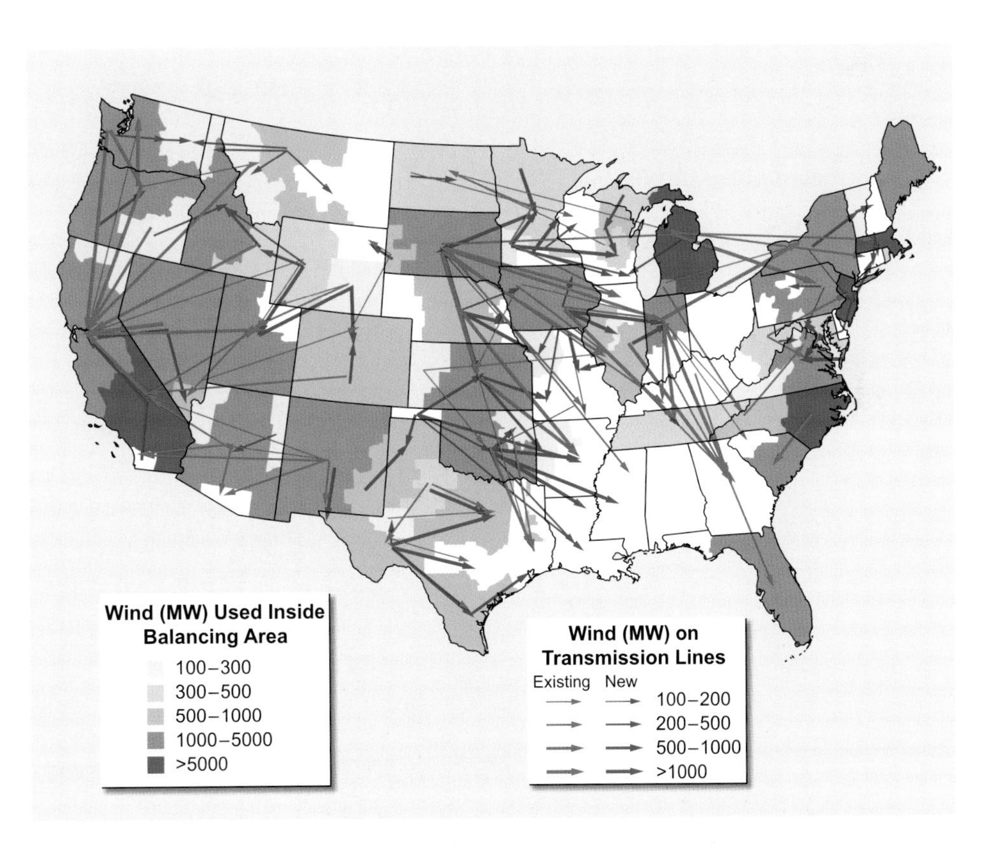

FIGURE 7.2 *Map indicating potential new transmission corridors for integrating 300 GW of wind power.*
Source: DOE, 2008.

though it recognized the barriers to installing new transmission in general. This estimate is lower than other estimates. Separately, American Electric Power (AEP) developed a conceptual interstate transmission plan for integrating more than 300 GW from wind power and for reducing existing transmission bottlenecks. AEP estimates such a system would include 19,000 miles of new high-voltage (765 kV) transmission lines and require investments on the order of $60 billion (AEP, 2007). The more recent Joint Coordinated System Plan (JCSP), discussed below in this section, estimated that integrating 20 percent wind into most of the eastern U.S. electricity system would require 15,000 miles of new extra-high-voltage lines at a cost of $80 billion (JCSP, 2009). Though these studies have differing assumptions resulting in varying estimates, they all indicate the magnitude of investment in transmission required to integrate large amounts of wind power into the electric grid.

Environmental and Energy Impacts

The 20 percent wind power scenario would cause significant land-use and atmospheric emissions impacts. The estimated land area needed to realize this scenario would be 50,000 km^2, which includes the land used directly for the turbines and other land requirements. Only about 2–5 percent of the land use would be for the turbines themselves, with the rest of the area between turbines that could be available for agricultural or other uses.

Figure 7.3 shows reductions of carbon dioxide (CO_2) emissions with 20 percent wind compared to the reference case. Atmospheric emissions of CO_2 and other pollutants would be significantly reduced. The scenario estimates that wind power would replace coal- and gas-fired electricity generation and reduce CO_2 emissions to 800 million tons per year in 2030. Also shown in Figure 7.3 is the trajectory required to reduce electricity sector CO_2 emissions by 80 percent, which is the overall target for reductions of greenhouse gas (GHG) emissions necessary to maintain CO_2 at or below 450 parts per million. Increasing wind power generation would also result in reductions of other atmospheric pollutants associated with fossil-fuel electricity generation, though there would be emissions from natural-gas-fired power plants needed for backup generation. However, the impact on NO_x and SO_2 emissions is less than what would be expected from assuming that electricity generation from fossil fuels is replaced with a non-carbon-emitting technology such as wind power. Because emissions of NO_x and SO_2 are subject to caps on emissions, reductions of emissions from wind-generated electricity might

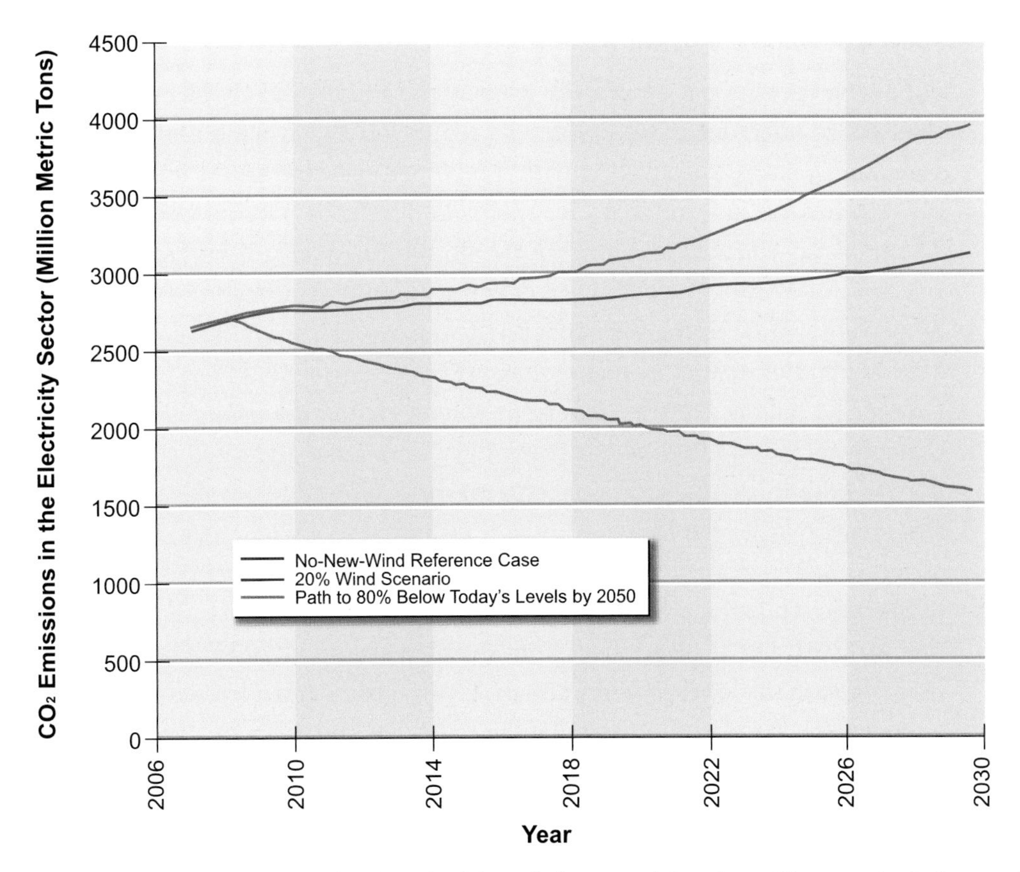

FIGURE 7.3 *Reductions in CO_2 emissions resulting from 20 percent wind scenario compared to the no-new-wind reference case. Also shown is the trajectory for reducing CO_2 emissions by 80 percent.*
Source: DOE, 2008.

be reallocated to other plants. Other air toxics emitted from coal and natural gas electricity generation are not capped and would be reduced in replacing fossil-fuel electricity generation with wind power.

The impact on the energy mix would be largest for natural gas, with the 20 percent wind scenario displacing about 50 percent of electric utility natural gas consumption compared to 18 percent of coal consumption in 2030 (DOE, 2008). The 20 percent wind scenario would also greatly reduce the need for imported liquefied natural gas. However, maintaining electricity system reliability would require additional capacity from natural gas combustion turbines that could respond to wind fluctuations in some combination with the transmission upgrades.

Joint Coordinated System Plan

Following the national 20 percent wind study, a multi-stakeholder group within the Eastern Interconnection prepared a report looking at wind integration issues from a regional perspective. As with the 20 percent wind study, it included multiple stakeholders in a collaborative that held numerous public workshop meetings. The Joint Coordinated System Plan (JCSP, 2009) looked at two scenarios, one a reference case with 5 percent market penetration by wind and the second with 20 percent wind. For the 5 percent wind scenario, the study estimated a need for 10,000 miles of new extra-high-voltage (EHV) transmission lines at an estimated cost of $50 billion. For the 20 percent wind scenario, the projected transmission requirement was 15,000 miles of new EHV lines at an estimated cost of $80 billion. In both cases, the additional transmission allowed renewable and baseload steam energy from the Midwest to be transmitted to a wider area. The study assumed that increased wind generation would primarily offset baseload steam production while requiring more production from fast-response, gas-fired combustion turbines. The JCSP study did not envision electricity storage as having a role in integrating this level of wind power. That report is intended to be part of an ongoing set of studies that examine the reliability and economic impacts of alternative combinations of supply- and demand-side resource technologies, densities and locations, and transmission infrastructure options. The group also plans to conduct sensitivity analyses to determine the implications of varying assumptions such as fuel and technology costs, load projections, plant retirements, and carbon regulation options and costs (JCSP, 2009).

Summary of High Wind Power Penetration Scenarios

It is clear that the high wind penetration scenarios outlined above represent a departure from present conditions. For manufacturers to make the investments needed to develop such capacities and supply chains, substantial capital and a stable policy environment would be required. These scenarios also would require significant land area for the spacing needed between wind turbines, though the actual area occupied by the turbines is a small portion of the land. Realizing the scenarios would entail substantial economic activity, including the addition of thousands of new manufacturing and construction jobs in the wind industry, and would provide significant carbon reductions. DOE's 20 percent wind study estimated a reduced demand for natural gas for electricity generation, though 20 percent wind would increase the need for the use of high-cost combustion-turbine natural gas capacity. The 20 percent wind scenarios of both the DOE and the JCSP demonstrate the need for substantial increases in transmission capacity. There are sufficient resources, technologies, and generally positive economics to increase wind power's contribution to the electricity sector. What these 20 percent wind penetration scenarios emphasize are the scale of the challenges and the benefits for the future.

High Solar Electricity Penetration Scenarios

A variety of scenarios discuss increased market penetration by solar photovoltaics (PV) and concentrating solar power (CSP). Examples range from the comparatively modest *Solar America Initiative* (SAI; DOE, 2007b) to the more optimistic *U.S. Photovoltaic Industry Roadmap* (PV Roadmap; SEIA, 2001, 2004) and the "Solar Grand Plan" (Zweibel et al., 2008). Another study examined a scenario for reaching 10 percent electricity generation from solar by 2025 (Pernick and Wilder, 2008). These scenarios consider issues similar to those addressed in the 20 percent wind power scenarios, such as the potential impacts of renewables' high market penetration on manufacturing, implementation, economics, and the environment. Further, solar electricity can provide insights into attributes of distributed energy sources. Because of the higher costs associated with solar energy, all scenarios consider the significant cost reductions that would have to occur to make solar electricity competitive with other electricity sources.

Distributed Solar Power—SAI and PV Roadmap Scenarios

The SAI and the PV Roadmap scenarios assume that 100–200 GW_p (W_p indicates peak power) of solar PV would be introduced by 2030 and that a majority of the newly installed generation would be distributed in residential, commercial, and industrial applications.[1] Tables 7.3 and 7.4 provide the assumptions used in these scenarios. As shown in Table 7.3, the SAI considered two scenarios: a low-penetration scenario assuming that a total of 5 GW_p of PV would be installed by 2015 and 70 GW by 2030, and a high-penetration scenario assuming that a total of 10 GW_p of PV would be installed by 2015 and 100 GW_p by 2030. In the PV Roadmap scenario, installed capacity would reach 200 GW_p by 2030, and 670 GW_p by 2050. In order for solar PV to be competitive with other electric-

[1]The SAI scenarios assume that all PV installations are distributed electricity sources, and the PV Roadmap assumes that 1/6 of installed capacity is grid (wholesale) generation.

TABLE 7.3 Electricity Capacity and Generation Under SAI and PV Roadmap Scenarios and Total U.S. Electricity Generation Estimated Under AEO 2008 Out to 2030

Scenario	Cumulative PV Capacity (GW)	PV Generation (TWh/yr)	U.S. Electricity Generation, from AEO 2008 (TWh/yr)	PV Percent of Total Generation, from AEO 2008
2015 SAI low	5	8.3	4485	0.2
2015 SAI high	10	16.6	4485	0.4
2015 PV roadmap	9.6	15	4485	n/a
2030 SAI low	70	116.5	5235	2.2
2030 SAI high	100	166.4	5235	3.2
2030 PV roadmap	200	410	5235	7.8
2050 PV roadmap	670	1400	n/a	n/a

Note: AEO 2008, Annual Energy Outlook 2008 (EIA, 2008b); n/a, not available; PV, photovoltaics; SAI, Solar America Initiative.
Source: Data from Grover (2007) and SEIA (2004).

ity sources, both scenarios assumed that the installed system cost of PV would decrease significantly. For example, in both SAI scenarios, costs would decrease from the 2005 value of $8/W_p$ to $3.3/W_p$ in 2015 and $2.5/W_p$ in 2030 (Grover, 2007). The PV Roadmap assumed costs would decrease to $3/W_p$ in 2020 and $1.9/W_p$ in 2050 (SEIA, 2004). Table 7.3 compares the estimated PV electricity generation for these scenarios and total U.S. electricity generation under the Energy Information Agency's reference case AEO 2008 (EIA, 2008b). Assuming a capacity factor of 19 percent, the SAI scenarios would represent between 2 and 3 percent of the total 2030 electricity generation in the AEO 2008 case. The electric-

TABLE 7.4 Annual Installations, System Costs, and Performance for Solar America Initiative High-Penetration and Photovoltaics Roadmap Scenarios

	2015	2030	2050
Annual installed capacity			
SAI high (GW)	2.74	10.4	n/a
PV Roadmap (GW)	2.3	19	31
System costs			
SAI high ($/W)	3.3	2.5	n/a
PV Roadmap ($/W)	3.68	2.33	1.93

Note: n/a, not available; PV, photovoltaics; SAI, Solar America Initiative.
Source: Data from Grover (2007) and SEIA (2004).

ity generation under the PV Roadmap scenario would represent almost 8 percent of the estimate from AEO 2008.

Providing solar PV as distributed electricity generation, as opposed to wholesale generation to the grid, has several advantages. Reducing the need to integrate this portion of PV-generated electricity into the transmission grid would reduce the costs of developing and maintaining transmission facilities. And localized use of electricity would eliminate the losses that occur during transmission. Distributed PV is also easier to site and eliminates land-use impacts. Because available solar energy tends to peak in the afternoon, solar PV delivers electricity directly to residences and businesses close to the time of peak electricity demand. A 19 percent capacity factor has been estimated for a 4.6 MW PV array operated by Tucson Electric Power that was sited to maximize sun exposure (Curtright and Apt, 2008). However, a goal of 19 percent capacity would be high for residential, commercial, and industrial applications where installation of solar panels would have to use existing rooflines and orientations.

Both the SAI scenarios and the PV Roadmap require substantial increases in manufacturing capacities. As shown in Table 7.4, annual U.S. installations of PV over the 2005–2030 time period would grow at a 20 percent rate to 2.7 GW_p in 2015 and 10 GW_p in 2030 under the SAI high scenario. Annual U.S. installations of PV over the 2005-2030 time period for the PV Roadmap would grow at a 22 percent rate to 19 GW_p in 2030. Both cases would result in large increases in manufacturing capabilities. Bradford (2008) estimated that 2007 global PV production was 3.7 GW and grew at an annual average rate of more 45 percent from 2001 to 2007. But global demand also continues to be strong. For example, Bradford (2008) projected that demand for PV in Europe would increase to 4.5 GW_p by 2010.

These scenarios pose significant materials, employment, and capital needs. The primary concern regarding PV materials would be the availability of sufficient polysilicon supplies to produce crystalline silicon PV cells. Global polysilicon supplies were tight in 2007, but there is evidence that supply conditions should improve in 2008 and later (Prometheus Institute, 2007). Though using thin-film technologies would require fewer materials, some shortages might occur; for example, technologies using copper indium gallium selenide (CIGS) could be limited by the amount of naturally occurring indium. The SAI scenarios in 2030 would produce about 120,000 jobs in the manufacturing and installation of PV systems and require $26 billion for manufacturing and installation costs (Grover, 2007). The PV Roadmap estimated that its scenario would produce 260,000 jobs for the manufacturing and installation of PV systems in 2030 and 350,000 jobs in 2050.

Environmental impacts under the SAI scenarios include reductions in atmospheric emissions of CO_2 and other pollutants, as well as potential waste-generation impacts associated with PV manufacturing. SAI estimated reductions in CO_2 and other atmospheric emissions, assuming that PV generation of electricity would replace fossil-fuel generation on a one-to-one basis, and that 75 percent of the fuel displaced would be natural gas and 25 percent would be coal (Grover, 2007). Using this assumption, the SAI low scenario would reduce annual CO_2 emissions by almost 70 million tons in 2030, while its high scenario would reduce annual CO_2 emissions by almost 100 million tons in 2030. These reductions would be about 2–3 percent of estimated annual CO_2 emissions for the electricity sector under AEO 2008 (EIA, 2008b). As with wind power, the impact on NO_x and SO_2 emissions would be less than what would come from replacing fossil fuels with solar PV for the generation of electricity, because of cap and trade policies. Emissions of both NO_x and SO_2 are subject to caps on emissions and thus credit for reductions of emissions from solar electricity generation might be reallocated to existing fossil-fuel plants. Other air toxics emitted from coal and natural gas electricity generation are not capped, however, and would be reduced by replacing fossil-fuel electricity generation with solar PV. As noted in Chapter 5, other impacts related to solar PV not incorporated in this scenario include the waste generation associated with its production and the energy payback time, or the number of years before the PV system becomes a net energy producer.

Grid Solar—CSP and Grand Plan Scenarios

Expanding CSP in California

Expanding the market for concentrating solar power (CSP) represents an approach to providing solar electricity to the electricity grid. Stoddard et al. (2006) described scenarios for increased cumulative market penetration of CSP at two different levels, 2100 MW or 4000 MW, by the year 2020, and the associated economic, energy, and environmental impacts of CSP in California. The report concluded that the size of the resource in California would offer even greater potential. Plants between 100 and 200 MW in size with parabolic trough technology and 6 hours of storage were assumed. It was estimated that with the solar resource in California, each 1000 MW of CSP would produce 3600 GWh/yr. The report concluded that the levelized costs of energy from the CSP plants would make them competitive with natural gas combustion turbines by 2015. The 4000 MW scenario estimated reductions in CO_2 by 7.6 million tons per year, compared with natural gas combined cycle plant, and provision of 3000 permanent jobs associated with the operation of the plants.

The Solar Grand Plan

The "Solar Grand Plan" scenario proposed meeting approximately 70 percent of electricity demand by 2050 through the development of large-scale solar PV farms and CSP plants (Zweibel et al., 2008). The plan also envisioned an extensive direct current (DC) transmission system and compressed air storage facilities distributed throughout the country to enable solar electricity to provide baseload capabilities nationally. It assumed that system costs for thin-film cadmium telluride would fall to $\$1.20/W_p$ and that efficiencies would increase to 14 percent. Almost 3,000 GW_p of capacity would be built, covering 30,000 square miles in the southwestern United States. This scenario assumed that only 10 percent of the generation would come from distributed PV installations. Another 560 GW of capacity would use CSP technologies, which would require 16,000 square miles of land area, also in the southwestern United States. Electricity would be delivered nationally over 100,000–500,000 miles of high-voltage DC transmission lines and partially stored in compressed-air storage facilities to provide power for turbines to generate year-round power. The scenario called for 400 storage facilities with a total capacity of more than 500 billion cubic feet (for information on compressed-air storage, see Chapter 3). It was estimated that a cumulative \$420 billion subsidy from the federal government would be required for the overhaul of the energy infrastructure. Under the plan's scenario, it was projected that U.S. CO_2 emissions would be reduced by 3.6 billions tons per year in 2050, meaning that CO_2 emissions in 2050 would be 62 percent lower than CO_2 emissions in 2005.

The Solar Grand Plan would require large cost reductions, efficiency improvements, and the development of massive storage and transmission infrastructure. The land requirements alone, more than 46,000 square miles, are enormous. One limiting factor would be whether sufficient tellurium exists for manufacturing solar cells at the scale necessary. Approximately 30,000 square miles of CdTe cell area would be used to reach this level of electricity generation, and a typical cell width of 2×10^{-6} meters would require slightly less than the total resource base shown in Figure 6.2 and more than the resource base estimated by the U.S. Geological Survey (USGS, 2007). The USGS resource base is estimated from the only economical source of tellurium, which is a by-product of producing copper, lead, and bismuth. Estimates from both the USGS and Feltrin and Freundlich (2008) have indicated that such a scenario would require most if not all of the world's tellurium production.

Reaching 10 Percent Solar by 2025 Scenario

Capacity Requirements

The 10 percent solar study by Pernick and Wilder (2008) examined conditions that would allow combined PV and CSP electricity generation to reach 486 TWh by 2025, approximately 10 percent of the estimated 4,858 TWh of total electricity generation (Pernick and Wilder, 2008).[2] To attain this level of generation, installed PV capacity would have to rise from less than 1 GW_p in 2007 to more than 210 GW_p in 2025 at an annual average growth rate greater than 30 percent, and CSP would have to rise from less than 0.5 GW_p to more than 40 GW_p over the same period at an annual average growth rate of almost 30 percent (Table 7.5). To achieve this growth rate would require annual installation of almost 50 GW_p of PV and almost 7 GW_p of CSP in 2025. Though this level of installation is quite high, it is smaller than the 60 GW_p of natural gas electricity generation installed in the United States in 2002.

Costs and Capital Requirements

Pernick and Wilder (2008) estimated that the installed price for PV would decline to \$1.48–1.82/W_p by 2025 and that the price for CSP would decline to \$0.88/W_p (Table 7.6), based on the assumption that costs decline by 18 percent for every doubling of capacity. With this decline in costs, solar PV would reach cost parity with conventional retail electricity rates throughout much of the United States by around 2015. Figure 7.4 compares projections of retail electricity rates for various cities in the United States with the cost of electricity from PV. It should be noted that the real retail price of electricity has not increased since 1960 (EIA, 2008c). The report estimated that to reach these levels of market penetration would require investment of \$26–33 billion per year with a total cost of \$450–560 bil-

---

[2]The total electricity generation estimate was derived from the EIA base case AEO 2008 and has been reduced by what the authors assumed would occur from energy efficiency improvements from the Energy Independence and Security Act of 2007.

TABLE 7.5 Photovoltaic and Concentrated Solar Power Installation and Generation Under 10 Percent Solar Scenario

Year	Cumulative PV Installed Capacity (GW)	Annual Generation from PV (TWh)	Cumulative CSP Installed Capacity (GW)	Annual CSP Generation (TWh)	Total Projected Annual Electricity Generation (MWh)	Percent of Total Electricity Generation from PV and CSP (PV% and CSP% shown in parentheses)
2007	0.87	1.6	0.42	0.92	4120	0.06 (0.04 and 0.02)
2010	2.2	4.1	0.78	1.7	4220	0.14 (0.10 and 0.04)
2015	11.2	21	4.0	8.9	4400	0.67 (0.47 and 0.20)
2020	53	97	16.5	36	4610	2.89 (2.11 and 0.79)
2025	213	392	43	94	4860	10 (8.06 and 1.94)

Note: CSP, concentrating solar power; PV, photovoltaics.
Source: Adapted from material in Pernick and Wilder, 2008.

lion (in 2008 dollars). This estimated cost covered only the installed costs, not the costs for transmission upgrades. Rooftop PV installations on commercial roofs, connected directly to the distribution system where they are installed, do not require transmission system upgrades, similar to PV installations envisioned under Southern California Edison's initiative to install 250 MW_p of PV capacity (Southern California Edison, 2008). However, CSP requires high-quality solar resources found in the Southwest, which would require transmission capacity to move electricity to demand centers. This scenario did not estimate market penetration requirements (e.g., materials or employment requirements) beyond the estimates of annual and cumulative installations and did not estimate energy payback times and environmental impacts.

Summary

The 10 percent solar scenario would result in a dramatic increase in solar's contribution to electricity generation that would require aggressive growth rates (annual average growth rates of 30 percent and greater) lasting for almost two decades. It would require large cost reductions for both PV and CSP continuing over the same

TABLE 7.6 Crystalline Silicon PV, Thin-Film, and CSP Cost Assumptions in 10 Percent Solar Scenario

	Average Price per Watt Installed		
Year	For Crystalline Silicon PV (range of costs per kWh)	For Thin-Film and Low-Price and Low-Price Bulk-Purchase Crystalline Silicon PV (range of costs per kWh)	For CSP
2007	$7.00/W ($0.19–0.32/kWh)	$5.50/W ($0.15–0.25/kWh)	$3.50/W
2010	$5.59/W ($0.15–0.25/kWh)	$4.39/W ($0.12–0.20/kWh)	$2.78/W
2015	$3.85/W ($0.10–0.18/kWh)	$3.02/W ($0.08–0.14/kWh)	$1.89/W
2020	$2.65/W ($0.07–0.12/kWh)	$2.08/W ($0.06–0.10/kWh)	$1.29/W
2025	$1.82/W ($0.05–0.08/kWh)	$1.48/W ($0.04–0.07/kWh)	$0.88/W

Note: CSP, concentrating solar power; PV, photovoltaics.
Source: Adapted from material in Pernick and Wilder, 2008.

time period and approximately $500 billion (2008 dollars) in investment in manufacturing and installation capacity to meet this target. The 10 percent solar scenario would also require a much greater involvement of electric utilities in using solar electricity capacity, improvements to the electricity grid to integrate intermittent distributed electricity generation, and national standards for solar interconnections to allow solar to become a "plug-and-play" technology (Pernick and Wilder, 2008). The motivation for pursuing such a strategy includes taking advantage of the ability of solar to produce power during times of peak demand and to serve as a price hedge against escalating fuel costs and potential carbon costs.

Using Multiple Renewables to Reach 20 Percent of Total U.S. Electricity Generation

The scenarios discussed above show the potential for renewables to increase electricity generation and the scale and integration associated with rapid expansion of any single renewable. In this section the panel describes a projection combining multiple renewable technologies that could meet the goal of providing 20 percent

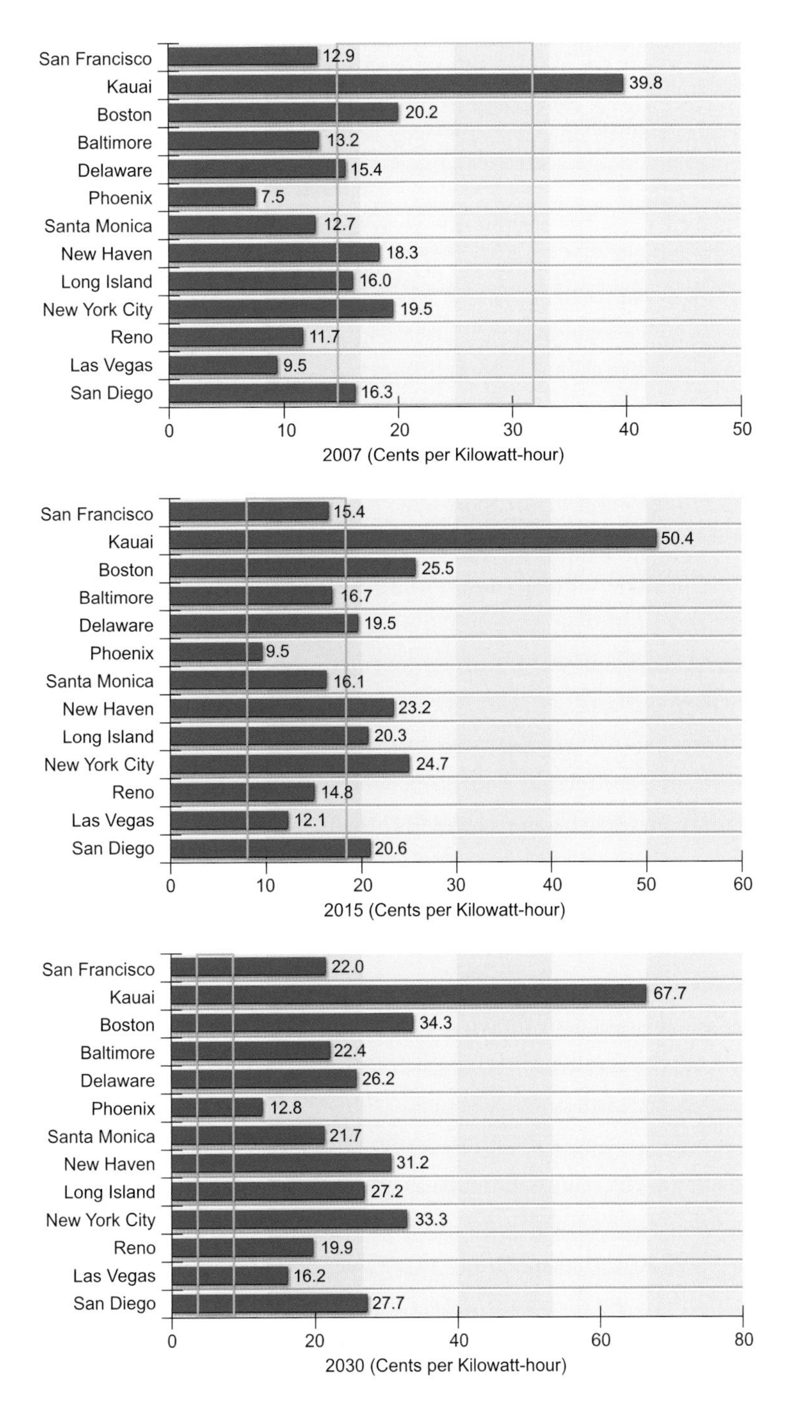

FIGURE 7.4 *Comparison of projected average retail electricity rates with projected high and low PV costs, indicated by yellow box outline, for 2007, 2015, and 2030. Source: Pernick and Wilder (2008), Clean Edge, Inc. (www.cleanedge.com).*

of total electricity generation by 2035 from new renewable electricity generation. This is not a scenario like those described above in this chapter because it does not consider material and deployment needs, capital requirements, or interactions with other electricity sources.

Assuming the use of multiple renewable resources and technologies to reach the 20 percent goal would address some of the scale and integration issues associated with meeting this level of electricity generation with a single renewable. Using an array of renewables could reduce the growth that would be required from individual sources and the scale-up challenges for manufacturers, materials, and human resources. Considering an array of renewables also might ease their integration into the electricity system, particularly for wind generation. Obtaining 20 percent of electricity generation from wind power as a single source will be a challenge, in that the 20 percent refers to an annual average, and wind power is intermittent. Because wind is not available all the time, it might have to represent much more than half the generation at times in order to reach the 20 percent annual average. Wind energy tends to be most abundant at night and in the spring and fall, when demand is low. Balancing wind with multiple renewable resources—including solar, which does not normally peak when wind does, and baseload power from geothermal and biopower—could mitigate the temporal variability in generation. As discussed in Chapter 2, using multiple renewable resources would take advantage of the geographical variability in the resource base. Relying on multiple renewable resources would not eliminate the need to expand transmission capacity or to make other improvements in the electricity infrastructure to enable the integration of renewables, nor would it reduce the magnitude of costs. However, it can provide other attributes, such as providing baseload generation and combining different intermittent renewables to reduce the temporal variability in generation.

Table 7.7 lists a set of renewables that, under the projection described here, would reach 20 percent of electricity generation by 2035. Achieving that goal would depend on wind power capacity additions of 9.5 GW per year, a slight increase over the 8.4 GW installed in 2008. Table 7.7 also shows solar growing to 70 GW by 2035, a much smaller gain than those projected in the high market penetration solar scenarios described above. It assumes that an additional 13 GW would come from conventional geothermal by 2035, which is consistent with the Western Governors' Association's estimated potential resource base in the western United States (WGA, 2006). It also assumes that an additional 13 GW would come from biomass. The mix of renewable resources shown in Table 7.7 is not

TABLE 7.7 Capacity and Generation from Multiple Renewable Resources Sufficient to Meet 20 Percent of Estimated U.S. Electricity Demand in 2035

	Generating Capacity (GW)	Capacity Factor	Electricity Generation (GWh)
Wind	252	0.35	786,429
Solar	70	0.15	91,980
Biomass	13	0.90	102,492
Geothermal	13	0.90	102,492

Note: The estimate of total electricity generation for 2035 comes from AEO 2009's estimate for 2030 (EIA, 2008d) projected out to 2035 using a 0.9 percent growth rate.

presented as the optimal set to meet the target of obtaining 20 percent of total electricity generation from additions of renewable resources. This set is merely one mix that could be considered, given the available resource base, readiness of renewable electricity technologies, and what might be practicable for an aggressive but achievable expansion of market penetration.

SCENARIOS COUPLING RENEWABLES TO ENERGY MARKETS THROUGH CARBON POLICIES

Another set of scenarios examines how renewables interact with other sources of electricity, other sources of energy, and end-use energy demands. This approach incorporates scenarios developed with long-term energy–economic models. The long-term outlook for renewable electricity will depend largely on the ability of renewable electricity technologies to compete against fossil-fuel and nuclear electricity generation. Further, an important consideration would be the extent to which a policy might affect end-use electricity and energy demands. For example, a policy that might influence the price of carbon would likely induce investment in energy efficiency, in addition to making renewables more economical. Energy–economic models allow assessment of potential impacts of demographic, economic, and regulatory factors on renewable electricity within a framework that considers how such factors interrelate to other sources of electricity and end-use energy demands. Therefore, these models are important for understanding possible future pathways for renewable electricity penetration.

These models do not predict the future, and their results are not forecasts.

Rather, they provide a convenient framework for understanding the impacts of critical assumptions and variables using a simplified, internally consistent representation of the energy–economic system. This caveat is especially true for models that simulate the evolution of the energy system over a century or more. It needs to be remembered that these scenarios depend on underlying technology and behavioral assumptions, and results would change if other assumptions were employed.

CCSP Climate Stabilization Scenarios

One set of scenarios that explicitly couples the U.S. and the global energy sectors to the objective of reducing greenhouse gas (GHG) emission concentrations are the scenarios developed for the U.S. Climate Change Science Program (CCSP, 2007). Using three long-term energy–economic models linked to climate representations (listed below), the CCSP simulated changes in the energy system that would stabilize atmospheric CO_2 emissions at approximately 450, 550, 650, or 750 parts per million (ppm):

- Integrated Global Systems Model (IGSM) of the Massachusetts Institute of Technology's Joint Program on the Science and Policy of Global Change (Sokolov et al., 2005);
- Model for Evaluating the Regional and Global Effects (MERGE) of GHG reduction policies developed jointly at Stanford University and the Electric Power Research Institute (Manne and Richels, 2005); and
- MiniCAM Model of the Joint Global Change Research Institute, a partnership between the Pacific Northwest National Laboratory and the University of Maryland (Brenkert et al., 2003).

To reduce CO_2 emissions, these models apply different levels of carbon taxes to all sources of emissions. Imposing costs for carbon from greenhouse gas emissions made fossil-fuel generation increasingly less competitive compared to non-carbon-based energy sources and technologies that use fossil fuels along with CO_2 capture and storage. Carbon prices also induced energy efficiency improvements and reductions in demand. Each model estimates the fee on carbon that it would take to stabilize atmospheric GHG at concentrations of 450–750 ppm. For example, the carbon taxes projected to be necessary to stabilize GHG at concentrations of 550 ppm in 2050 are $35/ton carbon (MERGE), $70/ton (MiniCAM), and

$250/ton (IGSM). The report on these scenarios describes the differences in the models and discusses selected results to provide insight on the following questions (CCSP, 2007):

- What emissions trajectories over time are consistent with meeting the four alternative stabilization levels, and what are the key factors that shape them?
- What energy system characteristics are consistent with each of the four alternative stabilization levels, and how might these characteristics differ among stabilization levels?
- What are the possible economic consequences of meeting each of the four alternative stabilization levels?

The models showed variability both in their reference case representations of the future for renewables and in the responsiveness of renewables to increases in fossil-fuel costs. The IGSM model postulates less renewable energy supply than in the MiniCAM model in both the reference and climate-constrained scenarios for a variety of reasons, including differences in assumed technology availability and institutional settings. Figure 7.5 shows the results for these two models for the reference and 550-ppm stabilization scenarios for renewable electricity generation. The IGSM model, less optimistic than the MiniCAM model in terms of the future market penetration of renewables, projects almost no increase in renewables in response to application of carbon taxes. Figure 7.6 shows the results of the MiniCAM reference and 550-ppm stabilization scenario in the larger context of the U.S. primary energy mix. One response to carbon taxes would be demand reduction and efficiency improvements (labeled Energy Reduction from Reference in Figure 7.6), which would be substantially greater than the increase in renewable electricity production. Clarke et al. (2007a) contains a more detailed description of the scenarios whose projections are graphed in Figure 7.6, along with documentation of the assumptions underlying the models (Clarke et al., 2007a,b).

Scenarios Projecting Effects of the Lieberman-Warner Climate Security Legislation

The EIA often analyzes proposed legislation related to energy and electricity. Its report *Energy Market and Economic Impacts of S. 2191* (EIA, 2008a) was prepared in response to a request for analysis of the Lieberman-Warner Climate Secu-

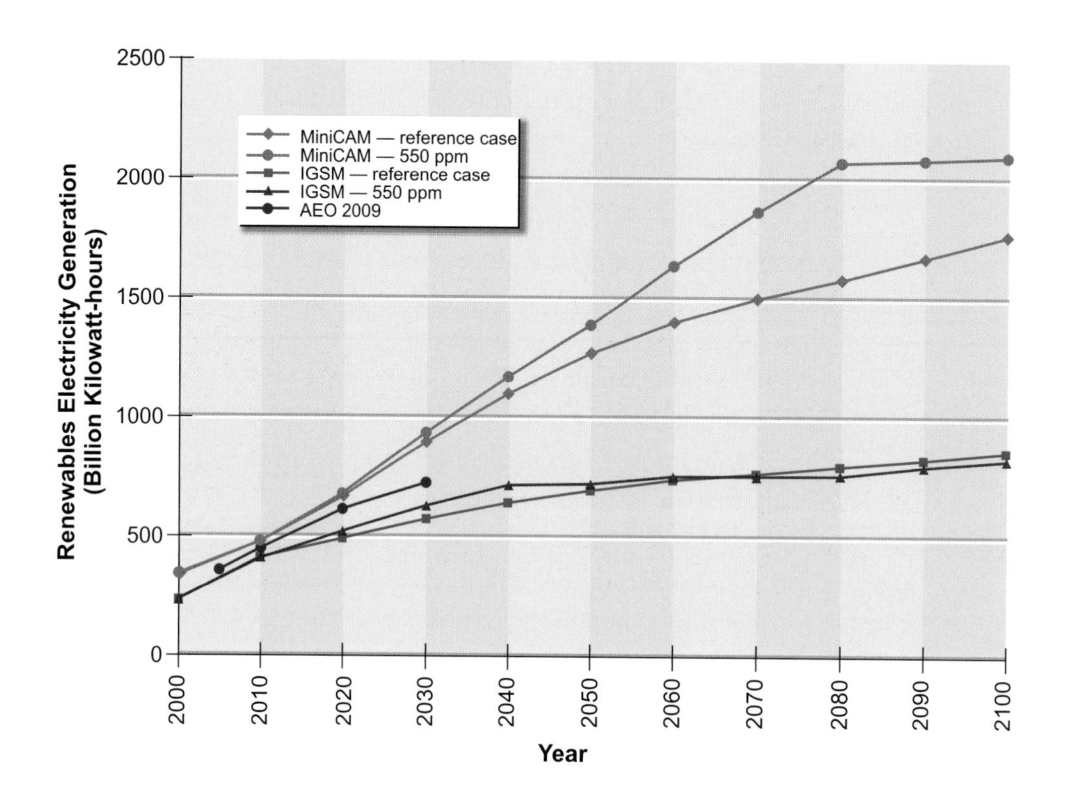

FIGURE 7.5 *Renewables electricity generation from reference case and 550 ppm stabilization scenarios for MiniCAM and IGSM models compared to AEO 2009.*
Source: CCSP (2007) and EIA (2008d).

rity Act of 2007 (the CSA). Regulating greenhouse gas emissions through market-based mechanisms, energy efficiency programs, and economic incentives, the CSA sets caps on annual emissions, primarily of CO_2, that decline from 5775 million metric tons of CO_2 equivalent in 2012 (7 percent below 2006 emission levels) to 1732 million metric tons in 2050 (72 percent below 2006 levels). The emission allowances created under the legislation are tradable and bankable.

Case Scenarios and Methods

The EIA projected the effects of CSA's provisions out to 2030 by modeling several scenarios with varying assumptions about future technology costs and how the legislation might be implemented. Here the panel focuses on two scenarios from the six offered in the EIA report and compares them to the AEO 2008 reference case (EIA, 2008b). The EIA's core CSA scenario assumed that key low-emission

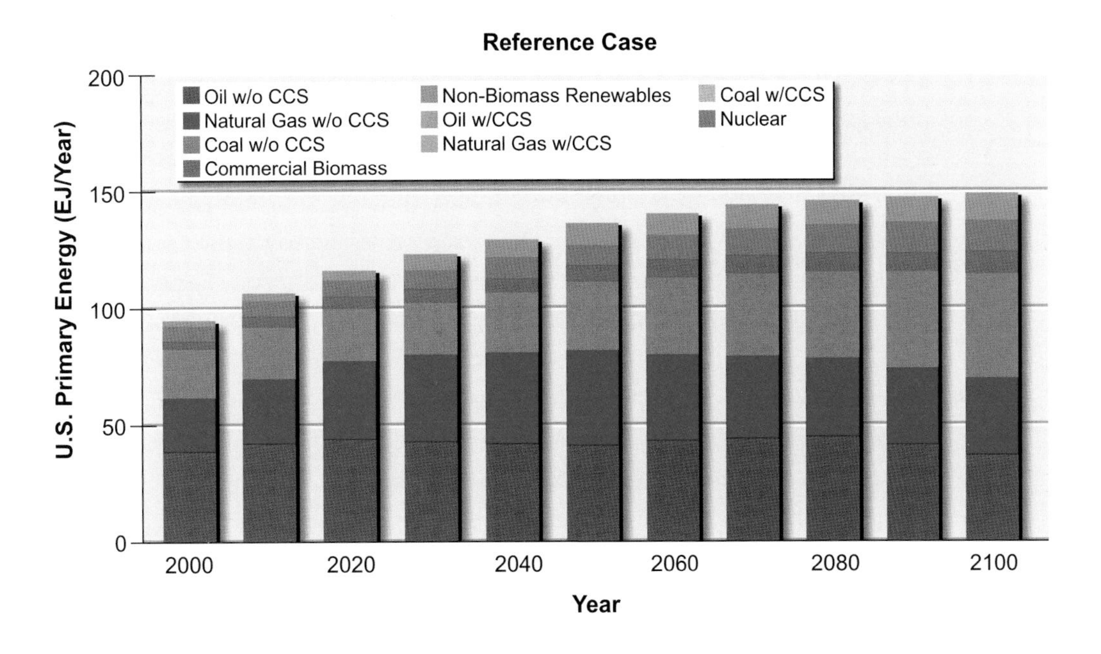

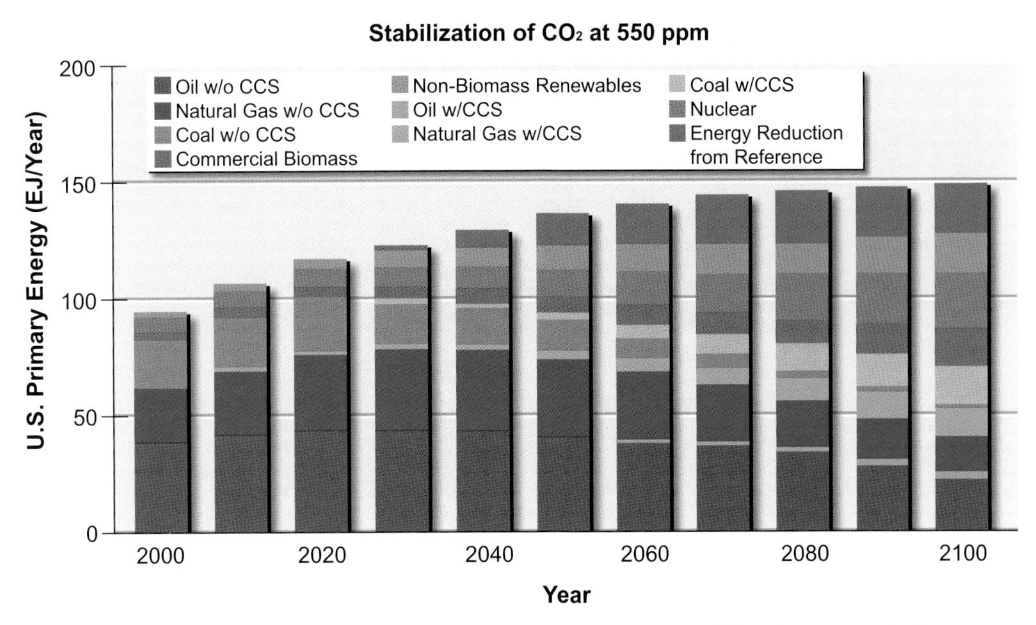

FIGURE 7.6 *MiniCAM reference case and 550 ppm stabilization scenario impacts on the U.S. energy mix, which includes efficiency improvements and demand reductions in the category labeled "Energy Reduction from Reference."*
Source: CCSP, 2007.

technologies, such as nuclear, fossil fuel with carbon capture and storage (CCS), and renewables, would be developed and deployed in the timeframe for emissions reduction set by the CSA without facing any major problems. The EIA's high-cost scenario used the basic assumptions of the core case, except that it applied a 50 percent higher cost of nuclear, fossil fuel with CCS, and biomass-generating technologies to reflect a more pessimistic perspective regarding the costs of these technologies and the feasibility of introducing them rapidly on a large scale.

The EIA used the National Energy Modeling System (NEMS) for its analysis of the CSA (EIA, 2003). NEMS calculated changes in energy-related CO_2 emissions for the various cases by adjusting the cost of fossil fuels and the GHG allowance prices variables that affect energy demand, the energy mix, and energy-related CO_2 emissions. The NEMS Macroeconomic Activity Module is used for analyzing the macroeconomic impacts of GHG reduction policies. This module solves for the energy–economy equilibrium by iteratively interrelating the energy supply, demand, and conversion modules of NEMS (EIA, 2003). Thus, NEMS is sensitive to energy prices, energy consumption, and allowance revenues, and it solves for the effects of policy such as that legislated in the CSA on macroeconomic and industry-level variables.

Energy Market and Electricity Mix

As expected, the projected greenhouse gas emissions in scenarios with emissions regulations are significantly lower than those in the reference case. The EIA's core CSA scenario described above would result in an 85–90 percent reduction of CO_2-equivalent emissions by 2030, and its high-cost case in a 50–60 percent reduction during the same timeframe. The majority of the emissions reduction would come from the electric power sector, a projection that is relevant to this panel's work. These reductions would be achieved by deployment of new nuclear, renewable, and fossil fuel with CCS facilities. Major determinants of the energy and economic impact of the CSA bill include the potential for and the timing of the development and commercial marketing of low-emissions electricity generation technologies. Another determinant is the degree to which companies might be able to purchase emission reduction credits overseas, a topic that is not discussed further here.

Figure 7.7 shows the impact of EIA's core and high-cost CSA scenarios on the overall electricity mix. With the regulation of greenhouse gas emissions in place, coal consumption, especially for electricity generation, would be significantly reduced by 2030. Many coal power plants without CCS would be forced

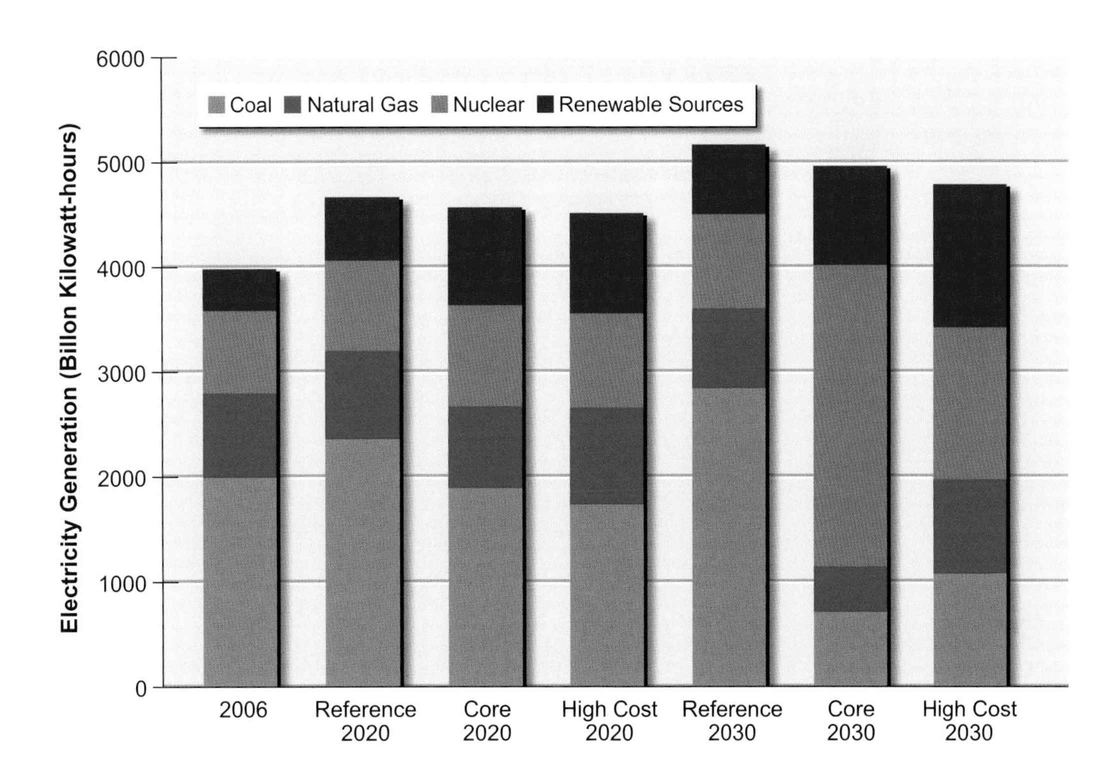

FIGURE 7.7 *Mix of electricity generation from EIA core and high-cost analysis of CSA bill compared to electricity mix in 2006 and to the AEO 2008 reference.*
Source: EIA, 2008b.

to retire early, because retrofitting with CCS technology is generally impracticable, and so is not simulated in the model. The energy-generation mix for the EIA's core CSA scenario would be composed of coal with CCS, nuclear, and renewable technologies, primarily wind and biomass. One important characteristic of the core case is the strong growth in nuclear power. If these low-emission technologies face trouble in deployment, as in the high-cost case, there would be a shift to electricity generation from natural gas to offset the reduction in coal generation.

The EIA estimated that renewable electricity generation would be significantly higher under the provisions of the CSA, with the vast majority of the increase from wind generation, followed by generation from biomass (EIA, 2008a). How each renewable energy resource would contribute to the total supply of electricity generated in the three scenarios (AEO 2008, core, and high cost) is shown in Table 7.8. The increase in total renewable generation is especially strong in the high-cost case. Table 7.9 shows the projected average annual growth rates

TABLE 7.8 Percent of Total U.S. Electricity Generated from Renewable Sources as Projected in Energy Information Administration Analysis of Three Scenarios

	2020			2030		
	Reference Case	Core Case	High-Cost Case	Reference Case	Core Case	High-Cost Case
Hydropower	6.87	7.18	7.37	6.23	6.63	7.13
Geothermal	0.55	0.98	1.21	0.65	1.14	1.45
Municipal waste	0.44	0.56	0.65	0.44	0.54	0.89
Biomass	1.79	5.54	5.30	1.72	3.74	4.58
Solar	0.059	0.06	0.061	0.066	0.068	0.095
Wind	2.33	5.76	6.73	2.57	5.63	13.94
Total renewable	12.0	20.1	21.3	11.6	17.8	28.1
Total non-hydropower renewable	5.13	12.92	13.93	5.37	11.17	20.97

Source: Data from EIA, 2008a.

TABLE 7.9 Average Annual Growth Rate (Percent) from 2005 to 2030 for Each Source of Renewable Electricity Generation

	Hydropower	Geothermal	Municipal Waste	Biomass	Solar	Wind
Reference case	0.49	3.05	1.88	9.45	18.51	8.78
Core case	0.57	5.38	2.93	18.02	18.51	12.03
High-cost case	0.71	6.34	5.08	21.94	19.4	15.85

Source: Data from EIA, 2008a.

of each renewable resource from 2005 to 2030. With GHG-emissions-regulating legislation in place, the NEMS model shows a sharp increase in the growth rate of biomass, solar, and wind generation, especially for the high-cost case, Wind generation would increase significantly, averaging annual growth at 16 percent in the high-cost case, and would grow to constitute 14 percent of the U.S. electricity mix by 2030. Despite this projected rapid growth, NEMS does not indicate a 20 percent contribution by wind energy to the U.S. electricity mix, as is projected in the 20 percent wind scenario discussed above in this chapter. Interestingly, despite the rapid growth rate for solar electricity in all cases, averaging 19 percent annually, solar would still contribute less than 1 percent of total U.S. electricity generation. These values are much smaller than the 10 percent solar generation described in the DOE study discussed above (DOE, 2008). Finally, the EIA estimates significant growth in the use of biomass for electricity generation: by 2030 biomass it would be used to generate 4–5 percent of the U.S. electricity supply.

Summary of Macroeconomic Impacts and Model Uncertainties

EIA's estimates of the macroeconomic impacts of the CSA include an increase in energy prices for consumers, especially in the cost of electricity, with increases of 11–64 percent, mainly as a result of high GHG allowance prices. Also projected by EIA is a reduction of total electricity consumption (5–11 percent). The large increases in energy costs would reduce economic output, lessen purchasing power, and lower aggregate demand for goods and services. In the core CSA case, the gross domestic product would fall by approximately 0.2 percent and would fall by approximately 0.8 percent in the high-cost case.

Many major uncertainties are associated with the EIA projections. It is difficult to foresee how existing technologies might evolve or what new technologies might emerge as market conditions change, particularly when those changes are fairly dramatic. To meet greenhouse gas emission reduction targets, future electricity providers will have to rely on technologies that today play a relatively small role or have not been built in the United States in some time. The actual cost of implementing legislation such as the CSA would depend on unknowns such as future reductions in the cost of renewable technologies, the potential for successful commercialization of CCS, and future costs for nuclear power—all of which cannot be predicted by the model.

FINDINGS

Shown in bold below are the most critical elements of the panel's findings, based on its examination of previously produced scenarios, regarding the future expansion of renewable electricity and factors affecting renewables expansion and integration into the U.S. electricity supply system.

Scale of Deployment

An understanding of the scale of deployment necessary for renewable resources to make a material contribution to U.S. electricity generation is critical to assessing the potential for renewable electricity generation. Large increases over current levels of manufacturing, employment, investment, and installation will be required

for non-hydropower renewable resources to move from single-digit- to double-digit-percentage contributions to U.S. electricity generation.

The scenarios described in this chapter indicate some of the characteristics and impacts associated with accelerating the integration of more renewable generation in the U.S. electricity market. Wind power, an intermittent source of electricity, would be the largest contributor in the near term. DOE (2008) shows that 20 percent of U.S. electricity generation could be obtained from wind and integrated into the nation's electricity system. Follow-up studies such as JCSP (2009) assess the impacts of 20 percent wind at a regional level. Solar PV and CSP could also contribute to attaining additional renewable electricity generation by 2035. Solar electricity is the only renewable resource that has a sufficiently large resource base to supply a majority of the electricity demands of the United States. Today's prices prevent solar electricity from being a widespread economic option at this time. However, the ability of solar PV to produce electricity at the point of consumption means that it competes with the higher retail price of electricity as opposed to the wholesale price of electricity. Solar CSP can provide utility-scale solar power at lower costs than solar PV, though it is limited to favored sites in the U.S. Southwest that have abundant direct solar radiation. Additional contributions could come from biopower and conventional geothermal resources, which can provide baseload power. Thus, if renewables were to contribute an additional 20 percent or more of all U.S. electricity generation by 2035, the largest portion of new renewable electricity generation would come from wind power, but other renewables would also contribute to making this goal a reasonable possibility.

The numbers from the 20 percent wind penetration study (DOE, 2008) demonstrate the challenges and opportunities. To reach the 20 percent target would require installing 100,000 wind turbines; incurring $100 billion worth of additional capital investments and transmission upgrades; and requiring 140,000 jobs be filled. Achieving this goal could reduce CO_2 emissions by 800 million metric tons. The high solar market penetration scenarios also present challenges associated with scaling up this resource. The 10 percent solar study (Pernick and Wilder, 2008) would require that annual installation of PV increase to almost 50 GW in 2025 and installation of CSP to almost 7 GW, with prices for installed PV declining to $1.48–1.82/W and prices for installed CSP declining to $0.88/W in the same timeframe. The cost estimates for reaching the 10 percent solar goal are $26–33 billion per year, with a total cost of $450–560 billion.

In the panel's opinion, increasing manufacturing and installation capacity, employment, and financing to levels required to meet the goals for greatly

increased solar or wind penetration goals is doable. However, to do so would require aggressive growth rates, a large increase in manufacturing and installation capacity, and a large infusion of capital. The magnitude of the challenges is clear from the scale of such efforts.

Integration of Renewable Electricity

The cost of new transmission and upgrades to the distribution system will be important factors when integrating increasing amounts of renewable electricity. The nation's electricity grid needs major improvements regardless of whether renewable electricity generation is increased. Such improvements would increase the reliability of the electricity transmission system and would reduce the losses incurred with all electricity sources. However, because a substantial fraction of new renewable electricity generation capacity would come from intermittent z distant sources, increases in transmission capacity and other grid improvements are critical for significant penetration of renewable electricity sources. According to the Department of Energy's study postulating 20 percent wind penetration, transmission could be the greatest obstacle to reaching the 20 percent wind generation level. **Transmission improvements can bring new renewable resources into the electricity system, provide geographical diversity in the generation base, and allow improved access to regional wholesale electricity markets.** These benefits can also generally contribute positively to the reliability, stability, and security of the grid. **Improvements in the system's distribution of electricity are needed to maximize the benefits of two-way electricity flow and to implement time-of-day pricing. Such improvements would more efficiently integrate distributed renewable electricity sources, such as solar photovoltaics sited at residential and commercial units. A significant increase in renewable sources of power in the electricity system would also require fast-responding backup generation and/or storage capacity, such as that provided by natural gas combustion turbines, hydropower, or storage technologies.** Higher levels of penetration of intermittent renewables (above about 20 percent) would require batteries, compressed air energy storage, or other methods of storing energy such as conversion of excess generated electricity to chemical fuels. Improved meteorological forecasting could also facilitate increased integration of solar and wind power. Hence, though improvements in the grid and related technologies are necessary and valuable for other objectives, significant integration of renewable electricity will not occur without increases in transmission capacity as well as other grid management improvements.

Timeframes for Renewable Technologies

For the time period from the present to 2020, there are no current technological constraints for wind, solar photovoltaics and concentrating solar power, conventional geothermal, and biomass technologies to accelerate deployment. The primary current barriers are the cost-competitiveness of the existing technologies relative to most other sources of electricity (with no costs assigned to carbon emissions or other currently unpriced externalities), the lack of sufficient transmission capacity to move electricity generated from renewable resources to distant demand centers, and the lack of sustained policies. Expanded research and development is needed to realize continued improvements and further cost reductions for these technologies. Along with favorable policies, such improvements can greatly enhance renewable electricity's competitiveness and its level of deployment. Action now will set the stage for greater, more cost-effective penetration of renewable electricity in later time periods. **It is reasonable to envision that, collectively, non-hydropower renewable electricity could begin to provide a material contribution (i.e., reaching a level of 10 percent level or more with trends toward continued growth) to the nation's electricity generation in the period up to 2020 with such accelerated deployment.** Combined with hydropower, total renewable electricity could approach a contribution of 20 percent of U.S. electricity by the year 2020.

In the period from 2020 to 2035, it is reasonable to envision that continued and even further accelerated deployment could potentially result in non-hydroelectric renewables providing, collectively, 20 percent or more of domestic electricity generation by 2035. In the third timeframe, beyond 2035, continued development of renewable electricity technologies could potentially provide lower costs and result in further increases in the percentage of renewable electricity generated from renewable resources. However, achieving a predominant (i.e., >50 percent) level of renewable electricity penetration will require new scientific advances (e.g., in solar photovoltaics, other renewable electricity technologies, and storage technologies) and dramatic changes in how we generate, transmit, and use electricity. Scientific advances are anticipated to improve the cost, scalability, and performance of all renewable energy generation technologies. Moreover, some combination of intelligent, two-way electric grids; scalable and cost-effective methods for large-scale and distributed storage (either direct electricity energy storage or generation of chemical fuels); widespread implementation of rapidly dispatchable fossil-based electricity technologies; and greatly improved technologies for cost-effective long-distance electricity transmission will be required. Significant, sustained, and greatly expanded R&D focused on these technologies is also necessary if this vision is to be realized by 2035 and beyond.

REFERENCES

AEP (American Electric Power). 2007. Interstate transmission vision for wind integration. AEP white paper. Columbus, Ohio.

Bradford, T. 2008. Solar energy market update 2008—PV and CSP. Presentation at Solar Market Outlook: A Day of Data, February 19, 2008. New York.

Brenkert, A., S. Smith, S. Kim, and H. Pitcher. 2003. Model Documentation for the MiniCAM. PNNL-14337. Richland, Wash.: Pacific Northwest National Laboratory.

CCSP (U.S. Climate Change Science Program). 2007. Scenarios of Greenhouse Gas Emissions and Atmospheric Concentrations; and Review of Integrated Scenario Development and Application. Washington, D.C.

Clarke, L., J. Edmonds, H. Jacoby, H. Pitcher, J. Reilly, and R. Richels. 2007a. Scenarios of Greenhouse Gas Emissions and Atmospheric Concentrations. Sub-report 2.1A of Synthesis and Assessment Product 2.1. U.S. Climate Change Science Program and the Subcommittee on Global Change Research, Office of Biological and Environmental Research. Washington, D.C.: U.S. Department of Energy.

Clarke, L., J. Edmonds, S. Kim, J. Lurz, H. Pitcher, S. Smith, and M. Wise. 2007b. Documentation for the MiniCAM CCSP Scenarios. PNNL-16735. Richland, Wash.: Battelle Pacific Northwest Laboratory.

Commission of the European Communities. 2007. Renewable Energy Road Map Renewable Energies in the 21st Century: Building a More Sustainable Future. Communication. Report to the Council and European Parliament. Brussels.

Curtright, A., and J. Apt. 2008. The character of power output from utility-scale photovoltaic systems. Progress in Photovoltaics 16(3):241-247.

DeMeo, E., W. Grant, M.R. Milligan, and M.J. Schuerger. 2005. Wind plant integration. IEEE Power & Energy Magazine 3(6).

DOE (U.S. Department of Energy). 2007a. Annual Report on U.S. Wind Power Installation, Cost, and Performance Trends. Office of Energy Efficiency and Renewable Energy. Washington, D.C.

DOE. 2007b. Solar America Initiative—A Plan for the Integrated Research, Development, and Market Transformation of Solar Energy Technologies. Washington, D.C.

DOE. 2008. 20% Wind Energy by 2030: Increasing Wind Energy's Contribution to U.S. Electricity Supply. Office of Energy Efficiency and Renewable Energy. Washington, D.C.

EIA (Energy Information Administration). 2003. The National Energy Modeling System: An Overview 2003. Washington, D.C.: U.S. Department of Energy, EIA.

EIA. 2008a. Energy Market and Economic Impacts of S. 2191, the Lieberman-Warner Climate Security Act of 2007. Washington, D.C.: U.S. Department of Energy, EIA.

EIA. 2008b. Annual Energy Outlook 2008: With Projections to 2030. DOE/EIA-0383(2008). Washington, D.C.: U.S. Department of Energy, EIA.

EIA. 2008c. Annual Energy Review 2007. Washington, D.C.: U.S. Department of Energy, EIA.

EIA. 2008d. Annual Energy Outlook 2009 Early Release. U.S. Department of Energy DOE/EIA-0383(2009). Washington, D.C.: U.S. Department of Energy, EIA.

Feltrin, A., and A. Freundlich. 2008. Material considerations for terawatt level deployment of photovoltaics. Renewable Energy 33:180-185.

GE Energy. 2005. The Effects of Integrating Wind Power on Transmission System Planning. Prepared for the New York State Energy Research and Development Authority. New York.

Goldberg, M., K. Sinclair, and M. Milligan. 2004. Job and Economic Development Impact (JEDI) model: A user-friendly tool to calculate economic impacts from wind projects. Presented at the Global Windpower Conference, Chicago, Ill. Washington, D.C.: American Wind Energy Association.

Grover, S. 2007. Energy, Economic, and Environmental Benefits of the Solar America Initiative. Subcontractor Report NREL/SR-640-41998. National Renewable Energy Laboratory. Washington, D.C.: U.S. Department of Energy.

JCSP (Joint Coordinated System Plan). 2009. Joint Coordinated System Plan 2008. Available at http://www.jcspstudy.org/.

Manne, A., and R. Richels. 2005. MERGE—A model for global climate change. Energy and Environment (R. Loulou, J. Waaub, and G. Zaccour, eds.). New York: Springer.

Parsons, B. 2006. Grid impacts of wind power variability: Recent assessments from a variety of utilities in the United States. Presentation at the European Wind Energy Conference, March 2006. Athens, Greece. Brussels: European Wind Energy Association.

Pernick, R., and C. Wilder. 2008. Utility Solar Assessment (U.S.A.) Study Reaching Ten Percent by 2025. San Francisco: Clean Edge, Inc. Available at http://www.cleanedge.com/reports/reports-solarUSA2008.php.

Prometheus Institute: For Sustainable Development. 2007. U.S. Solar Industry: The Year in Review 2006. SEIA/Prometheus Institute Joint Report. Cambridge, Mass.

SEIA (Solar Energy Industries Association). 2001. Solar Electric Power—The U.S. Photovoltaic Industry Roadmap. Washington, D.C.

SEIA. 2004. Our Solar Power Future—The U.S Photovoltaic Industry Roadmap Through 2030 and Beyond. Washington, D.C.

Smith, J.C., and B. Parsons. 2007. What does 20 percent look like? IEEE Power & Energy Magazine 5:22-33.

Sokolov, A., C. Schlosser, S. Dutkiewicz, S. Paltsev, D. Kicklighter, H. Jacoby, R. Prinn, C. Forest, J. Reilly, C. Wang, B. Felzer, M. Sarofim, J. Scott, P. Stone, J. Melillo, and J. Cohen. 2005. The MIT Integrated Global System Model (IGSM) Version 2: Model Description and Baseline Evaluation. Report 124. MIT Joint Program on the Science and Policy of Global Change. Cambridge, Mass.

Southern California Edison. 2008. Southern California Edison launches nation's largest solar panel installation. News release. Rosemead, Calif.: Edison International.

Stoddard, L., J. Abiecunas, and R. O'Connell. 2006. Economic, Energy, and Environmental Benefits of Concentrating Solar Power in California. Golden, Colo.: National Renewable Energy Laboratory.

USGS (U.S. Geological Survey). 2007. Mineral Commodity Summaries. Reston, Va.: USGS Minerals Information.

UWIG (Utility Wind Integration Group). 2006. Grid Impacts of Wind Power Variability: Recent Assessments from a Variety of Utilities in the United States. Reston, Va. Available at http://www.uwig.org/opimpactsdocs.html.

WGA (Western Governors' Association). 2006. Clean and Diversified Energy Initiative: Geothermal Task Force Report. Washington, D.C.

Wiley, L. 2007. Utility scale wind turbine manufacturing requirements. Presentation at National Wind Coordinating Collaborative's Wind Energy and Economic Development Forum, April 24, 2007. Lansing, Mich.

Wiser, R. 2008. The development, deployment, and policy context of renewable electricity: A focus on wind. Presentation at the fourth meeting of the Panel on Electricity from Renewable Resources, March 11, 2008. Washington, D.C.

Zavadil, R., J. King, L. Xiadong, M. Ahlstrom, B. Lee, D. Moon, C. Finley, L. Alnes, L. Jones, F. Hudry, M. Monstream, S. Lai, and J. Smith. 2004. Wind Integration Study—Final Report. Xcel Energy and the Minnesota Department of Commerce, EnerNex Corporation and Wind Logics, Inc. Available at http://www.uwig.org/XcelMNDOCStudyReport.pdf.

Zweibel, K., J. Mason, and V. Fthenakis. 2008. A solar grand plan. Scientific American 298(1):64-73.

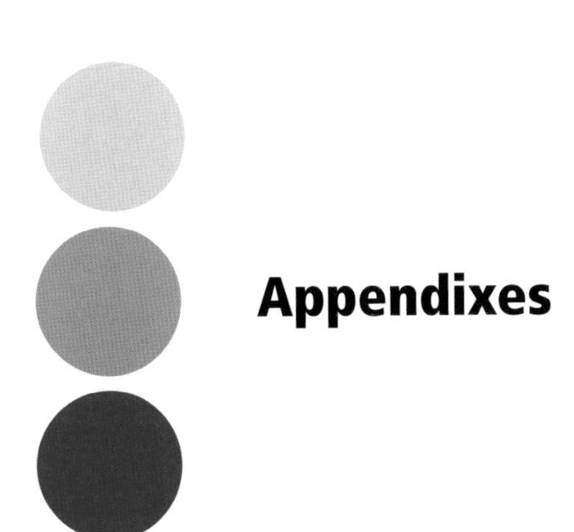

Appendixes

America's Energy Future Project

In 2007, the National Academies initiated the America's Energy Future (AEF) project (Figure A.1) to facilitate a productive national policy debate about the nation's energy future. The Phase I study, headed by the Committee on America's Energy Future and supported by the three separately constituted panels whose members are listed in this appendix, will serve as the foundation for a Phase II portfolio of subsequent studies at the Academies and elsewhere, to be focused on strategic, tactical, and policy issues, such as energy research and development priorities, strategic energy technology development, policy analysis, and many related subjects.

A key objective of the AEF project is to facilitate a productive national policy debate about the nation's energy future.

COMMITTEE ON AMERICA'S ENERGY FUTURE

HAROLD T. SHAPIRO, Princeton University, *Chair*
MARK S. WRIGHTON, Washington University in St. Louis, *Vice Chair*
JOHN F. AHEARNE, Sigma Xi and Duke University
ALLEN J. BARD, University of Texas at Austin
JAN BEYEA, Consulting in the Public Interest
WILLIAM F. BRINKMAN, Princeton University
DOUGLAS M. CHAPIN, MPR Associates
STEVEN CHU,[1] Lawrence Berkeley National Laboratory

[1]Resigned from the committee on January 21, 2009.

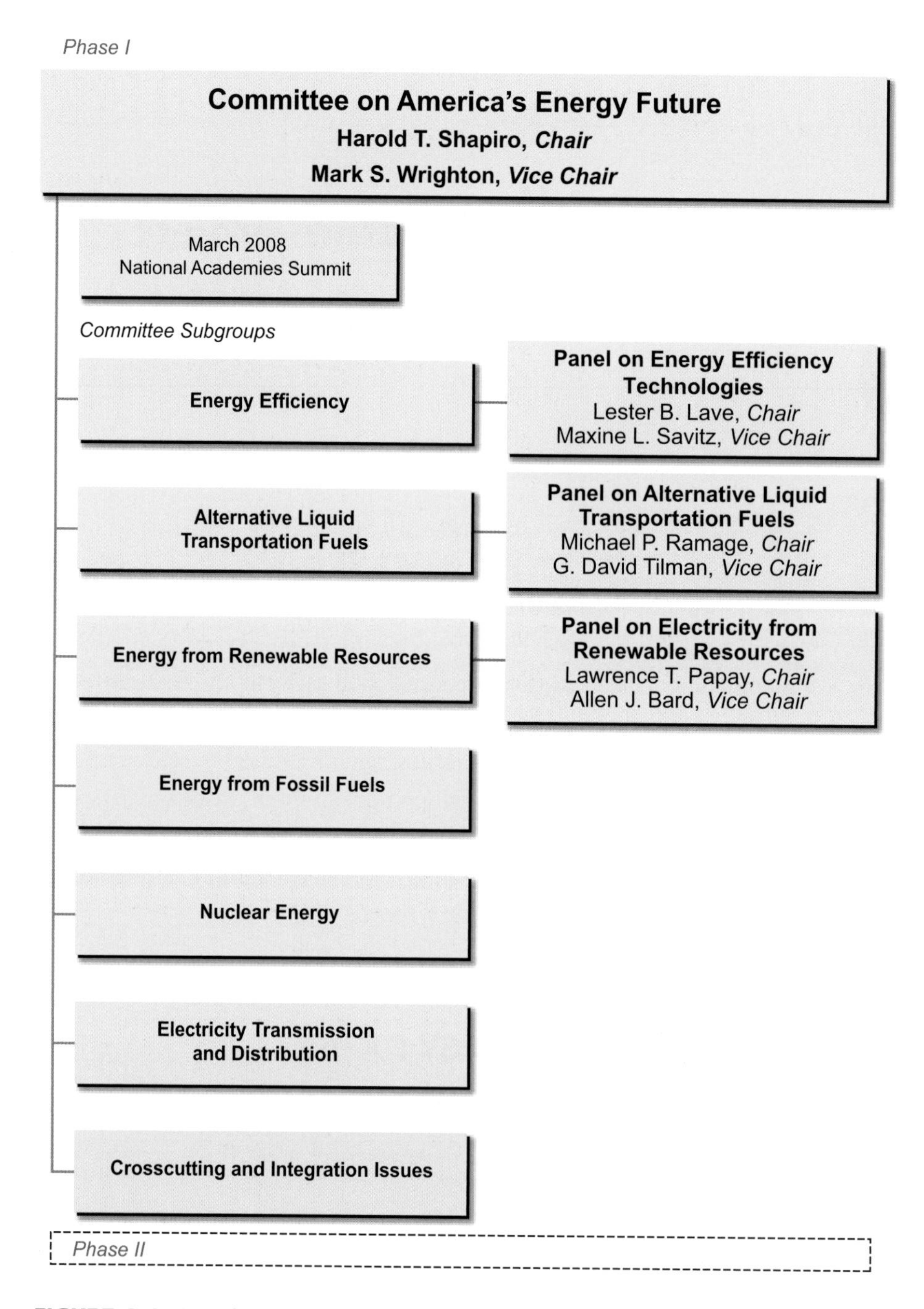

FIGURE A.1 *America's Energy Future Project.*

CHRISTINE A. EHLIG-ECONOMIDES, Texas A&M University
ROBERT W. FRI, Resources for the Future
CHARLES H. GOODMAN, Southern Company (retired)
JOHN B. HEYWOOD, Massachusetts Institute of Technology
LESTER B. LAVE, Carnegie Mellon University
JAMES J. MARKOWSKY, American Electric Power Service Corp. (retired)
RICHARD A. MESERVE, Carnegie Institution for Science
WARREN F. MILLER, JR., Texas A&M University
FRANKLIN M. ("Lynn") ORR, JR., Stanford University
LAWRENCE T. PAPAY, PQR LLC
ARISTIDES A.N. PATRINOS, Synthetic Genomics, Inc.
MICHAEL P. RAMAGE, ExxonMobil (retired)
MAXINE L. SAVITZ, Honeywell, Inc. (retired)
ROBERT H. SOCOLOW, Princeton University
JAMES L. SWEENEY, Stanford University
G. DAVID TILMAN, University of Minnesota, St. Paul
C. MICHAEL WALTON, University of Texas at Austin

PANEL ON ENERGY EFFICIENCY

LESTER B. LAVE, Carnegie Mellon University, *Chair*
MAXINE L. SAVITZ, Honeywell, Inc. (retired), *Vice Chair*
R. STEPHEN BERRY, University of Chicago
MARILYN A. BROWN, Georgia Institute of Technology
LINDA R. COHEN, University of California, Irvine
MAGNUS G. CRAFORD, LumiLeds Lighting
PAUL A. DeCOTIS, Long Island Power Authority
JAMES DeGRAFFENREIDT, JR., WGL Holdings, Inc.
HOWARD GELLER, Southwest Energy Efficiency Project
DAVID B. GOLDSTEIN, Natural Resources Defense Council
ALEXANDER MacLACHLAN, E.I. du Pont de Nemours and Company (retired)
WILLIAM F. POWERS, Ford Motor Company (retired)
ARTHUR H. ROSENFELD, E.O. Lawrence Berkeley National Laboratory
DANIEL SPERLING, University of California, Davis

PANEL ON ALTERNATIVE LIQUID TRANSPORTATION FUELS

MICHAEL P. RAMAGE, ExxonMobil Research and Engineering Company (retired), *Chair*
G. DAVID TILMAN, University of Minnesota, St. Paul, *Vice Chair*
DAVID GRAY, Noblis, Inc.
ROBERT D. HALL, Amoco Corporation (retired)
EDWARD A. HILER, Texas A&M University (retired)
W.S. WINSTON HO, Ohio State University
DOUGLAS R. KARLEN, U.S. Department of Agriculture, Agricultural Research Service
JAMES R. KATZER, ExxonMobil Research and Engineering Company (retired)
MICHAEL R. LADISCH, Purdue University and Mascoma Corporation
JOHN A. MIRANOWSKI, Iowa State University
MICHAEL OPPENHEIMER, Princeton University
RONALD F. PROBSTEIN, Massachusetts Institute of Technology
HAROLD H. SCHOBERT, Pennsylvania State University
CHRISTOPHER R. SOMERVILLE, Energy Biosciences Institute
GREGORY STEPHANOPOULOS, Massachusetts Institute of Technology
JAMES L. SWEENEY, Stanford University

PANEL ON ELECTRICITY FROM RENEWABLE RESOURCES

LAWRENCE T. PAPAY, Science Applications International Corporation (retired), *Chair*
ALLEN J. BARD, University of Texas, Austin, *Vice Chair*
RAKESH AGRAWAL, Purdue University
WILLIAM L. CHAMEIDES, Duke University
JANE H. DAVIDSON, University of Minnesota, Minneapolis
J. MICHAEL DAVIS, Pacific Northwest National Laboratory
KELLY R. FLETCHER, General Electric
CHARLES F. GAY, Applied Materials, Inc.
CHARLES H. GOODMAN, Southern Company (retired)
SOSSINA M. HAILE, California Institute of Technology
NATHAN S. LEWIS, California Institute of Technology
KAREN L. PALMER, Resources for the Future

JEFFREY M. PETERSON, New York State Energy Research and Development Authority
KARL R. RABAGO, Austin Energy
CARL J. WEINBERG, Pacific Gas and Electric Company (retired)
KURT E. YEAGER, Galvin Electricity Initiative

AMERICA'S ENERGY FUTURE PROJECT DIRECTOR

PETER D. BLAIR, Executive Director, Division on Engineering and Physical Sciences

AMERICA'S ENERGY FUTURE PROJECT MANAGER

JAMES ZUCCHETTO, Director, Board on Energy and Environmental Systems (BEES)

AMERICA'S ENERGY FUTURE PROJECT STAFF

KEVIN D. CROWLEY, Director, Nuclear and Radiation Studies Board (NRSB), *Study Director*
DANA G. CAINES, Financial Manager, BEES
SARAH C. CASE, Program Officer, NRSB
ALAN T. CRANE, Senior Program Officer, BEES
GREG EYRING, Senior Program Officer, Air Force Studies Board
K. JOHN HOLMES, Senior Program Officer, BEES
LaNITA JONES, Administrative Coordinator, BEES
STEVEN MARCUS, Editorial Consultant
THOMAS R. MENZIES, Senior Program Officer, Transportation Research Board
EVONNE P.Y. TANG, Senior Program Officer, Board on Agriculture and Natural Resources
MADELINE G. WOODRUFF, Senior Program Officer, BEES
E. JONATHAN YANGER, Senior Program Assistant, BEES

B Panel Biographical Information

Lawrence T. Papay (NAE) is currently a consultant with a variety of clients in electric power and other energy areas. His expertise and knowledge range across a wide variety of electric system technologies, from production to transmission and distribution, utility management and systems, and end-use technologies. He has held positions including senior vice president for the Integrated Solutions Sector, Science Applications International Corporation, and senior vice president and general manager of Bechtel Technology and Consulting. He also held several positions at Southern California Edison, including senior vice president, vice president, general superintendent, and director of research and development (R&D), with responsibilities for areas such as bulk power generation, system planning, nuclear power, environmental operations, and development of the organization and plans for the company's R&D efforts. Dr. Papay's professional affiliations currently include or have included the Electric Power Research Institute's Research Advisory Committee; and the Atomic Industrial Forum; the U.S. Department of Energy's Energy Research Advisory Board, Laboratory Operation Board, and Environmental Management Advisory Board; the Department of Homeland Security S&T Advisory Committee; and chair of the California Council on Science and Technology and the Renewable Energy Institute. He is a member of the National Academy of Engineering. He received a B.S. degree in physics from Fordham University and S.M. and Sc.D. degrees in nuclear engineering from the Massachusetts Institute of Technology.

Allen J. Bard (NAS) is professor of chemistry and biochemistry and holds the Norman Hackerman/Welch Regents Chair in chemistry at the University of Texas,

Austin. His research interests include electro-organic chemistry, photo-electrochemistry, electrogenerated chemiluminescence, electroanalytical chemistry, and fuel cells. His interests include energy policy related to fossil fuels and renewable energy sources. He has published widely and is the winner of numerous honors and awards, including the Willard Gibbs Award, the Pauling Award, and the Priestley Metal. He was president of the International Union of Pure and Applied Chemists and served as editor-in-chief of the *Journal of the American Chemical Society* from 1982 to 2001. He is a member of the National Academy of Sciences. He has served on the National Research Council's (NRC's) Energy Engineering Board (EEB), and has also served as chair of the Board on Chemical Sciences and Technology and chair of the EEB Committee on Potential Applications of Concentrated Solar Photons. He received a Ph.D. degree in chemistry from Harvard University.

Rakesh Agrawal (NAE) is a Winthrop E. Stone Distinguished Professor in the School of Chemical Engineering at Purdue University. Previously, he was an Air Products Fellow at Air Products and Chemicals, Inc., where he worked from 1980 to 2004. A major thrust of his research is related to energy issues and includes novel processes for fabrication of low-cost solar cells, biomass and coal to liquid fuel conversion, hydrogen production from renewable resources, and energy systems analysis. His research interests include basic and applied research in gas separations, process development, synthesis of distillation column configurations, adsorption and membrane separation processes, novel separation processes, gas liquefaction processes, cryogenics, and thermodynamics. He holds more than 116 U.S. and 500 foreign patents. These patents are used in more than 100 chemical plants with a capital expenditure in excess of a billion dollars. He has authored 66 technical papers and given many lectures and presentations. He chaired the Separations Division and the Chemical Technology Operating Council of the American Institute of Chemical Engineers (AIChE) and also a Gordon Conference on Separations. He was a member of the NRC Committee on Alternatives and Strategies for Future Hydrogen Production and Use. He is currently a member of the AIChE's board of directors and also its Energy Commission. He is also a member of the NRC Board on Energy and Environmental Systems (BEES). He has received several awards, including the J & E Hall Gold Medal from the Institute of Refrigeration (U.K.); Presidential Citation for Outstanding Achievement from the University of Delaware; and from the AIChE the Gerhold, Excellence in Industrial Gases Technology, Institute Lecture, Chemical Engineering Practice, and Fuels and

Petrochemicals Division awards. Dr. Agrawal received a B.Tech. from the Indian Institute of Technologies in Kanpur, India; a M.Ch.E. from the University of Delaware; and an Sc.D. in chemical engineering from the Massachusetts Institute of Technology.

William Chameides (NAS) is the dean of the Nicholas School of the Environment and Earth Sciences at Duke University. He is the former chief scientist for Environmental Defense, and before that the Smithgall Chair and Regents Professor of Earth and Atmospheric Sciences at the Georgia Institute of Technology. His research interests include atmospheric chemistry, tropospheric gas-phase and aqueous-phase chemistry; air pollution; global chemical cycles; biospheric-atmospheric interaction; and global and regional environmental change. His NRC service includes chair of the Committee on Atmospheric Chemistry and the Committee on Ozone-Forming Potential of Reformulated Gasoline, and a member of the Committee on Tropospheric Ozone Formation and Measurement. Dr. Chameides is a member of the National Academy of Sciences and a former member of the NRC Board on Environmental Studies and Toxicology. He received a B.A. degree from the State University of New York at Binghamton and M.Ph. and Ph.D. degrees in geology and geophysics from Yale University.

Jane H. Davidson is professor of mechanical engineering at the University of Minnesota and director of the Solar Energy Laboratory. Her current areas of research include solar systems for residential buildings, efficiency in building envelopes, and solar thermochemical cycles to produce fuels. She is a past editor of the *Journal of Solar Energy Engineering* and chair of the American Society of Mechanical Engineers (ASME) Solar Energy Division. She has served as an elected member of the boards of the American Solar Energy Society (ASES) and the Solar Rating and Certification Corporation. Her efforts in research and engineering education have been recognized with the 2007 American Solar Energy Society Charles Greeley Abbot Award, the 2005 University of Minnesota Distinguished Women Scholar Award in Science and Engineering, the 2004 ASME John I. Yellott Award, and the 2000 John Tate Award for Excellence in Undergraduate Advising. She is a fellow of ASME and ASES. She has B.S. and M.S. degrees in engineering science and mechanics from the University of Tennessee and a Ph.D. degree in mechanical engineering from Duke University.

J. Michael Davis is associate laboratory director for Energy and Environment at Pacific Northwest National Laboratory (PNNL), where he is responsible for ensuring that PNNL's energy and environmental programs continue to deliver outstanding science and technology solutions to the most important energy and environmental issues facing the nation and the DOE. Mr. Davis is known nationally as a spokesperson for hydrogen, renewable energy, and energy efficiency policy and technology issues. He has provided leadership for energy-related businesses and organizations, including responsibilities as president and CEO, serving as assistant secretary of energy, as president of the Solar Energy Industries Association, and as chair of the National Hydrogen Association. Mr. Davis also served as an associate professor of mathematics and civil engineering at the U.S. Air Force Academy (USAFA) following service in Vietnam. He received his B.S. degree in civil engineering from the USAFA and his M.S. degree from the University of Illinois.

Kelly R. Fletcher is the energy business program manager at GE Global Research. During his 18-year career in the nuclear energy division of GE Energy, he held various technical and leadership positions, including responsibility for regulatory services, e-business, strategic marketing, business development, and quality. In 2005, Mr. Fletcher was appointed general manager of nuclear technology where he managed activities for GE's nuclear products and services related to new product introduction, R&D, and intellectual property management. In 2006, he was appointed to the position of advanced technology leader for sustainable energy at GE Global Research, in which he is responsible for technology and business development in key sustainable energy areas—advanced energy storage, hydrogen technologies, CO_2-free power generation, and concepts for advanced nuclear power plants. Mr. Fletcher received his B.S. and M.S. degrees in nuclear engineering from the University of California, Berkeley.

Charles F. Gay was named corporate vice president and general manager of the Solar Business Group at Applied Materials in 2006. An industry veteran with 30-plus years of experience in the solar industry, Dr. Gay is responsible for establishing and building Applied Materials' solar business. Dr. Gay is also a co-founder of the Greenstar Foundation, an organization that delivers solar power and Internet access for health, education, and microenterprise projects to small villages in the developing world. Dr. Gay began his career in 1975 designing solar power system components for communications satellites at Spectrolab, Inc., and later joined ARCO Solar, where he established the research and development program and led the commercialization of single-crystal silicon and thin-film technologies. In 1990, Dr. Gay became president and chief operating officer of Siemens Solar Industries. From 1994 to 1997 he served as director of the DOE's National Renewable Energy Laboratory, the world's leading laboratory for energy efficiency and renewable energy research and technology. In 1997, Dr. Gay served as president and chief executive officer of ASE Americas, Inc., and in 2001 became chair of the advisory board at SunPower Corporation. He holds numerous patents for solar cell and module construction and is the recipient of the Gold Medal for Achievement from the World Renewable Energy Congress. Dr. Gay has a doctorate degree in physical chemistry from the University of California, Riverside.

Charles H. Goodman had a long career in electric utility research and development with Southern Company—primarily with regard to developing and improving power generation technologies and in addressing their associated public policy issues. His many contributions span heat transfer, emissions controls, environmental science, and advanced generation technologies. Prior to retirement in 2007 he was the senior vice president for generation policy for Southern Company. His responsibilities included serving as chair of the board for the FutureGen Industrial Alliance. Prior to 2006 he held the position of senior vice president of research and environmental policy. In that position he served as the chief environmental officer for Southern Company. He also directed R&D, environmental policy, environmental research, and compliance strategy development efforts for Southern Company. He served for many years on the Electric Power Research Institute's (EPRI) Research Advisory Committee and was chair of its Environment Sector Council. He is a member of the National Research Council/National Academy of Sciences Board on Energy and Environmental Systems. He served on the NRC Committee on Programmatic Review of DOE's Office of Power Technology, which reviewed the suite of renewable energy R&D technology programs. He has chaired the Environmental Staff Committee of the Business Roundtable and was a member of the Environmental Protection Agency's Clean Air Act Advisory Committee. His responsibilities have included oversight of the Power Systems Development Facility in cooperation with the DOE. Dr. Goodman received his M.S. and Ph.D. degrees in mechanical engineering from Tulane University and his undergraduate degree from the University of Texas at Arlington. He is a fellow of ASME.

Sossina M. Haile is professor of materials science and of chemical engineering at the California Institute of Technology. She earned her Ph.D. in materials science and engineering from the Massachusetts Institute of Technology. As part of her studies, Dr. Haile spent 2 years at the Max Plank Institute for Solid State Research in Stuttgart, Germany, first as a Fulbright Fellow and then as a Humboldt Fellow. Before assuming her present position at Caltech in 1996, she was a member of the faculty at the University of Washington. Her research broadly encompasses solid state ionic materials and devices, with particular focus on fuel cells. She has established a new class of fuel cells based on solid acid electrolytes and demonstrated record power densities for solid oxide fuel cells. Dr. Haile has published more than 100 papers and holds several patents on these and related topics, and she has been an invited speaker at numerous national and international conferences. In 2008 she was awarded an American Competitiveness and Innovation Fellowship from the National Science Foundation in recognition of "her timely and transformative research in the energy field and her dedication to inclusive mentoring, education and outreach across many levels." Since 2005 Dr. Haile has been a member of the NRC National Materials Advisory Board.

Nathan S. Lewis is the George L. Argyros Professor of Chemistry at California Institute of Technology (Caltech). His research interests include light-induced electron transfer reactions, both at surfaces and in transition metal complexes, and the photochemistry of semiconductor-liquid interfaces. Dr. Lewis has been a faculty member at Caltech since 1988 and has served as professor since 1991. He also served as the principal investigator at the Beckman Institute Molecular Materials Resource Center at Caltech since 1992. From 1981 to 1986, he was a faculty member at Stanford University—an assistant professor from 1981 to 1985 and a tenured associate professor from 1986 to 1988. Dr. Lewis has been an Alfred P. Sloan Fellow, a Camille and Henry Dreyfus Teacher-Scholar, and a Presidential Young Investigator. He received the Fresenius Award in 1990, the ACS Award in Pure Chemistry in 1991, the Orton Memorial Lecture award in 2003, the Princeton Environmental Award in 2003, and the Michael Faraday Medal of the Royal Society of Electrochemistry in 2008. He is currently the editor-in-chief of the Royal Society of Chemistry journal, *Energy & Environmental Science*. Dr. Lewis has published more than 300 papers and supervised approximately 60 graduate students and postdoctoral associates. He received his Ph.D. degree in chemistry from the Massachusetts Institute of Technology.

Karen L. Palmer is the Darius Gaskins Senior Fellow at Resources for the Future (RFF) in Washington, D.C., and the director of RFF's Electricity and Environment Program. Dr. Palmer specializes in the economics of environmental regulation of the electricity sector and the cost-effectiveness of energy efficiency programs. Her most recent work has focused on renewable energy and controls of multi-pollutants and carbon emissions from electrical generating plants. She has done extensive work analyzing different aspects of policy design for the Regional Greenhouse Gas Initiative. She is co-author of the book *Alternating Currents: Electricity Markets and Public Policy*. Dr. Palmer previously served as an economist in the Office of Economic Policy at the Federal Energy Regulatory Commission. She received a Ph.D. degree in economics from Boston College.

Jeffrey M. Peterson is currently the program manager for the Energy Resources Group at the New York State Energy Research and Authority, the primary research program for renewable and natural resource development. The goal of the program is to develop cooperative initiatives to introduce new energy and environmental technologies into the marketplace. Research projects range from partnering with New York State businesses to develop new technologies to supply the worldwide market for renewable energy, implementing a workforce training program for renewable technology, and sharing the risk of establishing new business enterprises or models to meet customer demand for renewable energy. He received B.S. and M.S. degrees in wood science and technology from the University of Massachusetts and an M.S. degree in industrial administration from Union College.

Karl R. Rábago is vice president for Distributed Energy Services with Austin Energy. Formerly the director of government and regulatory affairs for AES Wind Generation, he has nearly 20 years' experience in the renewable energy and sustainability fields, having held positions in academia, business, government, and the not-for-profit sector. He has served as a deputy assistant secretary for the DOE, as a public utility commissioner for the State of Texas, and as a managing director and principal of the energy and resources team at Rocky Mountain Institute. Mr. Rábago chairs the board of the Center for Resource Solutions, which manages the Green-e Certification program for green power and renewable energy credit products. He has a bachelor of business administration degree in business management from Texas A&M University, a juris doctorate from the University of Texas, and he holds LL.M. degrees from Pace University School of Law (environmental law) and the U.S. Army Judge Advocate General's School (military law).

Carl J. Weinberg is the principal of Weinberg Associates, which he founded in 1993 after 19 years with Pacific Gas and Electric Company where he managed the energy research and development program. Weinberg Associates was formed with the primary objective of accelerating the introduction of renewable and distributed power systems. His expertise covers technical, regulatory, policy, and environmental perspectives related to energy use. Mr. Weinberg's most recent activities involve policy issues and their technical considerations in the restructuring of the utility industry, with particular emphasis on the concept of sustainability in a competitive framework, and the introduction of distributed resources. He serves on the boards and working level committees of numerous energy efficiency and renewable energy organizations in the public and private sectors. He was the chair of the review panel for California's Public Interest Energy Research Program.

Kurt E. Yeager is executive director of the Galvin Electricity Initiative and former president and chief executive officer of Electric Power Research Institute. Previously, he was the director of energy R&D planning for the EPA Office of Research. He also was with the MITRE Corporation as associate head of the Environmental Systems Department. Mr. Yeager was a distinguished graduate of the Air Force Nuclear Research Officer's Program while serving 7 years on active duty. He is a fellow of the ASME and its Industry Advisory Board, a trustee of the Committee for Economic Development, and he serves on the boards of the U.S. Energy Association and the National Coalition for Advanced Manufacturing. He has served on the executive board of the National Coal Council as well as several National Academy of Engineering committees and the Energy Research Advisory Board to the secretary of energy. Mr. Yeager received a bachelor's degree from Kenyon College.

C Presentations to the Panel

FIRST MEETING: SEPTEMBER 18-19, 2007
WASHINGTON, D.C.

Overview of U.S. Renewable Energy, *Steve Chalk, Deputy Assistant Secretary for Renewable Energy, U.S. Department of Energy (DOE)*

Solar Energy—Photovoltaics and Solar-Thermal Technologies, *Craig Cornelius, Acting Program Manager Solar Energy Technologies, DOE*

Wind Energy, *Steve Lindenberg, Acting Program Manager Wind and Hydropower Technologies, DOE*

Hydropower and Ocean Energy (Wave and Tidal), *Steve Lindenberg, Acting Program Manager Wind and Hydropower Technologies, DOE*

Geothermal Energy, *J. Michael Canty, Drilling Technology Manager, Geothermal Technologies, DOE*

Biomass for Electricity, *Jacques Beaudry-Losique, Program Manager Biomass, DOE*

Renewable Hydrogen's Potential for Electricity Generation, *JoAnn Milliken, Program Manager Hydrogen, Fuel Cells and Infrastructure Technologies, DOE*

Renewable Energy Interconnection and Storage, Technical Aspects, *Ben Kroposki, National Renewable Energy Laboratory*

Grid Integration—The DOE Perspective, *Pat Hoffman, Acting Chief Operating Officer, Office of Electricity Delivery and Energy Reliability, DOE*

Perspectives from the House Science and Technology Committee, *Christopher King and Adam Rosenberg, Staff, U.S. House of Representatives, House Science and Technology Committee*

SECOND MEETING: DECEMBER 6, 2007
WASHINGTON, D.C.

Electricity from Renewables: An NREL Perspective, *Dan Arvizu, National Renewables Energy Laboratory*

Texas Alphabet Soup: SB7, SB20, RPS, CREZ & Other Fun Acronyms, *Mike Grable, Electric Reliability Council of Texas*

Renewable Energy: Progress and Grid Impact, *Pedro Pizarro, Southern California Edison*

California and Renewable Energy, *Martha Krebs, California Energy Commission*

The Future of Geothermal Energy, *Jeff Tester, Massachusetts Institute of Technology*

The Process of Developing Wind Power Generators, *Pete Bierden, General Electric*

Renewable Energy Projections and Modeling for the AEO 2007, *Christopher Namovicz, Energy Information Agency*

Integration of Wind Resources into the Grid, *J. Charles Smith, The Utility Wind Integration Group*

Long-Term Scenarios of Renewable Electricity Generation, *Steve Smith, Pacific Northwest National Laboratory and University of Maryland*

THIRD MEETING: JANUARY 16, 2008
WASHINGTON, D.C.

Hydropower at FERC, *Ann Miles, Federal Energy Regulatory Commission*

Energy Storage for a Greener Grid, *Imre Gyuk, DOE*

New Program Directions at DOE, *Steve Chalk, DOE*

FOURTH MEETING: MARCH 11, 2008
WASHINGTON, D.C.

Electric Energy Storage Briefing, *Dan Rastler, Electric Power Research Institute*

The Development, Deployment, and Policy Context of Renewable Electricity Sources: A Focus on Wind, *Ryan Wiser, Lawrence Berkeley National Laboratory*

Carbon Lock-In: Barriers to the Deployment of Renewable Energy, *Marilyn Brown, Georgia Institute of Technology*

Description of State Renewables Portfolio Standards

TABLE D.1 Description of State Renewables Portfolio Standards

State	Amount	Year	Description
Arizona	15%	2025	The Arizona Corporation Commission introduced new renewable energy standards in 2006. Customers will face a slightly higher Environmental Portfolio Surcharge to offset the cost of compliance. If a utility does not meet the standard, the Commission may assess a penalty for non-compliance. The new rules also require a growing percentage of the total resource portfolio to come from distributed generation. Sources of energy that count toward the standard include electricity produced from qualifying biogas, hydropower, fuel cells that use only renewable fuels, geothermal, hybrid wind and solar, landfill gas, solar, and wind.
California	20%	2010	On September 26, 2006, Governor Schwarzenegger signed Senate Bill 107, which requires California's three major utilities—Pacific Gas & Electric, Southern Edison, and San Diego Gas & Electric—to produce at least 20 percent of their electricity using renewable sources by 2010. Sources of energy that count toward the standard include biomass, solar thermal, photovoltaic, wind, geothermal, fuel cells using renewable fuels, small hydroelectric, digester gas, municipal solid waste conversion, landfill gas, ocean wave, ocean thermal, and tidal current.

continued

TABLE D.1 Continued

State	Amount	Year	Description
Colorado	20%	2020	On March 27, 2007, Governor Bill Ritter signed House Bill 1281, which increased Colorado's previous renewable portfolio standard. Under the new standard, large investor-owned utilities are required to produce 20 percent of their energy from renewable resources by 2020, 4 percent of which must come from solar-electric technologies. HB 1281 requires municipal utilities and rural electric providers to provide 10 percent of their electricity from renewable sources by 2020. Sources of energy that count toward the standard include solar, wind, geothermal, biomass, and small hydroelectric.
Connecticut	23%	2020	On June 4, 2007, Governor M. Jodi Rell signed House Bill 7432, which expanded the state's previous renewable portfolio standard. HB 7432 requires that 27 percent of the state's electricity come from renewable sources by 2020. The law includes standards for three classes of renewables. By 2020, 20 percent of the renewables must be from Class I, 3 percent must be from Class I or II, and 4 percent must be from Class III. Class I sources include solar, wind, new sustainable biomass, landfill gas, fuel cells (using renewable or non-renewable fuels), ocean thermal power, wave or tidal power, low-emission advanced renewable energy conversion technologies, and new run-of-the-river hydropower facilities with a maximum capacity of five megawatts. Class II sources include trash-to-energy facilities, biomass facilities not included in Class I, and certain hydropower facilities. Class III sources include customer-sited combined heat and power systems with a minimum operating efficiency of 50 percent installed at commercial or industrial facilities on or after January 1, 2006; electricity savings from conservation and load management programs that started on or after January 1, 2006; and systems that recover waste heat or pressure from commercial and industrial processes installed on or after April 1, 2007.

TABLE D.1 Continued

State	Amount	Year	Description
District of Columbia	11%	2022	On January 19, 2005, the Council of the District of Columbia enacted Bill A15-755, creating a renewable portfolio standard that requires 11 percent of the electricity sold in the District to come from renewable sources by 2022. The standard includes two tiers. Tier-one renewable resources include solar, wind, biomass, landfill gas, wastewater-treatment gas, geothermal, ocean (mechanical and thermal), and fuel cells fueled by tier one resources. Tier-two renewable resources include hydropower (other than pumped-storage generation) and municipal solid waste. The standard calls for an additional 0.386 percent of the state's renewable energy to come from solar energy by 2022.
Delaware	20%	2019	On July 24, 2007, Governor Ruth Ann Minner signed Senate Bill 19, which expanded the state's previous renewable portfolio standard to require that 2 percent of the state's electricity supply come from solar photovoltaics by 2019, in addition to 18 percent from other renewable sources by the same date. Sources of energy that count toward the standard include wind, ocean tidal, ocean thermal, fuel cells powered by renewable fuels, hydroelectric facilities with a maximum capacity of 30 megawatts, sustainable biomass, anaerobic digestion, and landfill gas.
Hawaii	20%	2020	On June 2, 2004, Governor Linda Lingle enacted Senate Bill 2474, which requires the state's public utilities to provide 20 percent of their electricity from renewable sources by 2020. Sources of energy that count toward the standard include wind, solar, ocean thermal, wave, and biomass resources.
Iowa	105 MW		In 1983, Iowa enacted the Iowa Alternative Energy Production law. The law requires the state's two investor-owned utilities—MidAmerican Energy and Alliant Energy Interstate Power and Light—to contract for a combined total of 105 megawatts of their generation from renewable-energy resources, including small hydropower facilities. Sources of energy that count toward the standard include photovoltaics, landfill gas, wind, biomass, hydroelectric, municipal solid waste, and anaerobic digestion.

continued

TABLE D.1 Continued

State	Amount	Year	Description
Illinois	25%	2025	On August 28, 2007, Governor Rod Blagojevich of Illinois signed into law Public Act 095-0481, which sets a statewide Renewable Energy Standard and an Energy Efficiency Portfolio Standard. Under the RES, utilities in Illinois must produce a certain percentage of their power from renewable sources, starting with 2 percent in 2008 and increasing to 25 percent by 2025. Seventy-five percent of the electricity used to meet the renewable standard must come from wind power generation; other eligible electricity resources include solar, biomass, and existing hydropower sources. The law also includes an efficiency standard that requires utilities to implement cost-effective energy efficiency measures to reduce electric usage by 2 percent of demand by 2015.
Maine	10%	2017	On September 28, 1999, Maine's Public Utilities Commission adopted a renewable portfolio standard, requiring that 30 percent of Maine's power come from renewable sources by 2000. Sources of energy that count toward the standard include fuel cells, tidal power, solar, wind, geothermal, hydroelectric, biomass, and generators fueled by municipal solid waste in conjunction with recycling. In June 2006, the state adopted a renewable portfolio goal to increase new renewable energy capacity by 10 percent by 2017. "New" renewable energy sources include those placed into service after September 1, 2005. In 2007 the state updated the 2006 goal and made it a mandatory target. Resources that satisfy the new capacity requirement cannot also be used to satisfy the 30 percent portfolio requirement.
Maryland	9.5%	2022	On April 24, 2007, Governor Martin O'Malley signed Senate Bill 595, which expanded Maryland's existing renewable portfolio standard to require that 2 percent of the state's electricity supply come from solar sources by 2022, in addition to 7.5 percent from other renewable sources by the same date. Sources of energy that count toward the standard include wind, qualifying biomass, methane from the anaerobic decomposition of organic materials in a landfill or wastewater treatment plant, geothermal, ocean, including energy from waves, tides, currents, and thermal differences, a fuel cell that produces electricity from qualifying biomass or methane, and small hydroelectric power plants.

TABLE D.1 Continued

State	Amount	Year	Description
Massachusetts	4%	2009	In April 2002, the Massachusetts Division of Energy Resources (DOER) adopted a previously outlined renewable portfolio standard. The regulations require that 4 percent of the state's electricity supply come from new renewable sources by 2009. Sources that count toward the standard include solar, wind, ocean thermal, wave, tidal, fuel cells using renewable fuels, landfill gas, and low emission advanced technology biomass. The system must have been installed after December 31, 1997, for the source to qualify as "new." After 2009, the minimum renewable standard shall increase by 1 percent per year until the DOER suspends the annual increase. The minimum renewable standard may at no time decrease below the percentage in effect at the time a suspension is implemented.
Minnesota	25%	2025	On February 22, 2007, Governor Tim Pawlenty signed into law Senate Bill 4, which mandates that 25 percent of Minnesota's power come from renewable sources by 2025. Xcel Energy, which currently generates about half of the state's electricity, will be required to produce 30 percent of its power from renewable sources by 2020. Sources of energy that count toward the standard include solar, wind, small hydroelectric power plants, hydrogen generated from renewable resources, and biomass from qualifying resources.
Missouri	11%	2020	On June 25, 2007, Governor Matt Blunt signed into law Senate Bill 54, which created a renewable energy objective for the state. The bill requires every utility to make a "good-faith effort" to supply 11 percent of their electricity with renewable sources by 2020. Sources of energy that count toward the objective include solar, wind, hydropower, hydrogen from renewable resources, and biomass. Utilities can also earn credit toward the objective through energy efficiency measures that include utility and consumer efforts to reduce the consumption of electricity.

continued

TABLE D.1 Continued

State	Amount	Year	Description
Montana	15%	2015	On April 28, 2005, Governor Brian Schweitzer signed into law Senate Bill 415, the Montana Renewable Power Production and Rural Economic Development Act, which established a renewable energy portfolio standard for the state. SB 415 mandates that 15 percent of the state's energy come from renewable sources by 2015, and for each year thereafter. Sources of energy that count toward the standard include wind, solar, geothermal, existing hydroelectric projects, landfill or farm-based methane gas, wastewater-treatment gas, low-emission, nontoxic biomass, and fuel cells where hydrogen is produced with renewable fuels.
New Hampshire	16%	2025	On May 11, 2007, Governor John Lynch signed into law House Bill 873, the Renewable Energy Act, which establishes a renewable energy portfolio standard for the state. HB 873 mandates that 25 percent of the state's electricity come from renewable sources by 2025, a goal Governor Lynch had previously set for New Hampshire. Sources of energy that count toward the standard include wind, solar, geothermal, hydrogen derived from biomass fuels or methane gas, ocean thermal, wave, current, tidal energy, methane gas, eligible biomass technologies, and existing small hydroelectric sources.
New Jersey	22.5%	2021	On April 12, 2006, the New Jersey Board of Public Utilities (BPU) approved new regulations that expanded the state's renewable portfolio standard. The BPU decision requires utilities produce 22.5 percent of their electricity from renewable sources, at least 2 percent of which must come from solar sources. Sources of energy that count toward the remainder of the standard include solar, wind, wave, tidal, geothermal, methane gas captured from a landfill, fuel cells powered by renewable fuels, electricity generated by the combustion of gas from the anaerobic digestion of food waste and sewage sludge at a biomass generating facility, and hydropower.
New Mexico	20%	2020	On March 5, 2007, Governor Bill Richardson signed into law Senate Bill 418, which established a renewable portfolio for the state. SB 418 mandates that by 2020, 20 percent of an electric utility's power come from renewable sources. Sources of energy that count toward the standard include solar, wind, hydropower, geothermal, fuel cells that are not fossil fueled, and qualifying biomass resources.

TABLE D.1 Continued

State	Amount	Year	Description
Nevada	20%	2015	On June 7, 2005, the Nevada Governor Kenny Guinn signed into law Assembly Bill 3, expanding Nevada's previous renewable portfolio standard. The updated standard requires that 20 percent of the state's electricity come from renewable energy sources by 2015, and for each year thereafter. Of the 20 percent, not less than 5 percent must be generated from solar renewable energy systems. Utilities can also earn credit for up to 25 percent of the standard through energy efficiency measures. Sources of energy that count toward the standard include biomass, fuel cells, geothermal, solar, waterpower, and wind.
New York	24%	2013	On September 22, 2004, The New York Public Service Commission adopted a renewable portfolio standard. The standard requires that 25 percent of the state's electricity come from renewable sources by 2013. The standard identifies two tiers of eligible resources, a "Main Tier" and a "Customer-Sited Tier." The Main Tier is mandatory and is to account for 24 percent of the standard. Eligible sources include biogas, biomass, liquid biofuel, fuel cells, hydroelectric, solar, ocean or tidal power, and wind. The Customer-Sited Tier will make up the remaining 1 percent of renewable energy sales and is to come from voluntary green market programs. Sources of energy that count toward the Customer-Sited Tier include fuel cells, solar, and wind resources.

continued

TABLE D.1 Continued

State	Amount	Year	Description
North Carolina	12.5%	2021	On August 20, 2007, Governor Mike Easley of North Carolina signed into law S.L. 2007-397, which establishes a Renewable Energy and Energy Efficiency Portfolio Standard for the state. Under the law, by 2021 electric public utilities must meet 12.5 percent of retail electricity demand through renewable energy or energy efficiency measures, and electric membership corporations and municipalities that sell electric power in the state would have to meet a standard of 10 percent by 2018. Resources that can be used to meet the standard include solar energy, wind energy, hydropower, geothermal energy, ocean current or wave energy, biomass resources, and energy efficiency measures. The law also includes provisions to encourage the use of solar energy, swine and poultry wastes, as well as implementation of energy efficiency programs.
Oregon	25%	2025	On June 6, 2007, Governor Ted Kulongoski signed Senate Bill 838, adopting a renewable electricity portfolio standard for the state. SB 838 requires the state's largest utilities to meet 25 percent of their electric load with new renewable energy sources by 2025. Sources of energy that count toward the standard include wind, solar, wave, geothermal, biomass, [and] new hydro or efficiency upgrades to existing hydro facilities.
Pennsylvania	18%	2020	On December 16, 2004, Governor Edward Rendell signed into law Pennsylvania's Alternative Energy Portfolio Standard, requiring that qualified power sources provide 18.5 percent of Pennsylvania's electricity by 2020. There are two tiers of qualified sources that may be used to meet the standard. Tier-one sources must make up 8 percent of the portfolio, and include wind, solar, coalmine methane, small hydropower, geothermal, and biomass. Solar sources must provide 0.5 percent of generation by 2020. Tier-two sources make up the remaining 10 percent of the portfolio, and include waste coal, demand side management, large hydropower, municipal solid waste, and coal integrated gasification combined cycle.

TABLE D.1 Continued

State	Amount	Year	Description
Rhode Island	15%	2020	On June 29, 2004, Governor Donald Carcieri signed the Clean Energy Act, requiring state electricity retailers to derive at least 3 percent of the electricity they sell in state from renewable energy by December 31, 2006. The percentage of renewable energy required will then rise 1 percent per year through 2020, though the Rhode Island Public Utility Commission (PUC) is authorized to revise the schedule after 2013. Existing renewable resources may only contribute 2 percent of the required amount of renewables in any year; the rest must be from new renewable energy production. Sources of energy that count towards the standard include direct solar radiation, wind, movement or the latent heat of the ocean, the heat of the earth, small hydroelectric facilities, eligible biomass, and fuel cells using renewable resources.
Texas	5,880 MW	2015	On August 1, 2005, Governor Rick Perry signed a bill increasing the amount of renewable generation required in the state. The law requires that 5,880 MW of new renewable generation be built in the state by 2015, which will meet about 5 percent of the state's projected electricity demand. The legislation also sets a cumulative target of installing 10,000 MW of renewable generation capacity by 2025. In an effort to diversify the state's renewable generation portfolio, the measure also includes a requirement that the state must meet 500 MW of the 2025 target with non-wind renewable generation.

continued

TABLE D.1 Continued

State	Amount	Year	Description
Vermont	10%	2013	On June 14, 2005, Governor Jim Douglas signed a renewable portfolio standard into law, requiring renewable generation to equal incremental load growth between 2005 and 2012, but not requiring utilities to hold renewable energy credits equal to renewable generation. If utilities have not met this requirement, the state will instate an RPS [renewables portfolio standard] equal to the percentage of load growth between 2005 and 2012. If the state experiences 7 percent load growth, but utilities have not obtained 7 percent of their electricity from eligible renewables by 2012, the state will adopt an RPS of 7 percent. Sources of energy that count toward the standard include wind, solar, small hydropower methane from landfill gas, anaerobic digesters, and sewage-treatment facilities, while excluding municipal solid waste. Vermont utilities are permitted to build generation capacity out of state to comply with the mandate. On March 20, 2008, Governor Jim Douglas signed the Energy Efficiency and Affordability Act of 2008, which established a renewable energy goal for the state. The law sets a goal of producing 25 percent of the energy consumed in the state from renewable sources, particularly Vermont's farms and forests, by 2025.
Virginia	12%	2022	On April 11, 2007, Governor Tim Kaine signed Senate Bill 1416, which established a voluntary renewable portfolio goal. The standard sets a renewable energy target of 12 percent of base year sales by 2022. The standard targets are defined as percentages of 2007 (the "base year") electricity sales minus the average annual percentage of power supplied from nuclear generators between 2004 and 2006. A utility may participate in the voluntary RPS program if it demonstrates that it has a reasonable expectation of achieving the 12 percent target in 2022. Sources of energy that count toward the target include solar, wind, geothermal, hydropower, wave, tidal, and biomass energy. Wind and solar receive a double credit toward RPS goals.

TABLE D.1 Continued

State	Amount	Year	Description
Washington	15%	2020	On November 7, 2006, Washington state voters approved ballot initiative 937, setting renewable energy standards for utility companies in the state. The measure requires all utilities serving 25,000 people or more to produce 15 percent of their energy using renewable sources by 2020. Such sources include wind, solar, and tidal power as well as landfill-methane capture. Sources of energy that count toward the standard include water, wind, solar, geothermal, landfill gas, wave, ocean, tidal power, gas from sewage treatment facilities, biodiesel fuel that is not derived from crops raised on land cleared from old growth or first-growth forests, and qualifying biomass resources.
Wisconsin	10%	2015	On March 17, 2006, Governor Jim Doyle signed Senate Bill 459, the Energy Efficiency and Renewables Act, which increased the state's previous renewable portfolio standard. The revised standard requires utilities to produce 10 percent of their electricity from renewable energy sources by 2015. Sources of energy that count toward the standard include solar, wind, waterpower, biomass, geothermal technology, tidal or wave action, and fuel cell technology that uses qualified renewable fuels.

Sources: U.S. Department of Energy, Energy Efficiency and Renewable Energy website, available at http://www.eere.energy.gov/states/maps/renewable_portfolio_states.cfm, and the Pew Climate website, available at http://www.pewclimate.org/what_s_being_done/in_the_states/rps.cfm.

E

Attributes of Life-Cycle Assessment

While it is the intention of all life-cycle assessments (LCAs) to cover technologies from "cradle to grave" in a systematic way, there is significant variability in the assumptions, boundaries, and methodologies used in these assessments. Therefore, comparisons of LCAs should be done with caution; each is an approximation of a technology's actual impact. A major complication in comparing LCAs is that there is no set standard by which such analyses are carried out. There are two basic kinds of LCAs: economic input/output (EIO) and process analysis (PA). PA can be thought of as a "bottom-up" approach in which specific information for the energy and emissions associated with each component of the technology is determined and then combined to obtain a complete life-cycle impact. In addition to being very time intensive, the PA can be limited in its utility by the fact that there is often a lack of information concerning one or more components of the technology, and this can lead to truncation errors.

EIO, on the other hand, does not track individual components, but instead uses economy sector level data to quantify the relationship between energy and the materials and processes produced. As compared to PA, EIO can be thought of as being a "top-down" approach. While having the advantage of being less time and data intensive, the EIO is limited in its accuracy by its dependence on highly aggregated data that may or may not be appropriate for the specific process or material being considered.

Recent investigations suggest that EIO analyses tend to give higher energy use and emissions estimates than does PA, perhaps because PA is only able to consider the subset of processes for which data are available (Fthenakis and Kim, 2007; Odeh and Cockerill, 2008). A third method, sometimes referred to as the "hybrid" LCA, attempts to address this issue by combining aspects of both techniques: using the PA approach where specific data are available and the EIO method where such data are not available.

To provide a coherent and consistent framework for LCAs, various agencies have compiled life-cycle inventory (LCI) databases that list the materials and energy inputs required for various technologies. For example, ECLIPSE (Frankl et al., 2004) is a LCI assessment project developed in Europe that looks at emissions and resource consumption. Other PA-type LCI databases include DEAM (Ecobilan, 2001), Franklin (Franklin Associates, 2008), and Ecoinvent (Pre Consultants, 2007b). EIO analyses typically rely on economic databases that are compiled by governmental bodies. For example, in the United States, the Department of Commerce generates data on air emission and water and energy use for 485 commodity sectors from various data sources.

Beyond the choice of LCA method and LCI database, there are any number of other factors that can affect LCA results and cause discrepancies among analyses. Assumptions about power plant capacity (or lifetime output), plant life expectancy, and energy infrastructure influence LCA results. In general, when comparing installations of the same energy technology, those with longer plant life expectancies and greater electrical output to the grid will have lower lifetime emissions per unit of electricity.

The nature of the underlying energy structure that supports the manufacture, operation, and dismantling of a given facility are also quite important. For example, the construction of a wind turbine in Sweden where much of the energy is produced using renewable sources will generally have less embedded CO_2 emissions than the same turbine produced in the United States, where coal-fired power plants generate a significant fraction of the electricity available on the grid.

A further source of discrepancies for LCAs arises from the fact that these assessments are aimed at technologies that are often undergoing continuous modification and improvement. Comparisons cited here of an LCA for a given technology may differ because the technology under consideration evolved over the period from one LCA to another. This is especially true for solar and wind technologies where the ongoing rate of innovations is quite rapid. A further complication arises from the fact that some LCAs assess impacts for a hypothetical, future installation of the technology.

Another shortcoming of LCA is that it addresses only a single environmental impact. If one is in the position of choosing one technology over another, it would be desirable to have a more integrated understanding of the overall environmental impact of one technology over another. Environmental valuation methods attempt to do this by estimating the overall ecosystem impact of a technology using currency or "willingness to pay" as the unit to integrate across types of impacts. Impact categories included in environmental valuation methods typically relate to damages to humans, ecosystems, and resources. ExternE (European Commission, 1997) and Eco-indicator 99 (Pre Consultants, 2007a) are examples of environmental valuation methods used in Europe.

The main criticism of environmental valuation methods is of the step where disparate effects of LCAs are weighted and normalized into a single value per technology. Often the development of a single value is not adequate to capture the complexities of a technology, and metrics like "willingness to pay" can vary over time. For this reason we limit our discussion to LCA results.

REFERENCES

Ecobilan, P. 2001. TEAM/DEAM. 2001. Bethesda, Md.: PriceWaterhouseCoopers.

European Commission. 1997. External Costs of Electricity Generation in Greece. ExternE Project. Brussels.

Frankl, P., Corrado, A., and S. Lombardelli. 2004. Photovoltaic (PV) Systems. Final Report. ECLIPSE (Environmental and Ecological Life Cycle Inventories for present and future Power Systems in Europe). Brussels: European Commission. January.

Franklin Associates. 2008. Available at http://www.fal.com/lifecycle.htm.

Fthenakis, V.M., and H.C. Kim. 2007. Greenhouse-gas emissions from solar-electric and nuclear power: A life-cycle study. Energy Policy 35:2549-2557.

Odeh, N.A., and T.T. Cockerill. 2008. Life cycle GHG assessment of fossil fuel power plants with carbon capture and storage. Energy Policy 38:367-380.

Pre Consultants. 2007a. Available at http://www.pre.nl/eco-indicator99.

Pre Consultants. 2007b. Available at http://www.pre.nl/ecoinvent/.

F Atmospheric Emissions from Fossil-Fuel and Nuclear Electricity Generation

Because of its high C-to-H ratio, coal is potentially the highest emitter of the energy sources available when it comes to greenhouse gas emissions. Emissions for traditional pulverized-coal plants hover near or above 1000 g CO_2e/kWh, about 2 orders of magnitude larger than most estimates for renewables. However, the emissions can be significantly mitigated to as low as ~40 g CO_2e/kWh with different configurations and most notably with carbon capture and storage (CCS) technologies, assuming that CCS can be successfully implemented (see Figure F.1).

Estimates for CO_2 emissions from pulverized-coal plants currently deployed range from 960 and 1050 g CO_2e/kWh; these estimates include the average for the United States, the United Kingdom, and Japan, as well as the average for the United States operating under new source performance standards (NSPS). Modest reductions (757 to 879 g CO_2e/kWh) are projected for new technologies that increase plant efficiency; these include a future low-emission boiler system (LEBS) (Spath et al., 1999), a U.K. supercritical pulverized coal plant (Odeh and Cockerill, 2008), and a U.K. integrated gasification combined cycle (IGCC) plant (Odeh and Cockerill, 2008). Emissions from a coal co-fired with biomass waste residue facility are estimated at 681 g CO_2e/kWh (Spath and Mann, 2004).

The lower end of the range from 43 to 255 g CO_2e/kWh includes a variety of coal technologies with future CCS methods. The carbon capture methods discussed in the literature include absorption by monoethanolamine (MEA) and selexol. (MEA is a post-combustion CO_2 capture method for the traditional pulverized-coal and biomass co-fired plants and thus could be used with the existing fleet. Selexol is used to capture CO_2 prior to combustion in IGCC

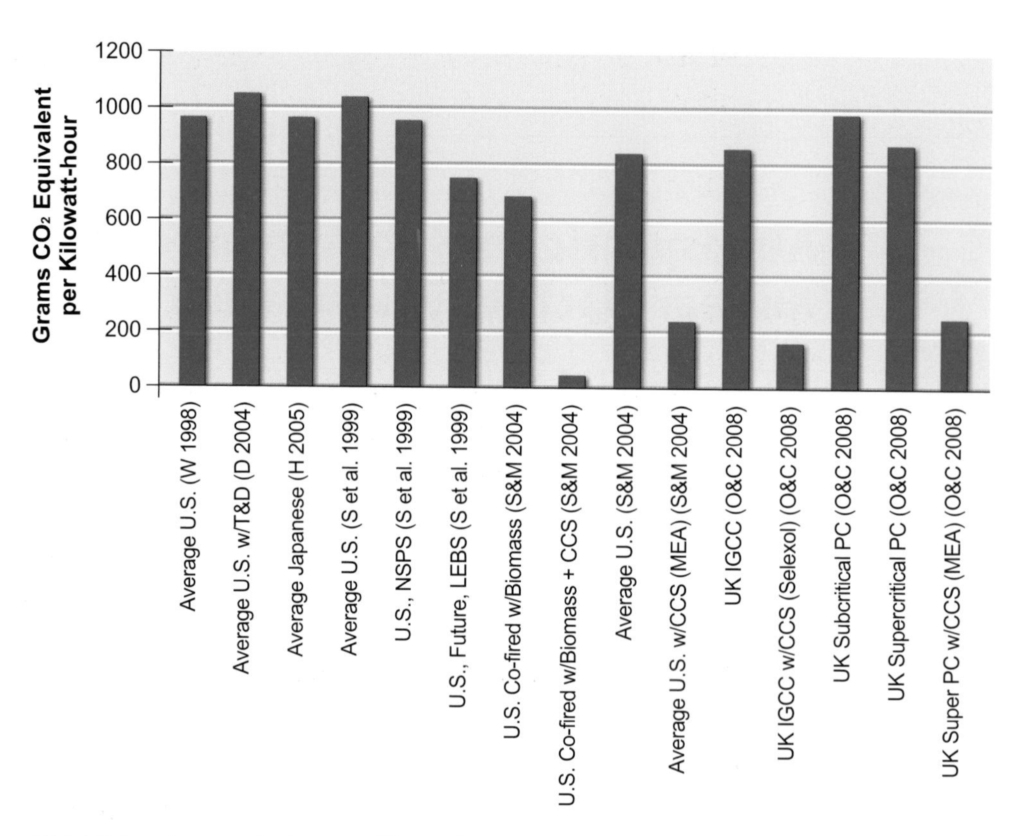

FIGURE F.1 *Range of life-cycle CO_2 equivalent emissions from various technologies for generating electricity from coal.*
Source: Based on data compiled from Denholm (2004), Hondo (2005), Odeh and Cockerill (2008), Spath and Mann (2004), Spath et al. (1999), and White (1998).

plants.) A hypothetical U.K. IGCC plant with carbon capture via selexol had a value of 167 g CO_2e/kWh (Odeh and Cockerill, 2008). Two hypothetical U.S. and U.K. average coal plants with carbon capture via MEA emit approximately 250 g CO_2e/kWh (Odeh and Cockerill, 2008; Spath and Mann, 2004). The lowest value of 43 g CO_2e/kWh is from a hypothetical coal plant co-fired with biomass residues (Spath and Mann, 2004). However, this estimate did not account for CO_2 emissions associated with the production, regeneration, or disposal of MEA.

Because MEA is highly reactive with SO_2, Odeh and Cockerill (2008) also evaluated a pulverized-coal plant with CCS, but without flue gas desulfurization (FGD) to evaluate how the life-cycle emissions of CO_2 would be affected by the interaction of the MEA with SO_2. Under this scenario significantly more MEA would be required, and because of the extra emissions associated with producing

MEA and other materials used in the capture process, they found that life-cycle CO_2e emissions doubled.

Natural Gas

Key factors affecting natural gas CO_2 emissions from natural gas facilities include plant efficiency and natural gas losses from production and distribution. Emissions for natural gas combined cycle (NGCC) plants had a small range of 469 to 518 g CO_2e/kWh. A higher value of 608 g CO_2e/kWh was reported for the only gas-fired plant evaluated (Hondo, 2005). CCS is not expected to have as large an impact on natural gas carbon emissions as it does for coal because upstream emissions are more significant in the natural gas fuel cycle. Two studies evaluated the future deployment of CCS with MEA for NGCC plants. Odeh and Cockerill (2008) found CCS reduced emissions from 488 to 200 g CO_2e/kWh and Spath and Mann (2004) found emissions dropped from 499 to 245 g CO_2e/kWh. The Spath and Mann (2004) result does not include CO_2 emissions associated with production, regeneration, or disposal of MEA.

Nuclear

For nuclear technologies, the studies reviewed report values from 3 to 10^6 g CO_2e/kWh, with all values except the low and high values clustered from 15 to 25 g CO_2e/kWh. The low value of 3 g CO_2e/kWh is from Vattenfall (2004) and the high value of 108 g CO_2e/kWh is from Storm van Leeuwen and Smith (2008). The Vattenfall study used PA methods to analyze two Swedish reactors where 80 percent of the fuel enrichment was performed by centrifuge. The reactors were assumed to operate at 85 percent capacity with a life expectancy of 40 years. The Storm van Leeuwen and Smith (2008) study used EIO methods to analyze a nuclear facility located outside of Sweden with fuel enrichment via gas diffusion and an 82 percent operational capacity over a life expectancy of 30 years. The nuclear subgroup of the America's Energy Future (AEF) Committee uses a narrower range of 24 to 55 g CO_2e/kWh. The narrower range was developed by the nuclear subgroup of the AEF Committee to represent the CO_2 emissions from the current fuel enrichment situation in the United States.

Fthenakis and Kim (2007) attribute most of the difference between low and high estimates on nuclear power to three factors: the energy mix of the country developing the plant, whether enrichment is via centrifugation or diffusion (diffusion tends to use 40 percent more electricity), and the type of LCA method used.

They found that EIO methods gave estimates 10–20 times higher than PA methods for side by side comparisons of nuclear plant construction. The difference between the results of these studies is a topic of interest in the nuclear power industry.

SO_2 EMISSIONS

Coal

Because of coal's high sulfur content, traditional pulverized-coal plants have the highest SO_2 emissions of all technologies considered here—approaching 7000 mg/kWh. Significantly lower emissions, however, are estimated for different coal-based configurations. The high-end values correspond to two cases from the United States: one case with SO_2 at 6700 mg/kWh based on average U.S. coal plant emissions in 1995, and a plant that complied with the NSPS with SO_2 at 2530 mg/kWh (Spath et al., 1999). The mid-range includes several cases from the United Kingdom that have SO_2 values of 1000–1250 mg/kWh (Berry et al., 1998; Odeh and Cockerill, 2008). Spath et al. (1999) analyzed a coal-fired plant with a low-emission boiler system (LEBS) that emitted SO_2 at 720 mg/kWh.

IGCC plants with or without CCS are estimated to emit SO_2 at between 200 mg and 330 mg/kWh. Odeh and Cockerill (2008) found that CCS to an IGCC plant caused a 10 percent increase in SO_2 emissions. On the other hand, the lowest SO_2 emissions were estimated for a supercritical coal plant with carbon capture via MEA. In this case SO_2 emissions were reduced from 1250 to 9 mg/kWh (Odeh and Cockerill, 2008), primarily by increasing SO_2 removal efficiency from 90 percent to 98 percent with flue gas desulfurization (FGD).

Natural Gas

Natural gas SO_2 emission data were reviewed for three studies that reported SO_2 emissions as negligible to 324 mg/kWh. Different methodological assumptions contribute to the very divergent results from the European and U.S. studies. The U.K. ExternE (Berry et al., 1998) study assumed SO_2 as negligible through the fuel cycle and the German ExternE study (European Commission, 1997) had a very small value for SO_2 of 3 mg/kWh from extraction only. In contrast, the U.S. study assigned a large value of 324 mg/kWh for SO_2 emissions, with more than 80 percent of the emissions from gas production and distribution and about 15 percent from construction and decommissioning of plant (Spath and Mann, 2000).

NO_x EMISSIONS

Coal

As is the case for other gaseous emissions, coal has potentially the greatest rate of NO_x emissions. Estimates range from 100 to 3,350 mg/kWh. These values represent a number of different configurations, including current average practices, future practices, different power plant designs, and with CCS technologies. The high end of the range includes a U.S. NSPS, an average U.S. case, and an average U.K. case. The average U.S. case emitted NO_x at 3350 mg/kWh (Spath et al., 1999), the U.S. NSPS case emitted NO_x at 2340 mg/kWh (Spath et al., 1999), and the average U.K. case emitted NO_x at 2200 mg/kWh (Berry et al., 1998). Hypothetical future cases from Berry et al. (1998) and Spath et al. (1999) had mid-range values for NO_x of 540–1000 mg/kWh. Emission results from Odeh and Cockerill (2008) suggest pulverized-coal plants with CCS via MEA will experience an increase in NO_x emissions. They found that carbon capture via MEA increased air emissions of NO_x from 410 to 590 mg/kWh for a supercritical pulverized-coal plant with SCR, FGD, and ESP. (NH_3 increases from 5 to 470 mg/kWh with CCS via MEA for coal because oxidation of MEA produces ammonia.)

The lowest NO_x emission values were from an IGCC. Without CCS, NO_x emissions were estimated at 120 mg/kWh; with CCS via selexol, NO_x emissions were estimated to decrease by 17 percent to 100 mg/kWh.

Natural Gas

NO_x emissions for NGCC plants were estimated to be considerable, ranging from 140 to 570 mg/kWh. The high value is from Spath and Mann (2000) for an average U.S. plant with SCR. The U.K. case with low NO_x burners had a value of 460 mg/kWh (Berry et al., 1998). The ExternE case in Germany reported a value of 277 mg/kWh for a plant with no NO_x controls (European Commission, 1997).

The lowest estimated emissions (140 mg/kWh) were from a study by Odeh and Cockerill (2008) of a plant equipped with SCR. They found that the addition of CCR using MEA increased NO_x emissions by 14 percent to 60 mg/kWh.

PARTICULATE MATTER EMISSIONS

Coal

Coal has a very wide range of particulate matter emission values. At the low end (4 mg/kWh), the estimated emission rate is as low as or lower than that of renewables. The high-end estimates, approaching 10,000 mg/kWh, are an order of magnitude or greater than for all other technologies. An emission rate for particulate matter of 9,210 mg/kWh was estimated for an average U.S. coal-fired plant, while a U.S. NSPS plant was estimated to emit PM at 9,780 mg/kWh (Spath et al., 1999). Future cases range from 4 to 160 mg/kWh and represent a variety of pollution controls and burner types. At the low end, a hypothetical IGCC plant configured with SO_2, NO_x, and PM removal systems emitted 4 mg/kWh (Odeh and Cockerill, 2008). CCS had no impact on the estimated PM emissions.

Natural Gas

Two LCA studies for PM emissions from natural gas facilities were found. They reported very different results. The U.S. study estimated a large emission rate of 133 mg/kWh for PM (Spath and Mann, 2000), whereas an ExternE study in Germany estimated a rate of 18 mg/kWh (European Commission, 1997). The difference in the PM emission results is due in part to differing methodological assumptions. In the U.S. study, Spath and Mann (2000) found that approximately equal percentages of PM were emitted from upstream processes and from the power plant itself, while the German ExternE study found negligible PM emissions from power generation (European Commission, 1997).

REFERENCES

Berry, J.E., M.R. Holland, P.R. Watkiss, R. Boyd, and W. Stephenson. 1998. Power Generation and the Environment—A U.K. Perspective. Brussels: European Commission. June.

Denholm, P.L. 2004, Environmental and Policy Analysis of Renewable Energy Enabling Technologies. Ph.D. dissertation, University of Wisconsin, Madison.

European Commission. 1997. ExternE National Implementation Germany. Brussels.

Fthenakis, V.M., and H.C. Kim. 2007. Greenhouse-gas emissions from solar-electric and nuclear power: A life-cycle study. Energy Policy 35:2549-2557.

Hondo, H. 2005. Life cycle GHG emission analysis of power generation systems: Japanese case. Energy 30:2042-2056.

Odeh, N.A., and T.T. Cockerill. 2008. Life cycle GHG assessment of fossil fuel power plants with carbon capture and storage. Energy Policy 38:367-380.

Spath, P., and M. Mann. 2000. Life Cycle Assessment of a Natural Gas Combined-Cycle Power Generation System. NREL/TP-570-27715. Golden, Colo.: National Renewable Energy Laboratory. September.

Spath, P, and M. Mann. 2004. Biomass Power and Conventional Fossil Systems with and without CO_2 Sequestration—Comparing the Energy Balance, Greenhouse Gas Emissions and Economics. NREL/TP-510-32575. Golden, Colo.: National Renewable Energy Laboratory. January.

Spath, P., M. Mann, and D. Kerr. 1999. Life Cycle Assessment of Coal-fired Power Production. NREL/TP-570-25119. Golden, Colo.: National Renewable Energy Laboratory. June.

Storm van Leeuwen, J.W. 2008. Nuclear power—The energy balance energy insecurity and greenhouse gases. Updated version of "Nuclear power—The energy balance" by J.W. Storm van Leeuwen and P. Smith. 2002. Available at http://www.stormsmith.nl/.

Vattenfall AB. 2004. Certified Environmental Product Declaration of Electricity from Vattenfall's Nordic Hydropower. Registration No. S-P-00088. Vattenfall AB Generation Nordic. Stockholm. February. Available at http://www.environdec.com/reg/088/.

White, S. 1998. Net Energy Payback and CO_2 Emissions from Helium-3 Fusion and Wind Electrical Power Plants. Ph.D. dissertation. UWFDM-1093. Fusion Technology Institute, University of Wisconsin, Madison.

Electricity from renewable resources